Université Joseph Fourier

Les Houches

Session LXXXII

Multiple aspects of DNA and RNA:

from Biophysics to Bioinformatics

Lecturers who contributed to this volume

Tom Duke
Jacques van Helden
Hidde de Jong
Jim Kadonaga
Alexei Khokhlov
Richard Lavery
John Marko
Alexander Samsonov
Maria Samsonova
Terence Strick
Denis Thieffry
Eric Westhof

ÉCOLE D'ÉTÉ DE PHYSIQUE DES HOUCHES

SESSION LXXXII, 2–27 AUGUST 2004

EURO SUMMER SCHOOL
NATO ADVANCED STUDY INSTITUTE
ÉCOLE THÉMATIQUE DU CNRS

MULTIPLE ASPECTS OF DNA AND RNA:

FROM BIOPHYSICS TO BIOINFORMATICS

Edited by

Didier Chatenay, Simona Cocco, Rémi Monasson,
Denis Thieffry and Jean Dalibard

ELSEVIER
2005

Amsterdam – Boston – Heidelberg – London – New York – Oxford
Paris – San Diego – San Francisco – Singapore – Sydney – Tokyo

ELSEVIER B.V.
Radarweg 29
P.O. Box 211, 1000 AE
Amsterdam,
The Netherlands

ELSEVIER Inc.
525 B Street, Suite 1900
San Diego, CA 92101-4495
USA

ELSEVIER Ltd
The Boulevard, Langford Lane
Kidlington, Oxford OX5 1GB
UK

ELSEVIER Ltd
84 Theobalds Road
London WC1X 8RR
UK

First edition 2005

Library of Congress Cataloging in Publication Data
A catalog record is available from the Library of Congress.

British Library Cataloguing in Publication Data
A catalogue record is available from the British Library.

ISBN: 0-444-52081-3
ISSN: 0924-8099

⊗ The paper used in this publication meets the requirements of ANSI/NISO Z39.48-1992 (Permanence of Paper).
Printed and bound by CPI Antony Rowe, Eastbourne

ÉCOLE DE PHYSIQUE DES HOUCHES

Service inter-universitaire commun
à l'Université Joseph Fourier de Grenoble
et à l'Institut National Polytechnique de Grenoble

Subventionné par le Ministère de l'Éducation Nationale,
de l'Enseignement Supérieur et de la Recherche,
le Centre National de la Recherche Scientifique,
le Commissariat à l'Énergie Atomique

Previous sessions

Publishers:
- Session VIII: Dunod, Wiley, Methuen
- Sessions IX and X: Herman, Wiley
- Session XI: Gordon and Breach, Presses Universitaires
- Sessions XII–XXV: Gordon and Breach
- Sessions XXVI–LXVIII: North Holland
- Session LXIX–LXXVIII: EDP Sciences, Springer
- Session LXXIX-LXXXI: Elsevier

Lecturers

DUKE Tom, Cavendish Lab., Madingley road, Cambridge CB3 0HE, UK

van HELDEN Jacques, Service de Conformation des Macromolécules Biologiques et de Bioinformatique, Univ. Libre de Bruxelles, CP 263, Campus Plaine, Boulevard du Triomphe, 1050 Bruxelles, Belgium

de JONG Hidde, INRIA Rhône Alpes, 655 avenue de l'Europe, 38330 Montbonnot St Martin, France

KADONAGA Jim, Section of Molecular Biology and Center for Molecular Genetics, Univ. of California, San Diego, La Jolla, CA 92093-0347 USA

KHOKHLOV Alexei, Head of the Chair of Polymer Physics and Crystalophysics, Physics dept., Moscow State Univ., Leninskie Gor, 117234 Moscow, Russia

LAVERY Richard, Lab. de Biochimie Théorique, Inst. de Biologie Physicochimique, 11 rue Pierre et Marie Curie, 75005 Paris, France

MARKO John, Dept. Physics, Univ. Illinois, 845 West Taylor Street, Chicago, IL 60607, USA

MUKAMEL David, Dept. Physics of Complex Systems, Weizmann Inst., 76100 Rehovot, Israel

NOVAK Bela, Group of Computational Molecular Biology, Dept. Agricultural Chemical Technology, Technical Univ., 1521 Budapest, Hungary

STRICK Terence, Cold Spring Harbor Laboratory, Cold Spring Harbor, NY 11724, USA

SAMSONOVA Maria, Dept. Computational Biology, Center for Advanced Studies, St Petersburg State Polytechnical Univ., 29 Polytechnicheskaya ul., 195251 St Petersburg, Russia

TADDEI François, Génétique Moléculaire Évolutive et Médicale, INSERM E9916, Fac. Médecine "Necker – Enfants malades", Univ. R. Descartes, 156 Rue de Vaugirard, 75730 Paris cedex 15, France

WESTHOF Eric, Structure des Macromolécules Biologiques et Mécanismes de Reconnaissance, Inst. de Biologie Moléculaire et Cellulaire, 15 rue Descartes, 67084 Strasbourg cedex, France

Short lectures and seminar speakers

CAVALLI Giacomo, Inst. Génétique Humaine, 141 rue de la Cardonille, 34396 Montpellier, France

FERNANDEZ Bastien, CPT/CNRS Luminy Case 907, 13288 Marseille cedex 09, France

GAUTHERET Daniel, TAGC INSERM, Luminy case 906, 13288 Marseille cedex 09, France

SAMSONOV Alexander, Theoretical Department, The Ioffe Physico-Technical Institute of the Russian Academy of Sciences, St. Petersburg, 194021 Russia

Organizers

CHATENAY Didier, LPS/ENS, 24 rue Lhomond, 75231 Paris cedex 05, France

COCCO Simona, LPS/ENS, 24 rue Lhomond, 75231 Paris cedex 05, France

KRICHEVSKY Oleg, Physics dept., Ben Gurion Univ., 84105 Beer Sheva, Israel

MONASSON Rémi, LPT/ENS, 24 rue Lhomond, 75231 Paris cedex 05, France

THIEFFRY Denis, LGPD/IBDM/CNRS, Campus Luminy, Case 907, 13288 Marseille cedex 9, France

DALIBARD Jean, LKB/ENS, 24 rue Lhomond, 75231 Paris cedex 05, France

Participants

BALDAZZI Valentina, Univ. of Roma "Tor Vergata", Dept. Physics, via della Ricerca Scientifica1, 00133 Roma, Italy

BAULIN Vladimir, LDFC, Univ. Louis Pasteur, 3, rue de l'Université, 67084 Strasbourg cedex, France

BOHINC Klemen, Univ. Ljubljana, College for Health Studies, Poljanska 26a, 1000 Ljubljan, Slovenia

CANARI Francesco, Univ. Roma "La Sapienza", Dip. di Chimica, Piazzale A. Moro, box 54-Roma 62, 00185 Roma, Italy

CHERTOVICH Alexander, Chair of Physics of Polymers and Crystals, Physics dept., Moscow State Univ., Moscow 119992, Russia

CLAUDET Cyril, Lab. Spectrométrie Physique, 140 rue de la Physique, BP 87, 38402 St Martin d'Hères, France

CUENDA Sara, Univ. Carlos III de Madrid, Dpto. De Matamaticas, Avda. Universidad, 30, 28911 Madrid, Spain

CORA Davide, Univ. Torino, Dept. Theoretical Physics, via Pietro Giuria 1, 10125 Torino, Italy

DEREMBLE Cyril, Lab. de Biochimie Théorique, Institut de Biologie Physico-Chimique, 11 rue Pierre et Marie Curie, 75005 Paris, France

DONSMARK Jesper, Dept. of Biophysics, Leiden Univ., Niels Bohr weg 2, NL 2333 CA Leiden, The Netherlands

DOUARCHE Nicolas, LDFC, 3 rue de l'Université, 67084 Strasbourg, France

GABAY Carmit, Ben Gurion Univ., Beer Sheva, 84105, Israel

GEURTS Pierre, Univ. Liège, Dept. Electrical and Computer Science, Sart Tilman B28, B-4000 Liège, Belgium

GOMEZ ULLATE David, Dip. Matematica, Univ. Bologna, Piazza Porta San Donato, 5, Bologna, 40126, Italy

GRIGORYAN Arsen, Dept. Molecular Physics, Yerevan State Univ., Al. Manoogian str.1, Yerevan 375025, Armenia

HORI Yuko, Dept. Physics, Brandeis Univ., MS 057, P.O. Box 549110, Waltham, MA 02454-9110 USA

IMPARATO Alberto, Dip. di Scienze Fisiche, Univ. di Napoli "Frederico II", Complesso Universitario di Monte San Angelo, 80126 Napoli, Italy

JOLI Flore, Groupe NMR, porte 349, LPBC-CSSB, Univ. Paris 13, 74 rue Marcel Cachin, 93017 Bobigny cedex, France

JUN Suckjoon, Physics dept., Simon Fraser Univ., Burnaby, B.C V5A 1S6, Canada; FOM Inst. AMOLF, Kruislaan 407, 1098 SJ Amsterdam, The Netherlands

KALOSAKAS George, MPI for Physics of Complex Systems, Nöthnitzer str 38, Dresden 01187, Germany

KHIZANISHVILI Ana, E. Andronikashvili Inst. of Physics of Georgia Academy of Sciences, 6 Tamarashvili str, 380077 Tbilisi, Georgia

KOSTER Daniel, Kavli Inst. of Nanoscience, Section Molecular Biophysics, Technical Univ. Delft, The Netherlands

MANNA Federico, Lab. Physique Mathématique et Théorique, Univ. Montpellier II, Sciences et Techniques du Languedoc, Case courrier 70, Place E. Bataillon, 34095 Montpellier cedex 05, France

MARCONE Boris, Dip. Fisica, Univ. degli Studi di Padova, via Marzolo 8, 35131 Padova, Italy

MARCONI Daniela, Dept. Physics, Viale Berti Pichat 6/2, Bologna, Italy

MARTIGNETTI Loredana, Univ. Di Torino, Dip. Fisica Teorica, VP Giuria, 1, 10125 Torino, Italy

MEGRAW Molly, Univ. Pennsylvania, Blockley Hall, Floor 14, 423 Guardian Drive, Philadelphia, PA 19104-6021, USA

MESSER Philipp, Inst. Theoretical Physics, Cologne Univ., Zülpicher str. 77, 50937 Köln, Germany

MOSCONI Francesco, Univ. Padova, Dept. Physics "G. Galilei", via F. Marzolo, 8, 35131 Padova, Italy

NORMANNO Davide, LENS, European Lab. For Non-Linear Spectroscopy, Via Nello Carrara 1, 50019 Sesto Fiorentino, Italy

PADINHATEERI Ranjith, Dept. Physics, Indian Inst. of Technology Madras, Chennai 600 036, Tamil Nadu, India

PAPAJ Grzegorz, Lab. of Bioinformatics and Protein Engineering, International Inst. of Molecular and Cell Biology, Trojdena 4, 02-109 Warsaw, Poland

PAPARCONE Raffaella, Dip. Chimica, Univ. "La Sapienza", P.le A. Moro 5, 00185 Roma, Italy

REYMOND Nancie, INSA Lyon, Lab. BF 21, Bâtiment Louis Pasteur, 20, av. A. Einstein, 69621 Villeurbanne cedex, France

ROSSETTO Vincent, MPI für Physik Komplexer Systeme, Nöthnitzer str 38, 01187 Dresden, Germany

ROSVALL Martin, Theoretical Physics, Umea Univ., SE-90187 Umea, Sweden

SACHIDANANDAM Ravi, Cold Spring Harbor Lab., 1 Bungtown Road, Freeman Building, Cold Spring Harbor, NY 11724, USA

SANKARARAMAN Sumithra, Dept. Physics, Univ. of Illinois, 845 w Taylor street, Chicago, IL 60607, USA

SHUSTERMAN Roman, Physics Dept, Ben Gurion Univ., Beer-Sheva, 84105, Israel

SKOKO Dunja, Univ. Illinois, 845 W. Tayor str., Physics MC 273, Chicago IL 60607, USA

TEIF Vladimir, Inst. Bioorganic Chemistry, Belarus National Academy of Sciences, Kuprevich street 5/2, 220141 Minsk, Belarus

TKACIK Gasper, 236 Carl Icahn Lab., Princeton Univ., Princeton, NJ 08544, USA

TOURLEIGH Yegor, Moscow Lomonosov State Univ., Dept. Biophysics, Leninskiye Gory, 1/12, Moscow, 119992, Russia

TRUSINA Ala, NORDITA, DK-2100 Copenhagen, Denmark

VAFABAKHSH Reza, IASBS, Zanjan, Post box 45195-159, Iran

VANDOOLAEGHE Wendy, Polymers and Colloids Group, Cavendish Lab., Univ. Cambridge, Madingley road, Cambridge CB3 0HE, UK

WEBER Jérémie, Inst. Curie, Lab. Physico-chimie, 11 Rue Pierre et Marie Curie, 75005 Paris, France

WILK Agnieszka, Molecular Biophysics Division, Inst. of Physics, Adam Mickiewicz Univ., Umultowska 85, 61-614 Poznan, Poland

YAN Koon-Kiu, Dept. Physics, Brookhaven National Lab., Upton, NY 11973, USA

Preface

In August 2004, the Ecole de Physique des Houches hosted a Summer School dedicated to biological, physical and computational aspects of nucleic acids. Central to vital processes, these biological molecules have been experimentally studied by molecular biologists for five decades since the discovery of the structure of DNA by J. Watson and F. Crick in 1953. Recent progresses, such as the development of DNA arrays, manipulations at the single molecule level, the availability of huge genomic databases, have foster the need for theoretical modeling. In particular, a global understanding of the structure and function of DNA and RNA require the concerted development and application of proper experimental and theoretical approaches, involving methods and tools from different disciplines, including physics. The aim of this Summer School was precisely to provide a comprehensive overview of these issues at the interface between physics, biology and information science.

The Summer School encompassed three main sections:

1) Biochemistry and Biology of DNA/RNA;

2) Biophysics: from Experiments to modeling and theory;

3) Bioinformatics.

The present book follows the same organization, and is mainly intended to advanced graduate students or young researchers willing to acquire a broad interdisciplinary understanding of the multiple aspects of DNA and RNA.

The first section comprises an introduction to biochemistry and biology of nucleic acids. The structure and function of DNA are reviewed in R. Lavery's chapter. The next contribution, by V. Fritsch and E. Westhof, concentrates on the folding properties of RNA molecules. The cellular processes involving these molecules are reviewed by J. Kadonaga, with special emphasis on the regulation of transcription of DNA. These chapters do not require any preliminary knowledge in the field, except that of elementary biology and chemistry.

The second section covers the biophysics of DNA and RNA, starting with basics in polymer physics with the contribution by R. Khokhlov. Advances in the understanding of electrophoresis, a technique of crucial importance in everyday molecular biology, are then exposed in T. Duke's contribution. Finally a large space is devoted to the presentation of recent experimental and theoretical progresses in the field of single molecule studies. T. Strick's contribution presents a detailed description of the various micro-manipulation techniques, and reviews recent experiments on the interactions between DNA and proteins (helicases, topoisomerases, etc.). The theoretical modeling of single molecules is presented by J. Marko, with a special attention paid to the elastic and topological properties of DNA.

The third section presents provides an overview of the main computational approaches to integrate, analyze and simulate molecular and genetic networks. First J. van Helden introduces a series of statistical and computational methods allowing the identification of short nucleic fragments putatively involved in the regulation of gene expression from sets of promoter sequences controlling coexpressed genes. Next the chapter by Samsonova *et al.* connects the issue of transcriptional regulation with that of the control of cell differentiation and pattern formation during embryonic development. This contribution ties the issues of data integration, image processing and dynamical modeling, focusing on a simple model organism (the fly). Finally, H. de Jong and D. Thieffry review a series of mathematical approaches to model the dynamical behaviour of complex genetic regulatory networks. This contribution includes brief descriptions and references to successful applications of these approaches, including the work of B. Novak, one of the teachers of the school, on the dynamical modeling of cell cycle in different model organisms, from yeast to mammals.

To complete the different chapters of this volume, the material corresponding to additional seminars and lectures, as well as to the public lecture in Les Houches by D. Chatenay can be download from the Summer School web page, at the url:

`http://w3houches.ujf-grenoble.fr/sessions_ete/ete-82/session-82.html`

The organization of the summer school and the publication of this volume could not be achieved without the invaluable contributions of the speakers and authors, all deserving warm thanks. We also express our gratitude to I. Lelièvre and B. Rousset for their diligent help in the organization of the School. Furthermore, we gratefully acknowledge the generous financial support from the CNRS, the NATO, and the European Union. Finally, we thank the attendees for the friendly and warm atmosphere that they were able to create during this 82th Session of the Les Houches Summer School.

D. Chatenay, S. Cocco, R. Monasson, D. Thieffry and J. Dalibard

CONTENTS

Contents

Course 9.   *A survey of gene circuit approach applied to*
              modelling of segment determination in fruit fly,
              by M.G. Samsonova, A.M. Samsonov,
              V.V. Gursky and C.E. Vanario-Alonso   *305*

Course 10. *Modeling, analysis, and simulation of genetic*
              regulatory networks: from differential equations
              to logical models, by Hidde de Jong and
              Denis Thieffry   *325*

Course 1

DNA STRUCTURE, DYNAMICS AND RECOGNITION

Richard Lavery

Laboratoire de Biochimie Théorique, CNRS UPR 9080
Institut de Biologie Physico-Chimique
13 rue Pierre et Marie Curie, Paris 75005, France

D. Chatenay, S. Cocco, R. Monasson, D. Thieffry and J. Dalibard, eds.
Les Houches, Session LXXXII, 2004
Multiple aspects of DNA and RNA: from Biophysics to Bioinformatics

Contents

1. Introduction to the DNA double helix

We live in the age of genomes and the DNA double helix which carries the genetic code has become a universally recognized icon. Today, the genomes of several hundred organisms have already been sequenced. Their size in base pairs (bp), the fundamental building block of the genetic code, is very variable, only 600,000 bp for *mycoplasma genitalium*, but roughly 3,300,000,000 bp for the human genome. The scale of these numbers means that "Mb", or mega-base pairs, has become the most common unit. Since the distance between two successive base pairs within DNA is roughly 0.34 nm (or 3.4 Å, where an Å is 10^{-10} m), the human genome is roughly 1 m long. Despite this length, the whole genome is packed into the nucleus of every one of our cells, a structure having a diameter on the order of 1 μm. DNA has a persistence length of roughly 500 Å (150 bp), which means that 1m of isolated DNA would form a random coil with a diameter of roughly 200 μm. This implies that something else must contribute to packing DNA into our cells, and this, as we will see later, is one of the roles of a wide variety of proteins which interact with the double helix (the protein-DNA complex found in the nucleus is termed chromatin).

Despite the very real complexity of cellular functioning, the sequencing of entire genomes means that it is becoming feasible, at least for the simplest organisms, to build lists of all the proteins encoded in the DNA message, to understand how the production of these proteins is controlled (initially through the subtle interplay of the proteins, known as transcription factors, controlling DNA→RNA transcription) and how these proteins interact with one another or act upon other molecules present within the cell, leading to energy storage, molecular synthesis, and so on. A number of projects are now targeting the construction of "minimal organisms" either in the laboratory (http://www.biomedcentral.com/news/20021122/05/) or within the computer (http://www.e-cell.org/). These efforts, which would have been impossible without genome sequencing, should bring us to a much deeper understanding of the true nature of life – a striking contrast to the definition I learnt in school, which was an acronym based simply on the observable characteristics of living organisms: "MERRING" (movement, excretion, reproduction, respiration, irritability, nutrition and growth).

Let's start the story of the DNA double helix by looking briefly at the history of its discovery. Although the Austrian monk, Gregor Mendel's work turned

5

out to be the true foundation of genetics, it was Freidrich Meischer who actually first isolated the carrier of the genetic message, DNA. He termed the molecule, or rather the molecular complex (somewhat degraded chromatin), he isolated "nuclein". Meischer, who had some difficulty publishing his work despite the fact that his boss Ernst Hoppe-Seyler was the founder of the first journal of biochemistry, showed great insight in imagining that a biological polymer (or biopolymer) might carry a message, coded in the linear organization of its monomeric building blocks. Nevertheless, it would take almost 60 years until this role was demonstrated for DNA. Part of the delay was due to the great organic chemist Phoebus Levene, did much of the pioneering work necessary to identify the molecular components of DNA (deoxyribonucleic acid) and how they were distinguished from RNA (ribonucleic acid). Unfortunately, he concluded that the four bases which occur along the DNA polymer, adenine (A), cytosine (C), guanine (G) and thymine (T) were probably organized in a tedious regular repeat: ACGTACGT.... This so-called tetranucleotide hypothesis implied that DNA could not be an information carrier, and led to proteins, which were known to have irregular sequences formed from 20 different amino acids, appearing to be much more attractive candidates as information storage molecules. With DNA research now seen as a side-track, it took many years before the pioneering work of Oswald Avery (ironically working at the Rockefeller Institute where Levine had formulated his tetranucleotide hypothesis) showed that injections of DNA led to genetic transformations in bacteria and put DNA back in the limelight.

Converting the chemical structure of DNA into a molecular conformation turned out to be a difficult task. A key step involved obtaining X-ray diffraction patterns for fibres, which could be easily pulled by inserting a glass rod into a solution of DNA. The earliest diffraction patters were obtained by William Astbury, but this work was perfected by Rosalind Franklin, working alongside John Kendrew in University College, London. While Rosalind Franklin attempted to solve the structure of DNA by laborious crystallographic techniques, Francis Crick and James Watson in Cambridge set about model building, using data from Francis Crick's earlier work on the theory of diffraction, which enabled them to identify the signature of a helical structure within the DNA fibre patterns. The first model of Watson and Crick was a disaster which meant that they had to continue their work in secret in order to avoid the wrath of William Bragg, the director of the Cavendish laboratory. They were not alone in making early mistakes, since Linus Pauling, one of the greatest structural chemists, also published a model of DNA (a triple helix with the bases on the outside) which was clearly incorrect. His mistake was to assume that the phosphate groups of DNA would be neutral and could hydrogen bond together. In fact, the phosphates are ionised in aqueous solution and therefore repel one another.

Table 1

DNA landmarks

1865	**Gregor Mendel** publishes his work on plant breeding with the notion of "genes" carrying transmissible characteristics
1869	"Nuclein" is isolated by Johann **Friedrich Miescher** in Tübingen in the laboratory of Ernst Hoppe-Seyler
1892	**Meischer** writes to his uncle "large biological molecules composed of small repeated chemical pieces could express a rich language in the same way as the letters of our alphabet"
1920	Recognition of the chemical difference between DNA and RNA **Phoebus Levene** proposes the "tetranucleotide hypothesis"
1938	**William Astbury** obtains the first diffraction patters of DNA fibres
1944	**Oswald Avery** (Rockefeller Institute) proves that DNA carries the genetic message by transforming bacteria
1950	**Erwin Chargaff** discovers [A]/[G] = [T]/[C]
1953	**James Watson** and **Francis Crick** propose the double helix as the structure of DNA based on the work of **Erwin Chargaff**, **Jerry Donohue**, **Rosalind Franklin** and **John Kendrew**
1980	**Richard Dickerson's** laboratory at UCLA publishes the first crystal structure of a DNA oligomer

The key to solving the problem came from the observation by Erwin Chargaff, that somehow the bases went together in pairs, A with T and G with C. When the organic chemist Jerry Donahue explained to Jim Watson that most textbooks of the day were wrong in showing the DNA bases as the so-called enol tautomers, rather than the keto form (which contains proton accepting carbonyl groups, rather than proton donating hydroxyl groups), the penny dropped and Watson was able to plug the base pairs together in the right way and discover that AT and GC pairs had exactly the same shape.

This implied that they could be build into the centre of a double helix which could then have a regular structure whatever the sequence of the bases.

This insight became one of the defining moments in 20th century science. The beauty of the double helix convinced everyone who saw it that it must be the right answer – not least because it clearly answered the question of how the genetic message could be copied. Due to base pairing, the two strands of the helix contained complementary messages: AATCAGTTGA... on one strand, lined up with TTAGTCAACT... on the other, separating the two strands and rebuilding the complementary message led to a new generation with two identical copies of the original molecule. This mechanism is described in the famous paragraph

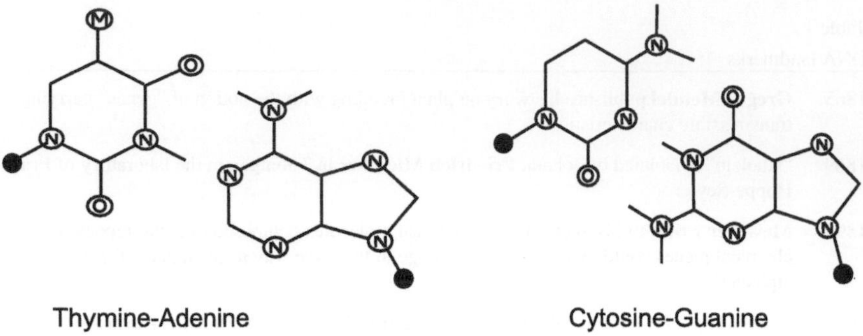

Thymine-Adenine   **Cytosine-Guanine**

Fig. 1. Base pairs.

of Watson and Crick's 1953 paper in Nature beginning "It has not escaped our notice . . .".

Despite the success of the double helix, it should be remarked there is actually not much information in a fibre diffraction pattern. The model of Watson and Crick was therefore very much a model. One problem it appeared to present was related to replicating the genetic message. In order to copy the strands of the double helix it is necessary to separate them. This is not trivial since the double helical structure implies that the two strands are interwound and cross over one another roughly every 10 base pairs. Separating the stands therefore requires unwinding them and not just pulling them apart. Given the duplication time seen in bacteria, simple calculations suggested that unwinding speeds would be so high that the corresponding centrifugal forces would lead to breaking chemical bonds. This independently led groups in New Zealand and in India to propose a so-called side-by-side model, where crossovers were avoided by making a double helix composed of alternate left- and right-handed segments. This model, which solved the centrifugal force problem, fitted the fibre diffraction data and was also compatible with early electron micrographs. The final solution came from Richard Dickerson's first single crystal structure of a DNA fragment (with the sequence CGCGAATTCGCG). It provided the first high-resolution view of DNA and it confirmed all the aspects of Watson and Crick's model. However, it also showed that a specific base sequence could locally deform the double helix. This slightly mars its beauty, but, as we shall see later, is an important factor in recognizing specific target sites within genomic DNA.

In order to understand DNA structure in more detail, we should now step back to its building blocks. Figure 2 shows the chemical constitution of DNA. The bases which we have already seen are divided in two families: adenine and

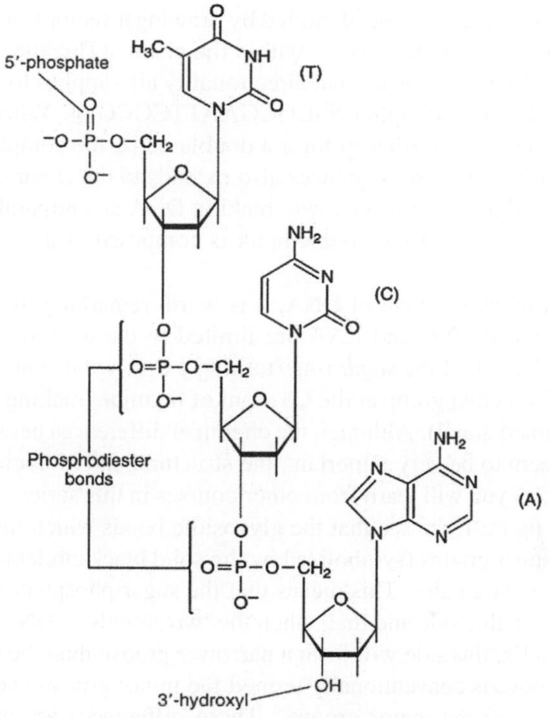

Fig. 2. DNA backbone.

guanine belong to the purine family (abbreviated as Pur or R) and have two conjugated rings; cytosine and thymine, which only have single rings, belong to the pyrimidine family (abbreviated as Pyr or Y). If, like a mathematician friend of mine, you can't remember which is which, just imagine yourself cleaning a DNA model with a "Rag" (pu**R**ine = **A**denine and **G**uanine). The bases are linked to a 5-membered sugar ring through the so-called glycosidic bond. This bond links the N9 atom of purine bases or the N1 atom of pyrimidine bases to the C1′ atom of the sugar. The sugar is called a deoxyribose. Together the sugar and the base constitute a nucleoside. (Before you get bored with chemical jargon, remember that when a base becomes a nucleoside it changes its name: adenosine, cytidine, guanidine and thymidine.) When we add a phosphate group (via a "phosphodiester bond") to the sugar ring a nucleoside becomes a nucleotide. Nucleotides can be linked together to form a polymer chain, since each sugar has two possible phosphate binding sites, the C3′ atom and the C5′ atom. Note that this implies that a polynucleotide chain has a direction. Conventionally, chains are written in

the $5'$-$3'$ direction, which can be identified by drawing a vector from the C5$'$ atom
to the C3$'$ atom of any sugar moiety within the chain. (This direction is downwards in figure 2.) This conventional directionality also applies to sequences and
thus CGCGAATTCGAG implies $5'$-CGCGAATTCGCG-$3'$. When two polynucleotide chains are put together to form a double helix, the complementarity already discussed for the base sequences also extends to the chain directions: one
chain goes up and the other goes down, making DNA an antiparallel double helix. Therefore each end of the double helix is composed of a $5'$ and a $3'$ strand
terminus.

Before continuing our visit of DNA, it is worth remarking that the chemical
differences between DNA and RNA are limited to the addition of a hydroxyl
group at the C2$'$ atom of the sugar ring (turning a deoxyribose into a ribose) and
the removal of a methyl group at the C5 atom of thymine (making it into another
pyrimidine, named uracil). Although the chemical differences between DNA and
RNA do not seem to be very important, the structural and biological differences
are significant, as you will learn from other courses in this series.

Looking at figure 1, we see that the glycosidic bonds which link the bases to
the backbone sugar groups (symbolized by the solid black circles) both lie on the
same side of the base pairs. This means that the sugar-phosphate backbones are
closer together on this side and that, when the two strands of DNA are wound up
into a double helix, this side will form a narrower groove than the opposing side.
The narrow groove is conventionally termed the minor groove and the opposing
groove is known as the major groove. These differences are quite important
when we think about other molecules, and notably proteins, binding to the double
helix, since there is more space to reach in and contact the bases on the major
groove side. The difference between the grooves becomes clear in the view of
the conventional B form of DNA shown in Fig. 3. The minor groove can be
seen in the centre of this figure. (In passing, the A and B notations for DNA
date from the fibre diffraction studies of Rosalind Franklin. A transition from the
"A" to the "B" form was seen as the relative humidity of the fibres increased.)
B-DNA has a diameter of roughly 20 Å. The base pairs are perpendicular to
the helical axis and separated by roughly 3.4 Å. There is a twist of roughly 34°
between successive base pairs, implying that there are roughly 10.5 bp per full
turn of the helix. Note that B-DNA is a right-handed double helix and, as already
mentioned, the strands run in opposite directions. To orient yourself, it is useful
to remember that, if we look into the minor groove of B-DNA, the $5'$-$3'$ direction
of the strand on the left will point upwards, while that of the strand on the right
will point downwards. If we look into the major groove, these directions will be
reversed.

If we study the DNA backbones in more detail, we will see that there are
six single bond rotations for each nucleotide along the sugar-phosphate pathway.

Fig. 3. B-DNA.

Table 2
Backbone dihedral angles

α :	O3′ – P – O5′ – C5′	g^-
β :	P – O5′ – C5′ – C4′	t
γ :	O5′ – C5′ – C4′ – C3′	g^+
δ :	C5′ – C4′ – C3′ – O3′	g^+
ε :	C4′ – C3′ – O3′ – P	t
ζ :	C3′ – O3′ – P – O5′	g^-
χ :	O4′ – C1′ – N1 – C2	g^- (Pyr)
	O4′ – C1′ – N9 – C4	g^- (Pur)

These dihedrals are denoted by the Greek letters α through ζ. A further single bond, the glycosidic bond χ, positions the base with respect to the sugar ring. Seven single bonds per nucleotide means that a single DNA chain is potentially very flexible and indeed studies of single chains show a persistence length equivalent to a single nucleotide unit. Much of this flexibility is lost in forming the double helix because of the pairing and stacking interactions involving the bases, but DNA still retains considerable conformational freedom, allowing for transitions between distinct conformational states (termed allomorphs) and also for considerable thermal fluctuation. Table 2 gives the definitions of the backbone dihedrals and shows there most common conformations in B-DNA. (The notations g^-, t and g^+, where g stands for *gauche* and t for *trans*, refer to dihedral angles corresponding to staggered conformations around 60°, 180° and −60° respectively.)

We must also discuss briefly the sugar rings themselves. The optimal valence angles of these 5-membered rings lead them to be most stable in non-planar "puckered" conformations. With respect to the mean plane formed by the ring atoms, these conformations have either two adjacent atoms out of plane, one on each side of the mean plane, or one atom out of-plane, leading to a so-called "envelope" conformation. There are a total of 10 major pucker states. These puckers are named after the atom which is most displaced from the ring plane and are termed "endo" if this atom lies on the same side as the C5′ exocyclic atom (or the nucleic acid base) and "exo" if the lie on the opposite side. Given the chemical environment of the sugar rings within DNA, two puckers turn out to be most stable. C2′-endo is preferred in B-DNA, while C3′-endo is preferred in A-DNA. (Shakespeare's famous "2B or not 2B" is a good way to remember which pucker goes with which form.) In publications concerning DNA structure, you will also see that sugar pucker described in terms of phase and amplitude. The variables come from the so-called pseudorotational representation of ring pucker, which treats the ring deformation as a sort of standing wave. The phase angle then characterizes the atom most displaced from the mean plane and the amplitude characterizes the extent of its displacement. The phase angles corresponding to the C3′-endo and C2′-endo forms are around 20° and 160° respectively. If you imagine these angles plotted on a 360° compass, you will understand why C3′-endo is sometimes referred to as a "north" pucker, while C2′-endo is " "south" pucker. The last thing its worth knowing about sugars is that the lowest energy route from south to north puckers goes though east direction, which corresponds to a O4′-endo pucker. This is due to steric hindrance which occurs between the C5′ atom and the base bound to C1′ if you try and push the O4′ atom below the mean plane (corresponding, as I'm sure you have already worked out, to an O4′-exo pucker).

I have already remarked that DNA can exist in more than one allomorphic form. Although the B form is the most common, transitions to other forms can take place as the result of physico-chemical changes (notably, differences in the solvent/salt environment) or as the result of physical constraints such as supercoiling and transitions are also be influenced by the base sequence. The A form of DNA, which was identified early on as occurring at low humidity, can be induced by changing to a water/alcohol mixture and is also favoured by high GC content. A-DNA is distinguished by a larger diameter than B-DNA (by roughly 4 Å), and a smaller rise (2.56 Å). The base pairs are also inclined with respect to the helical axis and, whereas in B-DNA the helical axis passes roughly through the centre of the base pairs, in A-DNA the base pairs are shifted almost 5 Å to-

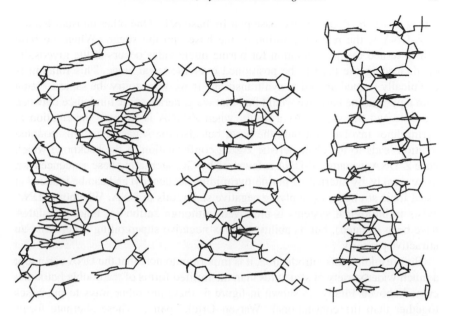

Fig. 4. A-, B- and Z-DNA structures.

wards the minor groove side. This means, as you can see on, the left of figure 4, that the minor groove becomes wide and shallow and the major groove is now deep and narrow. Despite these differences, the groove names derived from B-DNA (and based on the location of the glycosidic bonds) are maintained for the A form. In response to the change in position of the base pairs, there are also changes in the conformations of the backbones, the most important of which is a change in the sugar puckers from C2′-endo to C3′-endo. Note that A-DNA is still a right-handed, antiparallel helix. The B to A transition can occur rapidly and in a cooperative manner since it only requires minor conformational rearrangements. Local transitions to the A form of DNA probably occur within the cell and can certainly be induced by binding specific proteins.

Another important allomorphic form which you should know about is called Z-DNA. This form was originally detected by changes in the circular dichroism spectra for poly(dCG) sequences in high ionic concentrations. Z-DNA, shown on the right of figure 4, is notable in being a left-handed helix. It is less well known that, compared to B-DNA, the base pairs in Z-DNA have been turned through 180° around their long axis. The difficulty of carrying out such a rotation for a set of stacked base pairs explains why the transition from B to Z is much slower than from B to A, and that this transition generally starts at one point and works it way along the double helix, base pair by base pair. One other unusual feature of Z-DNA is the relative position of the base and the sugar. When the base pair is turned over, the rotation for purine nucleotides occurs at the glycosidic bond, leading the base to be positioned over the sugar ring. This rotation is chemically termed an anti to syn transition. If we were to try the same rotation with a pyrimidine base, we would find that we generate steric hindrance between the base and the sugar. As a result, when Z-DNA is formed, the rotation at pyrimidines involves not only the base, but also the sugar ring. This coupled rotation leads to the characteristic zigzag conformation of the backbone which gave Z-DNA its name. Z-DNA is favoured by GC alternating base sequences (or, more weakly, by alternating purine-pyrimidine sequences) and can be induced at usual ionic strengths by applying negative supercoils to DNA. Whether Z-DNA exists in biological systems is unproved (although antibodies binding Z-DNA have been isolated), but its ability to relax negative supercoiling is certainly an attractive property.

Before ending this introduction, it is important to note that the DNA can adopt a much wider variety of structures than the limited forms of the double helix discussed above. Firstly, as shown in figure 5, there are other ways to put bases together than the conventional "Watson–Crick" pairs. These alternate forms open the way for building more complex structures with three or even four DNA strands. Both of the latter possibilities, known respectively as triplex and quadruplex DNA, have biological and biotechnological interest. DNA triplexes occur naturally with the cell. They are also an attractive route towards artificial transcription control, since binding an appropriate single polynucleotide to a DNA duplex can inhibit transcription. This technique is known as the anti-gene strategy. The related technique of targeting a polynucleotide against a single stranded RNA to form a duplex is known as the anti-sense strategy. Since exogenous single strands are rapidly degraded within the cell, and also because the formation of duplex or triplex structures corresponds to a thermodynamic equilibrium, chemists have gone to considerable lengths to create modified nucleotides that will survive longer and bind better. If you are interested in this area, look up the work of Peter Nielsen on PNA (peptide nucleic acid) where the sugar-phosphate backbone has been replaced with modified peptide linkages, without damaging the possibility of base pairing with conventional polynucleotides.

A final structure which should be mentioned is the so-called Holliday junction (figure 6) which can be formed as a response to negative superhelical stress at inverted sequence repeats. Stress leads to local unpairing and the extrusion of two single strands which can then reform base pairs leading to a four helix junction. Although these junctions are conventionally drawn as square planar structures, they can fold up into a more compact tetrahedral form in the presence of Mg^{2+} ions.

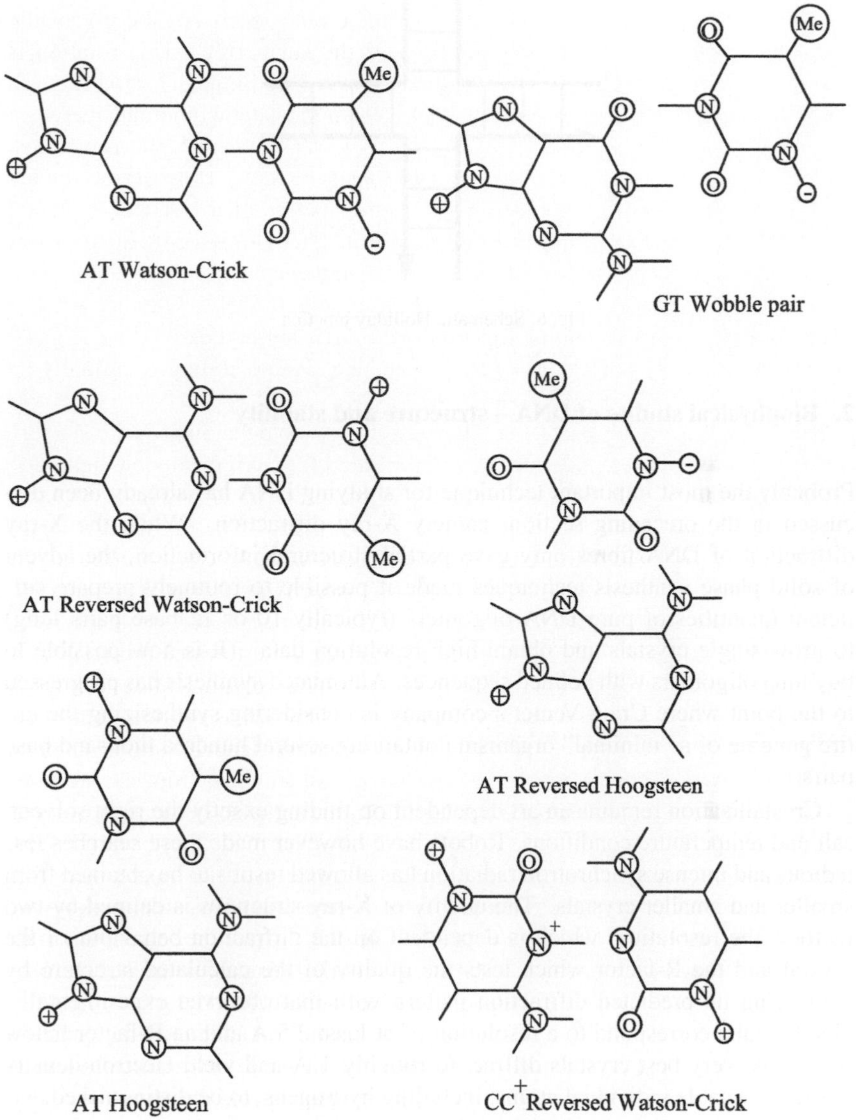

Fig. 5. Types of base pairing.

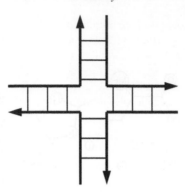

Fig. 6. Schematic Holliday junction.

2. Biophysical studies of DNA – structure and stability

Probably the most important technique for studying DNA has already been discussed in the preceding section, namely X-ray diffraction. While the X-ray diffraction of DNA fibres only gave partial structural information, the advent of solid phase synthesis techniques made it possible to routinely prepare sufficient quantities of pure DNA oligomers (typically 10 or 12 base pairs long) to grow single crystals and obtain high-resolution data. (It is now possible to buy long oligomers with defined sequences. Automated synthesis has progressed to the point where Craig Venter's company is considering synthesizing the entire genome of a "minimal" organism containing several hundred thousand base pairs.)

Crystallisation remains an art dependent on finding exactly the right solvent, salt and temperature conditions. Robots have however made these searches less tedious and intense synchrotron radiation has allowed results to be obtained from smaller and smaller crystals. The quality of X-ray structures is defined by two factors: the resolution, which is dependent on the diffraction behaviour of the crystal and the R-factor which tests the quality of the calculated structure by comparing its predicted diffraction pattern with that observed experimentally. Good results correspond to a resolution of at least 2.5 Å and an R-factor below 0.25. The very best crystals diffract to roughly 1 Å and yield electron density maps that enable individual atoms, including hydrogens, to be distinguished.

Although X-ray diffraction is undoubtedly the best technique available for determining the fine structure of DNA, the very process of forming crystals does run the risk of deforming the oligomers. Such deformations can notably affect helix bending and these problems have hindered the use of crystallography as a means for understanding sequence-induced structural changes in the double helix. The crystalline environment can also favour allomorphic transitions for some sequences (notably B→A) and other changes are also found more locally in the sugar-phosphate backbone conformations. Thus it is not clear that the crystalline conformation of an oligomer will necessarily reflect its conformation in solution, although a good match is found in many cases.

NMR spectroscopy (which studies atomic spin inversions) offers the possibility of studying DNA structure in solution and thus of avoiding crystal packing effects, however it does not provide the same quantity of information as X-ray diffraction. The two principal families of NMR experiments are termed COSY (COrrelation SpectroscopY) and NOESY (Nuclear Overhauser Effect SpectroscopY). COSY experiments give access to dihedral angles, while NOESY experiments provide through-space interatomic distances. (It remains difficult to directly exploit chemical shift data, although progress is being made in the quantum chemical calculations necessary to understand how local electronic structure influences the shielding of individual atomic spin centres.) Both COSY and NOESY methods require an initial analysis step where individual spectral peaks are assigned to specific chemical groups within specific nucleotides – using a stepwise procedure exploiting the chemical connectivity of the oligomer. Because peaks often overlap it is generally necessary to use pulse sequences which enable the spectra to be spread out over 2 or sometimes 3 dimensions. Since most NMR experiments study the spin inversions of hydrogen atoms, only a subset of the dihedral angles in the DNA backbone are accessible. Likewise, the distance dependence of NOESY coupling limits the detection of atomic interactions to distances below roughly 5 Å. This information is insufficient to obtain a 3D structure and means that model building, constrained to fit the available experimental data, is an indispensable part of NMR studies.

In contrast to X-ray diffraction, it is not possible to define a single resolution for NMR structures. Since a variety of models generally fit the spectral data (reflecting, in large part, the thermal motions of the oligomers in solution), NMR results are generally presented as an ensemble of 10 or 20 related conformations. While local elements of these structures (glycosidic angles, sugar puckers, base hydrogen bonding, . . .) are generally well defined, more subtle, long-range characteristics, such as axis bending, are more difficult to get at because they are sensitive to even small imprecisions in the data. A new NMR technique, termed RDC (residual dipolar coupling), offers a way of accessing longer interatomic distances and should result in considerable improvements for DNA. To date, only a few RDC studies have been carried out and it is too early to judge the quantitative improvement they represent.

As a result of both X-ray diffraction and NMR spectroscopy, a large number of DNA oligomer structures are now available in the PDB (Protein Data

Base, http://www.rcsb.org/pdb/) and NDB databases (Nucleic acid Data Base, http://ndbserver.rutgers.edu/). These databases are now grouped together under the RCSB (Research Collaboratory for Structural Bioinformatics) run by Helen Berman's group at Rutgers University. They also contain many DNA-ligand and DNA-protein complexes. Both databases provide tools for rapidly locating and for analyzing chosen structures and they represent a very valuable, and freely available, research resource.

Given the difficulties associated with both X-ray diffraction and NMR spectroscopy, lower resolution techniques can still play a significant role in DNA biophysical studies of DNA. Amongst these are UV (Ultra-Violet) and IR (Infra-Red) spectroscopy. UV radiation is sufficiently energetic to induce electronic transitions. Large molecules such as DNA do not give well-defined UV spectra, but rather large bands which are sensitive to changes in molecular disorder. This makes UV spectroscopy a useful tool for studying DNA melting, since the absorbance curve versus temperature (generally measured at a wavelength around 260 nm) gives access to T_m, the temperature at which 50% of the sampled has denatured from a double helix to disordered single strands. UV experiments on DNA fragments of known sequences show that higher GC content leads to a higher T_m. If you refer back to figure 1, you can see that this can be explained by the fact that GC base pairs are bound together by three hydrogen bonds, while AT pairs only have two. We will return to a more detailed discussion of double helix stability shortly.

A second use of UV radiation involves fluorescent resonance energy transfer (or FRET) between a fluorophore and a quencher group. An appropriate choice of these two groups enables short-range resonant interactions to inhibit the UV fluorescence of the fluorophore. If the two groups become spatially separated, this fluorescence is re-established. Attaching the two groups to appropriate positions in a DNA molecule enables FRET to be used as a powerful molecular ruler for distances up to the order of 10 Å.

Less energetic IR radiation is sensitive to bond vibrations and IR spectroscopy of DNA can therefore be used to define certain overall structural characteristics such as the dominant anti-syn state of the glycosidic bonds or the percentages of various sugar ring puckers. Raman FTIR (Fourier Transform Infra-Red spectroscopy) is the most common type of IR experiment since it avoids interference from water vibrations.

Another useful tool for studying DNA is CD (Circular dichroism). This type of spectroscopy measures how a molecule selectively absorbs left- or right-handed circularly polarised light. It is sensitive to molecular chirality and, in the case of DNA, was notably used to predict that the Z form was a left-handed double helix. It can provide data on the orientation of the base pairs with respect to the helical axis, but quantitative interpretation of the data is not easy since it

Table 3

Energetics of DNA

Stabilising factors:	Base pairing (electrostatics/LJ)
	Base stacking (hydrophobic)
	Ion binding (electrostatics)
	Solvation entropy
Destabilising factors:	Phosphate repulsion (electrostatics)
	Solvation enthalpy (electrostatics/LJ)
	DNA strand entropy

is still difficult to calculate the theoretical CD spectra corresponding to a given molecular conformation.

I should lastly mention the technique of neutron scattering (in either the elastic or inelastic regimes), which is gives access to information on dynamic fluctuations occurring in the picosecond to nanosecond timescale. This is a particularly interesting range for modellers since it corresponds to the timescale covered by molecular dynamics simulations. However, this technique has not been widely used for studying DNA. It should be remarked that one of its drawbacks is that it requires large quantities of the sample molecule.

I would now like to turn to a more general discussion of the stability of the DNA double helix. Many factors contribute to the equilibrium between the single and double stranded forms of DNA as shown in table 3. Amongst the stabilizing factors, most people immediately think of base pair hydrogen bonding. However, it should be remembered that in aqueous solution the bases are already interacting with water molecules before they come together to form base pairs. This balance means that a single base pair hydrogen bond contributes only roughly 1 kcal.mol^{-1} to stabilizing the double helix, compared to roughly 5 kcal.mol^{-1} if its formation enthalpy is measured in vacuum. (Note, in passing, that hydrogen bond stability results mainly from dipolar electrostatic interactions, but it also has a roughly 30% dispersion component, treated by the Lennard-Jones term in force fields.)

In fact, the double helix is more significantly stabilized by base stacking than by base pairing. This can be deduced from the fact that free bases in water stack on top of one another rather than forming hydrogen bonded pairs. Stacking enables the bases to remove the hydrophobic π-clouds of their conjugated rings from solution, without hindering access to the hydrophilic groups (involved in hydrogen bonding) around their peripheries. It should also be noted that stacking acts not only "vertically" between the successive bases in each strand of the double helix, but also "diagonally" between the neighbouring bases in opposing strands. This means that the total stacking between two successive base pairs is composed of four contributions, two vertical and two diagonal. Depending on the nature of the base pairs and the overall conformation of the double helix, the diagonal contributions can actually dominate the total stacking interaction. The next stabilizing factor is ion binding. Counter-ions are indispensable for overcoming the electrostatic repulsion between the anionic phosphates which are regularly spaced along each of the backbones of DNA. Their importance can be judged from the fact that the DNA double helix is not stable in pure water. Although water has a high dielectric constant ($\approx$80), this alone is not enough to overcome the phosphate-phosphate repulsion. The importance of electrostatics for DNA is also illustrated by the fact that its stability is enormously increased when one of its strands is replaced by the neutral "peptide nucleic acid" synthesized by Peter Nielsen (which was already mentioned earlier). The resulting hybrid is stable for hours in boiling water, in striking contrast to normal DNA. The last stabilizing factor to be considered is solvent entropy. This results from the fact that bringing together two single strands requires desolvation of their interacting faces. This releases a large number of previously bound water molecules into solution, with a consequent gain in entropy. Desolvation however also has a negative side, since removing the water molecules causes an enthalpy loss, a significantly destabilizing factor. A further disadvantage of forming a double helix is that loss of flexibility in the single strands (remember that the persistence length of DNA changes from roughly 4 Å to 500 Å upon forming a double helix) which represents an entropy penalty.

Each of the terms we have discussed can amount to hundreds of kcal.mol^{-1} for even a short DNA oligomer. It is a subtle balance between these large contributions which leads finally to a stable helix. A good way to get a feeling for the balances involved is provided by a thermodynamic study carried out in Ken Breslauer's group (see figure 7). The results of this study, which concern a 13 bp fragment of DNA with the sequence CGCATGAGTACGC, enable us to break down the helix-coil transition into two steps, the passage from a double-stranded helix (ds) into two helical single strands (s1 and s2) and then the passage of each of these strands to a disordered coil. The numbers given in figure 7 show that the double helix is stabilized by 20 kcal.mol^{-1}, but that this number actually hides opposing enthalpy and entropy changes which are roughly five times larger. If we now look at the single strands, their helical forms (h) are still stabilized enthalpically with respect to random coils (r) by roughly 30 kcal.mol^{-1} (almost exclusively due to base stacking). This term is however almost exactly compensated by the entropic loss associated with helical ordering. As a result, single strand ordering is easily disrupted by thermal agitation at room temperature.

Even for double helices, the overall balance of the enthalpic and entropic terms we have just discussed means that DNA (like most biological complexes) is only moderately stable at room temperature. This is an advantage in most cases, mak-

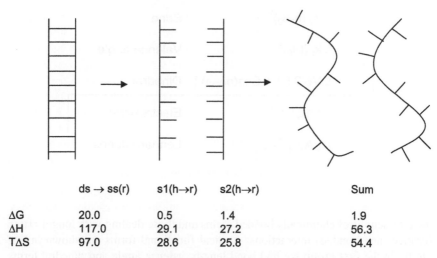

	ds → ss(r)	s1(h→r)	s2(h→r)	Sum
ΔG	20.0	0.5	1.4	1.9
ΔH	117.0	29.1	27.2	56.3
TΔS	97.0	28.6	25.8	54.4

Fig. 7. Double helix to coil transitions (values in kcal.mol^{-1}).

ing it easy to locally separate the two strands of a double helix in order to translate or replicate the genetic message. This ease can be increased further by negatively supercoiling, a strategy used by most living organisms. However, this moderate stability can become a liability when things get hot. This explains the fact that thermophiles need to keep their DNA positively supercoiled to avoid it unravelling.

3. DNA dynamics

Although experimental studies provide overall information on DNA dynamics (for example, through the atomic fluctuations probed by crystallographic B-factors), molecular simulations provide the only fully detailed information on dynamics and, in particular, offer the only hope at present of obtaining a comprehensive view of base sequence effects on dynamics. I would therefore like to spend a little time describing how simulations are carried out and what data they have provided so far.

The starting point for such simulations is an energy functional (or force field) which gives the conformational energy of a molecule as a function of the relative positions of its constituent atoms. Such force fields are empirical, although their terms are based on the various physical interactions which play a role in determining molecular structure. Their terms can be divided into those describing

R. Lavery

$$\Sigma\, k_b\, (b-b_0)^2 \qquad\qquad \text{Bond}$$

$$\Sigma\, k_t\, (t-t_0)^2 \qquad\qquad \text{Valence angle}$$

$$\Sigma\, V\phi/2\,[\,1 + Cos(n\phi-\omega)\,] \quad \text{Dihedral}$$

$$\Sigma\,[\,q_i q_j/\varepsilon r_{ij}\,] \qquad\qquad \text{Electrostatic}$$

$$\Sigma\,[\,A_{ij}/r_{ij}{}^{12} - B_{ij}/r_{ij}{}^6\,] \qquad \text{Lennard-Jones}$$

Fig. 8. Force field terms.

the interactions of chemically bonded atoms and those dealing with longer-range (termed, non-bonded) interactions. Typical functional forms are shown in figure 8. In the first group we find bond length, valence angle and dihedral terms which depend respectively on the relative positions of two, three or four linearly bonded atoms. The first two of these terms are generally quadratic functions, while the third is based on a cosine function. Each must be parameterized to reproduce the appropriate experimental (or quantum chemically calculated) value of the bond length, valence angle or dihedral (or set of symmetrically equivalent dihedrals) and be associated with an appropriate force constant determining the flexibility of the corresponding value. Such parameters need to be obtained for each chemically distinct class of bonds, valence angles or dihedrals, classes which are in turn determined by the "classes" of atom which compose them. Non-bonded interactions are generally limited to a Coulomb term describing the electrostatic interactions between fractional charges on each of the atoms and a Lennard-Jones term describing short-range interatomic repulsion and dispersion interactions. The fractional atomic charges within a molecule are generally determined by a fitting procedure aimed at reproducing the quantum chemically calculated electrostatic potential surrounding the molecule. (In the case of macromolecules, this procedure is actually applied to overlapping molecular fragments.) Lennard-Jones terms, including short range repulsion and dispersion, are obtained for interactions between all necessary atomic "classes", generally by fitting to data from the crystalline phases of small organic molecules. The reason for using the term atomic "classes" is that, given the empirical nature of force fields, it is not enough to have one set of parameters for a given atom type. It is necessary, for example, to distinguish aromatic and aliphatic carbon atoms, and amine and imine nitrogens, etc. Modern force fields typically contain around 50 atomic classes. For more details see the book by Leach cited at the end of this chapter.

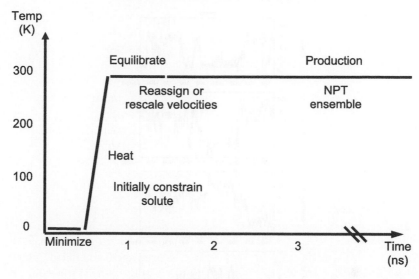

Fig. 9. MD protocol.

Carrying out a molecular dynamics simulation involves integrating Newton's equation of motion in time, using an appropriate force field to obtain the energy of the system and the forces acting on each atom. The integration has to be made in finite steps which must be smaller than the faster fluctuations occurring in the system. Since the vibrations of chemical bonds involving hydrogen have a characteristic time of a few femtoseconds (fs), the dynamics time step is generally set to 1-2 fs. Since DNA, like other biomacromolecules, is only stable in salty water, simulations must take into account a layer of water molecules containing counterions surrounding the solute molecule. This greatly increases the size of the system to be treated. As an example, a 15 base pair DNA double helix (containing roughly 1000 atoms) requires a solvent shell of roughly 5000 water molecules. The simulated system is contained within a box, which is typically a rectangular prism or a truncated octahedron. Artefacts linked to edge-effects are avoided by symmetrically reproducing the box in all directions. These so-called "periodic boundary conditions" mean that a solvent molecule or ion leaving the simulation box on one side will simultaneously re-enter the box on the opposite side.

Having set up the initial conformation of the system, and generally carried out energy minimization to ensure that there are no close atomic contacts, the system can be gently heated (by increasing the atomic velocities) until the desired simulation temperature is reached (figure 9). The so-called production phase of the simulation can then be carried out in various thermodynamic ensembles,

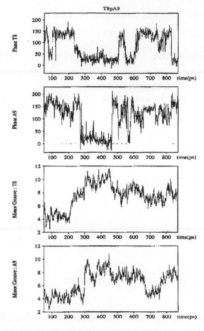

Fig. 10. MD time series.

the most common being the isothermal-isobaric ensemble (NPT), where both the temperature and the pressure of the simulated system are controlled by respectively modulating the atomic velocities and the size of the simulation cell. Given the computational cost of molecular dynamic simulations for macromolecular systems, production runs are typically limited to a few tens of nanoseconds.

What do we get out of such simulations? Firstly, it is possible to analyze in detail the way a DNA fragment fluctuates at room temperature. This can be done by making a movie of the simulation, superposing snapshots drawn from the simulation or studying time series of the backbone or helical parameters describing the fragment (figure 10 shows examples for sugar puckering and fluctuations in groove width). The results show that DNA undergoes dramatic fluctuations which considerably exceed those deduced by looking at an ensemble of DNA crystal structures. These fluctuations apply to both local backbone parameters (such as sugar puckers, phosphodiester dihedrals) and to helical parameters (such as rise and twist). They also show how base sequence can influence the structure and dynamics of the double helix and they have notably helped to understand how some sequences can induce bending of the helical axis (see below). By changing the solvent conditions it has also been possible to study spontaneous transitions between the A and B forms of DNA and it is also possible to study slower processes, such as base pair opening, by using constraints which force the simulation to sample what would otherwise be extremely rare events. I will return to this later.

An international group of laboratories (ABC, the Ascona B-DNA Consortium) has recently been formed to carry out enough DNA simulations to be able to get a comprehensive view of sequence effects. The initial stage of this project involved 15 ns simulations on a group of 39 oligomers. Each oligomer contained a tetranucleotide repeating sequence and these sequences were chosen so that the whole dataset would provide information on the 136 unique tetranucleotide sequences which can be formed from AT, TA, CG and GC base pairs. The project is far from finished, but the initial results have already led to some surprises, notably that the phosphodiester backbones can adopt finite set of conformational substates and that the barriers separating some of these substates are high enough to hinder proper sampling within a multi-nanosecond timescale. It has also been shown that ion distributions around DNA converge very slowly and that different monovalent ions have different influences on both backbone transitions and on the overall structure of DNA fragments.

4. Deformations of the double helix

Before discussing the major deformations that DNA undergoes, it is worth saying a few words about how to analyze its deformed conformations. Since DNA is a double helix it is useful to speak in terms of helical parameters. The names and geometrical sense of the various helical parameters is shown in figure 11.

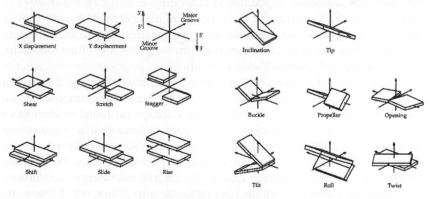

Fig. 11. Helical parameters.

The three horizontal groups of images in this figure correspond to three distinct groups of parameters: the first group positions the base pairs with respect to the helical axis, the second group positions one base with respect to another within a base pair and the third group positions one base pair with respect to its neighbouring base pair. For the latter two groups it is possible to generate a transformation matrix between the appropriate reference systems (bases or base pairs), from which the 3 translational and 3 rotational parameters can be deduced. For the first group however it is necessary to start by defining a helical axis. This is easy for a molecule with rigorous helical symmetry, but not so trivial when this symmetry is disturbed, which is often the case for DNA. In such cases, the generalized "axis" may be curved or kinked and indeed these characteristics are important in themselves in understanding the shape of a DNA fragment. It should also be remarked that the axis is useful for describing even regular DNA structures. This is not obvious for B-DNA, since the axis of the double helix passes almost through the middle of base pairs. However, for A-DNA, where the base pairs are strongly displaced towards the minor groove, the transformation matrix calculated between successive base pairs does not yield the standard rise and twist for the A conformation, namely 2.56 Å and 32.7°, but rather values which look closer to those of B-DNA, 3.44 Å and 30.7°. In the first case, the values correspond to a single rotation matrix coupled with a direct translation between the base pairs references, while in the latter they correspond to the special case where the rotation and translation involve the same unique screw axis.

We have devised an algorithm to analyse nucleic acid structures and obtain an optimal axis whether or not helical symmetry is respected. This algorithm, called CURVES, uses a least-squares procedure to distribute any distortion within the DNA molecule between curvature or dislocations in the axis and translational or rotational irregularity in the relative positions of successive bases with respect to their local axis segments. In addition to generating an axis (see the example in figure 12), CURVES provides all of the groups of parameters shown in figure 11 and also lists the backbone dihedrals and sugar puckers. If you are interested, you can obtain a free copy and a user guide from the web site of our laboratory (http://www.ibpc.fr/UPR9080/index_en.html).

Since we are on the subject of axis deformation it is worth noting that, almost 25 years ago, it was found that certain base sequences could spontaneously cause DNA to become curved. Particularly strong effects were found in kinetoplast DNA which contained segments of AT base pairs which repeated in phase with the helical twist of DNA. In general, 3-6 A's in one strand, optionally followed by T's and then either G's or C's to reach a total of 10–11 base pairs, led to strong curvature that could be observed directly by electron microscopy or indirectly by the fact that it slowed down the electrophoretic migration of DNA fragments through polyacrylamide gels or modulated the cyclization of DNA minicircles.

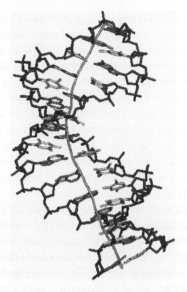

Fig. 12. Axis of bent DNA.

It was also found that so-called "A-tracts" were associated with reduced reactivity towards hydroxy radicals and with stabilized AT pairs, as measured by imino hydrogen exchange experiments. Independently of these studies, Ed Trifonov and his colleagues had noted that AA and TT dinucleotides had positional biases within nucleosome sequences, suggesting again that they were in some way linked to curvature. It is now known that spontaneously curved sequences are relatively common in genomic DNA and that they can play a role in controlling gene expression, notably by facilitating interactions between distant proteins bound to the double helix.

Understanding how A-tracts led to curvature turned out to be a difficult problem. Crystal structures of DNA fragments containing A-tracts were of limited use since the crystalline environment deformed the DNA and led either to the suppression of overall bends or to changes in bending direction that were incompatible with measurements made in solution. Initial models of bending assumed that AA (or TT) dinucleotides were associated with fixed roll angles (see figure 11) and acted like little wedges which caused uniform curvature when grouped together and repeated in phase with the rotation of the helix. That this idea was too simple was demonstrated in experiments by Paul Hagerman which showed that $(A_4T_4CG)_n$ was bent, while $(T_4A_4CG)_n$ was straight. If you look at roll in figure 11 you will see that turning an AA dinucleotide over to form a TT dinucleotide (i.e. changing the DNA stands to which the bases belong) does not change the sign of the roll. Hagerman's result shows that bending nevertheless, leading to the idea that a tilt angle could also be involved. Again looking at figure 11, you will see that inverting AA to make TT does change the sign of tilt. The story is however still more complicated since both the hydroxy radical and the hydrogen exchange experiments showed that A-tracts had a cooperative behaviour and that curvature built up rapidly beyond three successive AT pairs. It is now known that A-tracts adopt a modified form of the double helix, known as B'-DNA. This conformation is characterized by a narrow minor groove and high propeller twists. Curvature occurs not because of wedges between successive base pairs, but rather because of the junctions formed at either end of an A-tract between B'-DNA and conventional B-DNA. This is closer to the early model of curvature proposed by Don Crothers.

To come back to helical parameters for a moment, it is worth noting that A-tract curvature can nevertheless be described as if it was due to wedges between AA pairs. In fact, this model still serves to predict curvature from sequence, using the parameters developed by Trifononv, DeSantis and others. How can this be the case given what we know about B'-DNA? The answer is that looking at curvature as a succession of wedge angles or as the intrusion of a segment of DNA with a new conformation is equivalent to describing DNA with a set of local base pair to base pair transformation matrices or with screw transformations around a defined helical axis. These two ways of analyzing the problem are at the origin of much of the confusion which plagued studies of curvature during 20 years.

I would like to make one more remark about naturally curved DNA. If you have ever tried to coil up a garden hose, you will certainly have noticed that a long tube with intrinsic curvature is no longer easy to turn around its axis. The same thing happens with DNA. An intrinsic curvature, induced by specific local structures, fixes which side of the double helix will face in the direction of the curvature and which will face away. We have termed the variable which measures the rotational state of DNA with respect to the direction of curvature "rotational register" (figure 13). Since DNA has a high twist stiffness, curvature

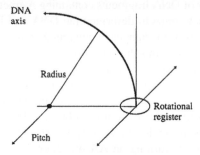

Fig. 13. Rotational register.

introduced at one position can have consequences many turns of the double helix further on and this implies that local curvature can be used to modulate which faces of DNA will contact one another at considerable distances. Note that only intrinsic curvature leads to anisotropy in rotational register. Referring back to the Hagerman sequences mentioned above, if $(T_4A_4CG)_n$ is forced to adopt the same radius of curvature that $(A_4T_4CG)_n$ adopts naturally, we will be able to detect which sequence is which by turning the double helix around its axis. $(T_4A_4CG)_n$ will turn freely while $(A_4T_4CG)_n$ will resist being turned away from its natural direction of curvature – think about it.

Let's continue with the subject of induced curvature. Curvature can be induced in the cellular environment when DNA is subjected to superhelical stress. Given the twist stiffness of DNA, attempting to wind up or unwind the double helix does not primarily lead to much change in twist. Since it is easier for the double helix to bend rather than twist, DNA undergoes buckling transitions leading to an interwound state formed of so-called plectonemes. You are already familiar with these structures if you use an old fashioned telephone with a cord connecting the body of the telephone to the handset. Always putting the handset down in the same way leads to supercoiling and plectoneme formation. The fundamental equation governing supercoiling is $L = W + T$, where L is the linking number, W is the writhe and T is the twist. If we limit ourselves for the moment to thinking about a closed circular DNA, then the linking number, which counts the number of times one DNA strand crosses the other, is a fixed integer, determined at the moment the DNA circle was closed. Twist is the total number of turns made by the double helix and writhe is the number of times the double helix crosses over itself in an out-of-plane supercoiled state. In a relaxed circular DNA, there will be no writhe and T will equal L. If we force the double helix to locally unwind, T decreases and positive writhe will be created. The same effect occurs if we remove several helical turns from the DNA before closure – leading to negative supercoiling. If we add several turns, leading to positive supercoiling, we create negative writhe. The first buckling transition of circular DNA leads to a figure of eight conformation. You can verify for yourself with a length of rubber tubing, that positive supercoiling leads to a left-handed "8" and negative supercoiling to a right-handed "8". If you add further turns, this handedness is naturally preserved, but more cross-overs are created.

However, if you play with your tubing you will also see that you can easily convert your cross-over structures into solendoidal forms. It is one of the great mysteries of life (at least for me) that, for a given type of supercoiling, the solendoidal form has the opposite handedness to that of the cross-over form (right-handed for positive supercoiling and left-handed for negative supercoiling). Biological texts generally discuss supercoiling in terms of superhelical density, σ. This is simply the number of extra turns added or removed, divided by the total

Fig. 14. Negative supercoiling.

number of turns in the relaxed system. In bacteria, σ is of the order of -0.05, that is one negative superturns per 20 turns of DNA. However, as already mentioned above, thermophiles generally contain positively supercoiled DNA. As you can imagine, it is very important for the cell to maintain the correct degree of super-coiling at all times, this is the job of the topoisomerases which cut either one or two DNA strands, modify the number of turns and then repair the cuts. These proteins are, not surprisingly, important drug targets.

While we are on the subject of supercoiling, I should add that although the examples discussed above refer to circular DNA, the same results apply to linear DNA segments whose ends are rotationally constrained. This occurs for long DNA loops within chromatin, and also for some single molecule experiments on DNA. These experiments, which enabled DNA mechanics to be investigated in a controlled environment, are now becoming a powerful means for studying protein-DNA interactions, but they have also led to a better understanding of what happens to DNA under extreme stress conditions. Single molecule experiments on DNA generally provide force-extension curves, but not structures. However, molecular modelling can help provide the missing data. Our laboratory has carried out a number of such studies, leading to proposals for the structures of two unusual forms of DNA: so-called S-DNA, which results from stretching the double helix and P-DNA which results from stretching combined with positive supercoiling. The detailed conformation of S-DNA seems to depend on which ends of the double helix are under traction. Its length is roughly 1.7 times that of normal DNA and this dramatic extension can be achieved either by unwinding the double helix and separating the base pairs along its axis, or by inclining the base pairs and reducing the helix diameter. P-DNA is even more surprising. It results

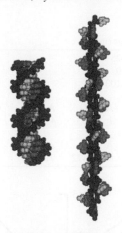

Fig. 15. P-DNA.

from very strong overtwisting, to the point where the rotation between successive base pairs passes from the usual value of 34° to almost 160°. Such extreme twists stretch of the phosphodiester backbones, break the Watson–Crick pairs and force the bases to the outside of the structure, while the backbones move towards the centre. The fact that this structure (shown in figure 15) has much in common with the incorrect structure Linus Pauling proposed for DNA explains our choice of the letter "P".

I now turn to a more localized type of DNA deformation, base pair opening. Although making P-DNA is one extreme way to open base pairs, they do in fact open spontaneously from time to time at room temperature. This can be verified by following the exchange of the imino protons of thymine (within AT pairs) or guanine (within GC pairs) with protons coming from surrounding solvent molecules. If you refer figure 1, the imino protons in question are the ones in the very centre of each base pair. Consequently, although these protons are chemically labile and exchange easily with the solvent, they are not accessible within double helical DNA. NMR experiments nevertheless show that proton exchange does take place on a millisecond timescale. As shown in table 4, exchange times depend on the nature of the base pair (GC being, not surprisingly, more stable than AT), but also on the local sequence. One example of this has already been mentioned and is illustrated in the table for specific oligomer sequences (data from the work of Maurice Guéron and Jean-Louis Leroy) – A-tracts lead to the formation of B′-DNA which greatly stabilizes the AT pairs involved and slows down exchange. Modelling can again help in understanding what is going on during such exchange. Despite the fact that proton exchange is very slow compared to the duration of typical molecular simulations, introducing restraints which enable

Table 4

Proton exchange times (ms)

GC	15–25
AT	5–10
A-tract	50–100

C	G	C	A	A	G	A	A	G	C	G	
	*	4	1	1	23	4	5	4	*	*	

C	G	C	A	A	A	A	A	A	C	C	G
	*	12	3	54	122	91	84	28	11	*	*

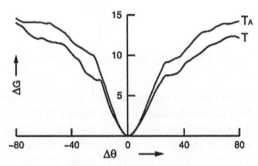

Fig. 16. Opening free energy.

dynamics to be sampled along a chosen conformational pathway can overcome the problem. We have achieved this by developing a geometrical restraint which enabled us to move a chosen base out of its normal position, towards the minor or major grooves, leaving the rest of the double helix to react freely. Using the technique of umbrella sampling, it is also possible to obtain an estimate of the free energy changes along such conformational pathways. The results confirm the experimental finding that base pairs open individually, producing little perturbation of the surrounding helix. (I remark that recent results from Irina Russu's group show that two adjacent pairs can open together within certain base sequences.) Opening begins with elastic deformation of the Watson–Crick pairs which move in a coupled way within a more or less harmonic potential. Once one base rotates enough to start breaking hydrogen bonds (about 30°) it uncouples from its partner (which returns to normal stacked position) and the free energy increases more or less linearly. Although opening is possible into either of the grooves of the double helix, the larger purine bases understandably show a preference for opening into the major groove which is less sterically hindered. The results in figure 16 (obtained by Emmanuel Giudice in our group) show the free energy for breaking an AT pair by moving thymine into the minor groove (on the left of the figure) or the major groove (on the right). The lower curve corresponds to an AT pair in an alternating (AT)n sequence, while the upper curve corresponds to an AT pair within an A-tract. The observed increase in opening free energy correlates with the experimentally observed slowing down of hydrogen exchange in this case.

5. DNA recognition

The survival and transmission of DNA's genetic message depends on interactions with a wide variety of other molecules, some essential and some harmful. Let's begin by looking at interactions with small molecules, generally termed ligands. DNA-binding ligands come in four basic families depending on how they interact with the double helix: covalent, coordinating, intercalating and groove binding. Figure 17 shows examples of the first class. Covalent ligands, as their name suggests, interact with DNA by forming one or more covalent bonds. These bonds can involve the bases, as with the infamous nitrogen mustards (used in 1st World War mustard gas) which bridge across two adjacent bases, or backbone sites, as with a variety of methylating agents such as ethylnitrosourea. A wide variety of both carcinogens and anti-tumour agents also belong to this category. Benzo[a]pyrene, a polycyclic aromatic hydrocarbon (PAH), one of the dangerous components of cigarette smoke is a good example. Within the cell, this molecule undergoes chemical attack on its most exposed ring to form an epoxide which reacts with water to form a diol. A second attack then forms another epoxide which again reacts, but this time with DNA forming a covalent bond with the exposed amino group of a guanine – which, unless repaired in time, can lead to unpleasant

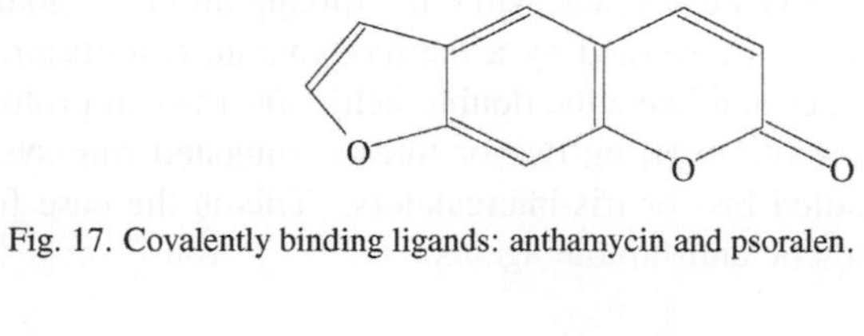

Fig. 17. Covalently binding ligands: anthamycin and psoralen.

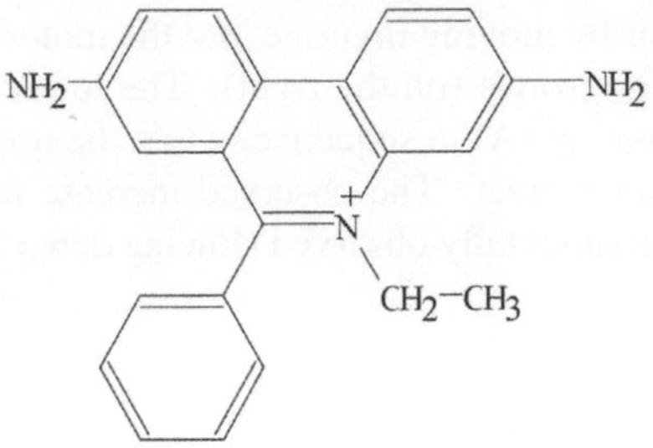

Fig. 18. Intercalator: ethidum.

consequences. (It is interesting to note in passing that the so-called KL-theory proposed in the 60's by Alberte and Bernard Pullman, well before the mode of action of PAH's were understood, successfully predicted the carcinogenicity of benzo[a]pyrene because it detected the reactivity of the primary bonds involved in the essential metabolic pathway, rather than detecting the reactivity directly related to interaction with DNA.)

The second family of ligands involve the formation of coordination bonds between transition metals and DNA, often with the bases. The best known example is called cis-platinum (cis-Pt: $(NH_3)_2$-Pt-$(Cl)_2$) which is still a commonly used anticancer agent. This compound reacts with DNA at several sites. The biologically most important adducts involve interactions with the major groove imino nitrogens (N7) of guanine and can lead to bidentate bonding to two neighbouring guanines within the same strand or bridging between the opposing stands. This leads to considerable DNA bending which can in turn influence subsequent protein binding. You can read more about this from Lippard group at http://web.mit.edu/chemistry/lippardlab/.

The third ligand family, the intercalators, constitute a large group of molecules containing conjugated ring systems that geometrically and physico-chemically mimic the base pairs. The hydrophobic nature of their π-electron clouds leads them to force their way into the double helix, stacking in between adjacent base pairs. This in turn leads to helical unwinding and stretching. It is a "rule" that in the presence of a saturating concentration of intercalator, only one inter-base pair site in two will be occupied. Referring back to our discussion of superhelicity, note that adding an intercalator to a relaxed circular DNA will create positive writhe (since the twist decreases, while the linking number remains fixed). Because intercalation is governed by a thermodynamic equilibrium, intercalators will regularly enter and leave the double helix. In order to prolong these interactions, it is possible to string two or three conjugated ring systems together, leading to so-called bis- or tris-intercalators. This is the case for a variety or natural and synthetic anti-tumour agents.

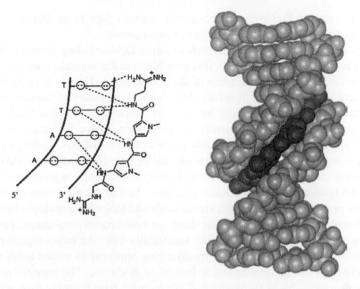

Fig. 19. Groove binder: netropsin and its DNA complex.

The last family of DNA-binding ligands is the groove binders. As their name suggests, they interact with the grooves of the double helix, generally with the minor groove, but without forming covalent bonds. This family is again formed of both natural and synthetic ligands and comprises both antibiotics and anti-tumour agents. Netropsin and dystamycin are two well known examples of groove binders. They are both built up from a linear string of rings and intervening linkers. In addition they have one or two positively charged end groups. The overall dimensions of these ligands allow them to easily fit within the minor groove of the double helix. It was initially thought that their charged end groups would interact with the negative phosphates.

However, early calculations from our laboratory showed that superposition effects led to the most negative electrostatic potentials of DNA occurring in the grooves. This suggested that cationic groove-binding ligands would lie entirely within the grooves, without direct phosphate interactions, and this was effectively found to be the case. Calculations also showed that AT-rich sequences led to more negative electrostatic potentials in the minor groove and thus explained the observed preference of cationic ligands for these sequences. Many attempts were made to change this preference, building new ligands from a chemical "tool kit" of rings, linkers and end groups. Success in this area was limited until it was shown that it was possible to fit two ligands side-by-side in the minor groove. This doubled the number of potentially hydrogen bonding groups and made it possible to design highly sequence-specific ligands (see Peter Dervan's work) which are now being used in gene control experiments.

Let's now move from small ligands to larger DNA-binding proteins. Before looking at these proteins in detail, it is possible to make several general remarks, based on the overall characteristics of the double helix. Firstly, in order to bind to a specific base sequence, it seems clear that it is necessary to contact the base pairs and to form direct specific interactions (steric fit, hydrogen bonds, etc.). Although this is not the whole story of recognition, as we will see below, it is nevertheless an important factor. Early models of protein-DNA binding assumed that there would be a code, somewhat like the genetic code, which would enable recognition to be broken down into fixed binary interactions: a given amino acid side chain binding to a given nucleic acid base. This assumption turned out to be false, the problem being that both amino acids and bases have multiple hydrogen bond donors and acceptors and that there are consequently no unique partners, in the way that Watson–Crick pairing specifically links the bases together. Proteins have overcome this problem by arranging appropriate amino acids within relatively rigid motifs, commonly α-helices or β-sheets. The specific recognition of a series of bases (typically 4–7 consecutive base pairs) is then achieved by an array of amino acid side chains. Naturally, it is still necessary to place this "recognition array" carefully with respect to the DNA groove and this is generally achieved by a set of salt bridges between the positively charged amino acids lysine or arginine and the negatively charged phosphate groups of DNA.

Even if recognition passes through an amino acid array, it is still necessary for the side chains to contact the bases. If we look at the overall structure of the double helix, this is clearly easier through the major groove. An α-helix will fit easily into the major groove, but not into the minor (this point was made my Paul Doty in the very article to be published in the *Journal of Molecular Biology*). In fact, there is another more subtle reason for preferring the major groove, which is linked to the hydrogen bonding possibilities of the bases. If you look at figure 20, the hydrogen bond acceptors and donor have been indicated by arrows and the letters A and D respectively. If you now look at the minor groove side of the base pairs (the bottom edges in the figure), you see that the letters are almost symmetry with respect to the so-called pseudo-dyad axis through the base pairs (corresponding to the vertical axis of the figure). This means that if we reverse and AT pair to make a TA pair, the hydrogen bonding profile is almost unchanged. The same is true if we change GC into CG. This is a problem if we want to use hydrogen bonding to specifically recognize a single base pair. If you now look at the major groove, you will see it does not pose the same problem since there is no symmetry around the pseudo-dyad axis. There are thus both steric and chemical reasons for DNA-binding proteins to prefer the major groove, and, indeed, most of them do.

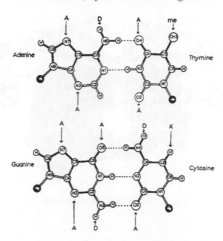

Fig. 20. Base pair hydrogen bonding.

One final generality concerning DNA-binding proteins concerns the number of base pairs they actually recognize. We can work this out by thinking about the number of sequences that can be built from a fixed number of base pairs. This number is simply 4^N for N pairs: 1 pair implies 4 possibilities (AT, TA, CG or GC), 2 pairs imply 16 possibilities and so on. This number increases rapidly, 10 pairs already allow 1 million sequences and by 15 pairs we have reached a billion sequences. Given that the latter number is roughly the size of the human genome, a given 15 base pair sequence would be expected to arise only once by chance. This means that a protein capable of specifically binding to such a sequence would effectively have a single site within the genome. Since this is exactly the degree of uniqueness which is required for gene control, we can deduce that highly specific proteins will recognize something of the order of 15 base pairs and this turns out to be the case. (Incidentally, this also implies that bioengineers attempting to control gene expression have to synthesize molecules having a similar "footprint" on DNA.)

The actual number of base pairs for known DNA-binding proteins is something between 5 and 25 base pairs. Shorter binding sites obviously corresponding to lower specificity (for example, many nucleases cut DNA strands after binding to 6-base pair sites). One should also note that longer binding sites are often recognized by multimeric (i.e. non-covalently linked) or multidomain (i.e. co-valently linked) proteins, where each domain actually recognizes of the order of 4–7 base pairs. Building up protein monomers, often under control of small signalling molecules, is an efficient way to recognize longer sequences and to profit

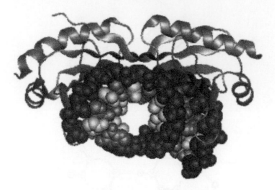

Fig. 21. TATA box binding protein.

from the combinatorial possibilities of plugging monomers together in different ways.

In fact, we are not finished with the complexities of protein-DNA interactions. As the number of crystal structures of protein-DNA complexes increased, it became clear that, in some cases, there were simply not enough specific contacts between amino acids and bases to account for the observed specificity. An example of this is shown in figure 21. The TATA box binding protein (TBP), which plays an important role in DNA transcription binds relatively specifically to a 7-base pair sequence, but only makes specific hydrogen bonds with two of the central base pairs. It was also noted that many such complexes contained significantly deformed DNA. Proteins often induce changes in groove width and this was particularly true for the small class of proteins (including TBP) which interact with the normally narrow minor groove.

Many also cause axis bending and most lead to some local deformations of helical or backbone parameters around the base pairs which are recognized. (This is certainly the case for TBP where, as figure 21 shows, DNA has an opened minor groove and overall bending of roughly 100°.) Lack of direct interactions, combined with DNA deformation led to the idea that part of the recognition might be indirect, that is, linked to the ease of deforming DNA. Given what we have seen about the energetic factors which stabilize DNA, such indirect recognition is not unreasonable. Base stacking and pairing both play an important role in the constitution of the double helix and both will be modified when we change the base sequence. The bases also have more subtle sequence specific interactions with the backbones. All these effects modulate not only the structure of the double helix, but also its local mechanical and dynamical properties. Although the concept of indirect recognition has been around for some time, it has been difficult to quantify since it requires more than a simple visual inspection of protein-DNA

Table 5

Factors in protein-DNA binding

Stabilising:	Hydrogen bonding (aa-base, aa-backbone)
	Partial intercalation of aromatic aa side chains
	Salt bridges (lysine or arginine to phosphate)
	Ion release
	Solvation entropy
Destabilizing:	DNA and protein deformation
	DNA and protein entropy
	Solvation enthalpy

complexes. For this reason we have carried out theoretical studies in which we measured both the protein-DNA interaction energy and the DNA deformation energy for a variety of complexes and for all possible base sequences. The results of this study suggest that indirect interactions play a significant role in nearly all protein-DNA complexes. As you might expect, an extreme case like TBP is almost entirely dominated by indirect recognition, but in many other complexes, even those showing apparently little DNA deformation, 30–50% of the recognition can often be attributed to the protein probing the local mechanical properties of the double helix. If you are interested in learning more about this, have a look at the recent publications of Guillaume Paillard and Cyril Deremble.

Before leaving the question of energetics, one should note that the formation protein-DNA complexes, like the formation of the DNA double helix itself, again involves many different contributions. Table 5 lists the main ones. Some are common to our discussion of the DNA double helix (see table 3), while others are specific to protein-DNA interactions. However both processes share with other biological interactions the fact that the observed free energies are subtle balances between many, much larger contributions.

We should now say something about the different families of proteins which bind to DNA. However, since I cannot use colour images in this chapter, and these structures are often very complex, I will limit myself to providing you with a table of the main families, containing one example of each drawn from the protein data bank. High performance graphic programs are now available free of charge for both Linux and Windows based computers and can be downloaded from the web. VMD (http://www.ks.uiuc.edu/Research/vmd/) and Pymol (http://pymol.sourceforge.net/) are probably amongst the most popular ones. So having read patiently through this chapter, I now encourage you to play a more active role. Go and get a program (if you do not already have one). Download the pdb files and get a real look at how DNA and proteins interact. I hope this will make you even more excited with the double helix than I have been able to do with this text.

Table 6

DNA-binding protein families

Family	Name	PDB code
Helix-turn-helix	Cyclic AMP activator protein	1BER
Zinc finger	ZIF268	1AAY
Minor groove binding	TATA box binding protein	1CDW
Leucine zipper	GCN4-AFT/CREB	2DGC
β-sheet	Arc repressor	1PAR
Enzymatic proteins	BamH1 endonuclease	2BAM
DNA packaging	Nucleosome	1AOI

References

Principles of Nucleic Acid Structure, W. Saenger, 1984 Springer-Verlag - *an old book, but still an excellent source of detailed information on DNA structure.*

DNA Structure and Function, R.R. Sinden, 1994 Academic Press - *a more recent book in the line of Saenger, good on DNA bending.*

Nucleic Acid Structure, Ed. S. Neidle, 1999 Oxford University Press - *a multi-author volume covering many aspects of DNA structure.*

Biochemistry, D. Voet and J.G. Voet, 1999, J. Wiley & sons - *good introduction to biological processes involving DNA.*

The Eighth Day of Creation, H.F. Judson, 1996 Cold Spring Harbour Press - *a very interesting history of the development of molecular biology.*

The Double Helix, J.D. Watson, 1999, Penguin Books - *indispensable reading for DNAists.*

Rosalind Franklin: the dark lady of DNA, Brenda Maddox, 2003, Harper Collins - *for correcting any misleading impressions from the previous book.*

Introduction to Protein Structure, C. Branden and J. Tooze, 1998, Garland Publishing - *good section on protein-DNA interactions.*

Molecular Modelling. Principles and Applications, A.R. Leach, 2001, Prentice Hall - *probably the best guide to molecular simulation techniques.*

Course 2

INTRODUCTION TO NON-WATSON–CRICK BASE PAIRS AND RNA FOLDING

V. Fritsch and E. Westhof

Institut de Biologie Moléculaire et Cellulaire du CNRS,
Université Louis Pasteur,
15 rue R.Descartes, F-67084 Strasbourg, France

E-mail: E.Westhof@ibmc.u-strasbg.fr
http://www-ibmc.u-strasbg.fr/upr9002/westhof/

D. Chatenay, S. Cocco, R. Monasson, D. Thieffry and J. Dalibard, eds.
Les Houches, Session LXXXII, 2004
Multiple aspects of DNA and RNA: from Biophysics to Bioinformatics

41

Contents

43

Contents

Ribonucleic acid (RNA) molecules are highly negatively charged polymers, the only polymers known to date able to both code genetic information and to perform chemical catalysis. Because of the negative charge on each nucleotide phosphate, electrostatics is central for folding, binding, and catalysis of RNA molecules. Experimental and theoretical studies have revealed the hierarchical folding of catalytical RNA molecules (ribozymes). The pairings of the secondary structure join first proximate regions in sequence, followed by parallel packing or end-to-end stacking of contiguous helices. Those preformed helical domains associate into bundles of helices to constitute the compact tertiary structure maintained via interactions between tertiary architectural motifs. This architectural hierarchy is coupled with an electrostatic hierarchy whereby RNA folding occurs first with an electrostatic collapse to compact states with most of the secondary structure elements induced by non-specific ion binding. Later, there is a cooperative transition to native states with all tertiary contacts induced by specific ion binding, especially magnesium ions. Thus, RNA molecules exhibit complex structures in which a large fraction of the bases engage in non-Watson–Crick base pairing, forming motifs that mediate long-range RNA–RNA interactions and create binding sites for proteins and small molecule ligands.

1. Definitions

The repeating unit in RNA is called nucleotide. Each nucleotide is composed of a purine (R) or pyrimidine (Y) base, a ribose sugar ring and a phosphate group (Figure 1 and Figure 2). The standard bases in RNA are guanine, adenine, cytosine and uracil (Figure 2). In addition to these standard bases, some other modified bases are found within nucleic acids (Figure 3). Funtionnal RNA, which are implied in protein synthesis as ribosomal RNA (rRNA) or transfert RNA (tRNA), often have such modified bases. A nucleotide without a phosphate group is named nucleoside. In a polynucleotide strand, nucleosides are joined together in a linear manner through phosphodiester linkages between 3' and 5' positions of two successive sugars (Figure 4). By convention, the sequence of a nucleic acid single strand is always written with the 5' end at the left and the 3' end at the right given the 5' → 3' direction. A single oligonucleotide strand is conventionally numbered from 5' to 3' end (Figure 4). Purine and pyrimidine

45

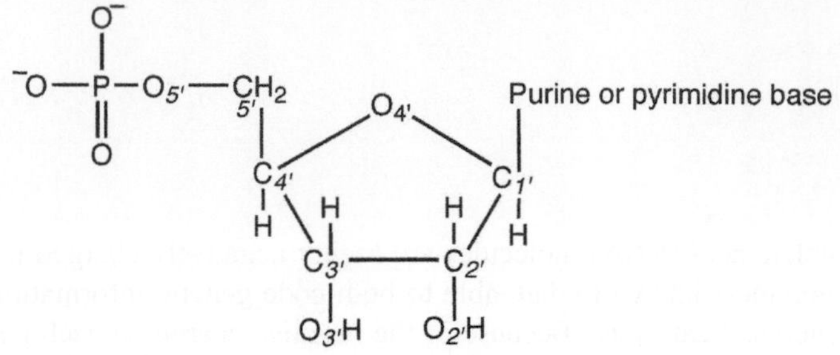

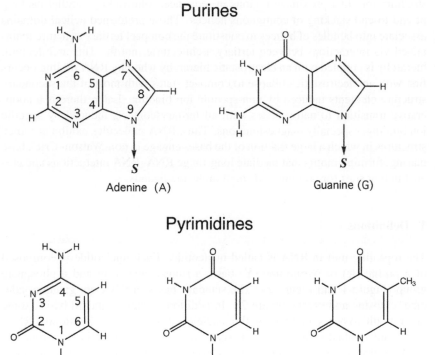

Fig. 1. A nucleotide has three characteristic components: a purine or pyrimidine base, a ribose sugar ring and a phosphate group. The numbering convention for each atom is given.

Purines

Adenine (A)

Guanine (G)

Pyrimidines

Cytosine (C)

Uracil (U)

Thymine (T)

Fig. 2. Purine and pyrimidine bases of nucleic acids (thymine is mainly found in DNA). The arrow shows the point of attachment to the sugar. This bond is called glycosidic bond.

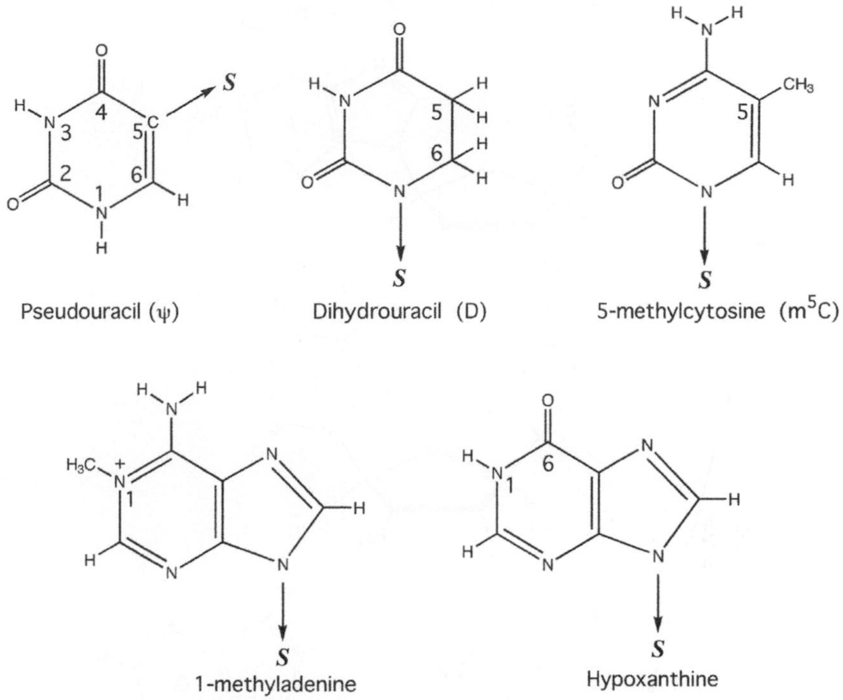

Fig. 3. Some modified bases of nucleic acids.

standard bases are planar heterocyclic molecules. They can interact together by forming hydrogen bonds. An hydrogen bond is a special kind of dipole–dipole interaction that can occur when a hydrogen atom bonded to an electronegative atom (as oxygen or nitrogen atoms) is close to another electronegative atom with nonbonding electrons. The electronegative atom with nonbonding electrons is called H-bond acceptor site while the electronegative atom which is covalently linked to the hydrogen atom is called H-bond donor site.

The model of DNA double helix proposed by James Watson and Francis Crick in 1953 consisted of two strands of nucleic acids with the sugar-phosphate backbone on the outside and the bases on the inside. The chains are held together by base-stacking interactions and hydrogen bonds between bases of both strands. In double DNA helix, a guanine is base-paired with a cytosine and an adenine with a thymine (canonical base pairs). In RNA, thymine is replaced by uracil (Figure 5). The position of both sugars related to the hydrogen bonds allows to precise the base-pairing mode. In the initial double helix model, all the base-pairing were

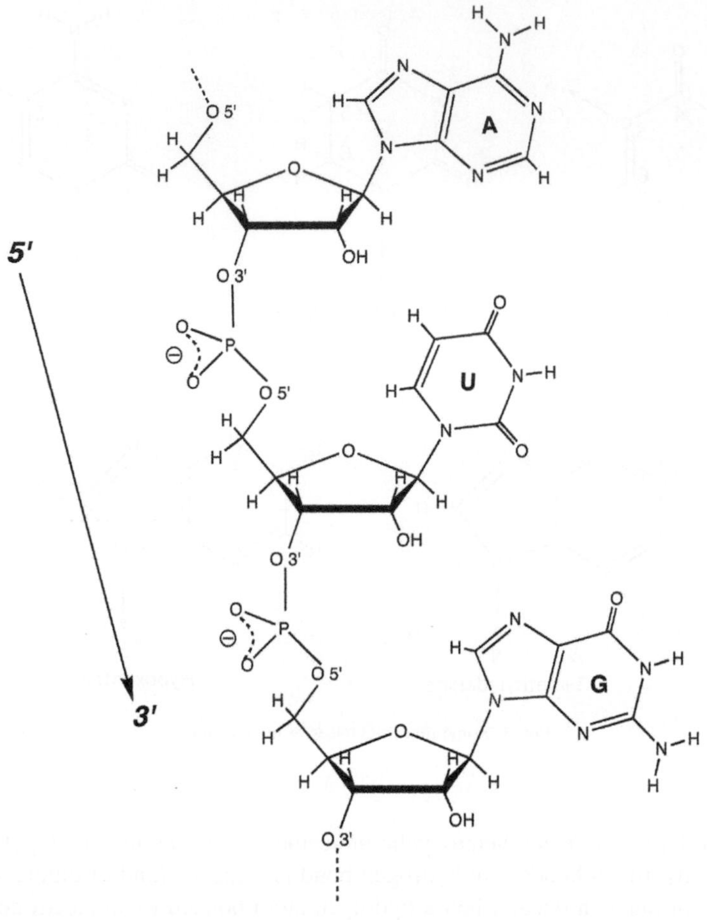

Fig. 4. Nucleotide units are joined together through phosphodiester linkages between 3′ and 5′ positions of sugars. The 5′ → 3′ direction is indicated.

proposed with a *cis* orientation of glycosidic bonds (both sugars are attached to the bases on the same side of the base pair). Such base-pairing in which A is base-paired to U or T and G to C with a *cis* sugars orientation are called *cis* Watson–Crick/Watson–Crick pairs. These base pairs are isosteric: the C1′–C1′ distance and the relative orientation of the glycosidic bonds are the same for each base pair. Such canonical Watson–Crick base pairs, however, represent only one of many possible base–base interactions. In fact, RNA purine and pyrimidine bases present three edges for hydrogen bonding interactions (Figure 6): the Watson–

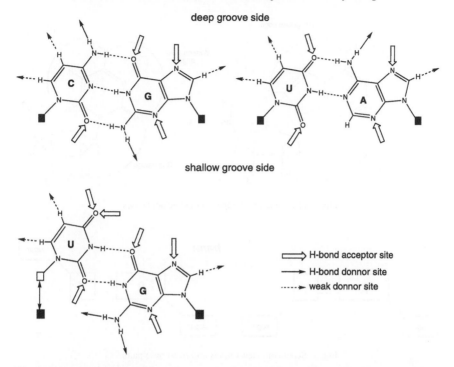

Fig. 5. Hydrogen bonding patterns between complementary bases. Isostericity between each base pair can be observed (top). C1′ atoms are shown by black squares. The G–U "wobble" pair is often observed in RNA structure (bottom). This base pair is not isosteric to the complementary *cis* Watson–Crick/Watson–Crick base pairs: to be isosteric, the C1′ atom of U (white square) have to be on the black square position. All the possible H-bond interaction sites around these base pairs are also indicated.

Crick edge, the Hoogsteen edge (for purine) or the C–H edge (for pyrimidine), and the Sugar edge (which includes the sugar 2′-hydroxyl group). The analysis of different RNA structures shows that base pairs with *trans* glycosidic bonds orientation are also observed in nucleic acid structures (Figure 7). RNA is a flexible molecule. Rotations are possible around a number of sugar-phosphate backbone bonds within a nucleotide and between nucleotides. A ribose sugar can adopt different puckering modes. Rotations of the base around glycosidic bond are also observed. Because of steric constraints between the base and the sugar-phosphate backbone, mainly two conformations are allowed: the *anti* and the *syn* conformations (Figure 8).

A triangle can be used to represent a nucleotide (Figure 9). For each triangle, the labels for the edges are as follow. The sides adjacent to the right angle repre-

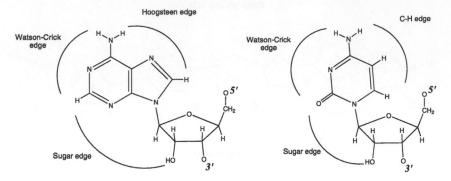

Fig. 6. Identification of edges in nucleic acid bases.

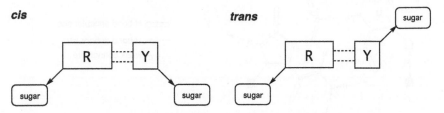

Fig. 7. Schematic views of *cis* and *trans* base pairs.

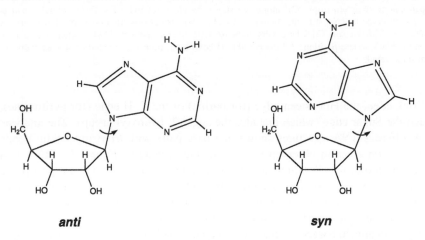

Fig. 8. *Anti* and *syn* conformations are obtained by the base rotation around the glycosidic bond.

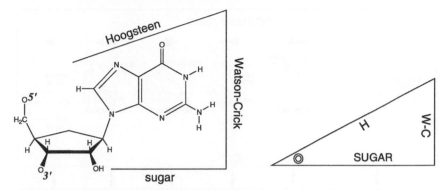

Fig. 9. Chemical structure of a purine nucleotide illustrating the three edges available for base-to-base interaction (left). Representation of an RNA base as a triangle, with edges labeled as in Figure 6 (right). For more details, see [1, 2].

sent the Watson–Crick and Sugar edges of each base while the hypotenuse of the triangle represents the Hoogsteen edge. A circle or a cross in the corner where the Hoogsteen and Sugar edges meet indicates the orientation of the sugar-phosphate backbone relative to the plane of the page: a circle means that the $5' \rightarrow 3'$ direction come from back to front while a cross means that it goes from front to back (when the base is drawn in the plane of the page). Two strands can be parallel or antiparallel. Although Hoogsteen edge applies only to purines, it is also used to refer to the C–H edge of pyrimidines as the atoms involved are normally found in the deep groove of the A-type helix.

A given edge of one base can potentially interact in a plane with any one of the three edges of a second base, and can do so in either the *cis* or *trans* orientation of the glycosidic bonds. The twelve possible, distinct edge-to-edge base-pairing geometries (or families) are illustrated in Figure 10 using the triangle representation for the bases. The upper row illustrates the six distinct *cis* pairings and the lower row the six *trans* pairings, each one positioned below the corresponding *cis* pair. Each pairing geometry is designated by stating the interacting edges of the two bases (Watson–Crick, Hoogsteen, or Sugar edge) and the relative glycosidic bond orientation, *cis* or *trans*. A historically based priority rule is invoked for listing the bases in a pair: Watson–Crick edge > Hoogsteen edge > Sugar edge. The twelve base pair geometries are listed in Table 1, with the local strand orientations in the default *anti* configurations of the bases with respect to the sugars [3].

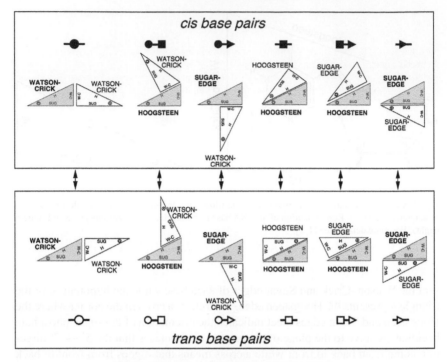

Fig. 10. The twelve possible base-pairing geometries.

2. The annotation of non-Watson–Crick base pairs and of RNA motifs

The rapidly growing number of three-dimensional RNA structures at atomic resolution requires that databases contain the annotation of such base pairs. The annotation facilitates the recognition of isosteric relationships among base pairs belonging to the same geometric family, and thus facilitates the recognition of recurrent 3D motifs from comparison of homologous sequences. Graphical conventions for accurately and unambiguously representing RNA motifs in secondary structure diagrams and in electronic databases have been defined. The annotation facilitates the 2D representation of complex 3D structures since conventions have been suggested for presenting the essential 3D features of RNA structures in a visually accessible and appealing 2D format. These include: (1) all canonical and non-Watson–Crick pairs, (2) changes in strand polarity in the folding of the RNA, (3) the occurrence of *syn* bases and (4) essential stacking interactions. The nomenclature and classification were devised in order to facilitate the organization of the vast amount of new structural data so that, when properly

Table 1

The twelve geometric families of nucleic acid base pairs with symbols for annotating secondary structure diagrams [1]. The local strand orientation is given in the last column, assuming that all bases are in the default *anti* conformation; a *syn* orientation for one base would imply a reversal of orientation; for the global orientation, the stereochemistry at the phosphate groups has to be considered. In the very rare case that both bases are *syn*, the strand orientations revert to those given in the table [3].

No.	Glycosidic Bond Orientation	Interacting Edges	Symbol	Default Local Strand Orientation
1	Cis	Watson–Crick/Watson–Crick		Anti-parallel
2	Trans	Watson–Crick/Watson–Crick		Parallel
3	Cis	Watson–Crick / Hoogsteen		Parallel
4	Trans	Watson–Crick / Hoogsteen		Anti-parallel
5	Cis	Watson–Crick / Sugar edge		Anti-parallel
6	Trans	Watson–Crick / Sugar edge		Parallel
7	Cis	Hoogsteen / Hoogsteen		Anti-parallel
8	Trans	Hoogsteen / Hoogsteen		Parallel
9	Cis	Hoogsteen / Sugar edge		Parallel
10	Trans	Hoogsteen / Sugar edge		Anti-parallel
11	Cis	Sugar edge / Sugar edge		Anti-parallel
12	Trans	Sugar edge / Sugar edge		Parallel

stored, comparisons with homologous sequences and retrieval of motifs would be rapid and accurate.

Annotation of 2D diagrams

Accurate and unambiguous annotation of RNA motifs on standard 2D drawings allows one to communicate succinctly the essential features of a motif. This, in turn, facilitates recognition of shared 3D tertiary motifs and foldings. What are the essential elements of such drawings, which can furthermore be coded easily and used for computer aided motif identification? Such diagrams should indicate:

1. The classical secondary structure (contiguous canonical pairs forming A-form double-stranded helices maintained by Watson–Crick and wobble pairs);

2. All non-Watson–Crick pairs and the geometric family to which they belong, designated using unique symbols;

3. All points in the covalent chain at which the strand polarity reverses direction;

4. Key base stacking interactions, to the degree possible without overly cluttering the picture;

5. Sequential numbering of nucleotides (5′ to 3′) to aid in tracing the covalent chain;

6. Which nucleotides adopt the less usual *syn* conformation about the glycosidic bond.

Nucleotides can be indicated by single black, capital letters (A, G, C, or U) as usual. Bold or red colored fonts are suggested to indicate which bases are in the less usual *syn* configuration of the glycosidic bond. To designate canonical Watson–Crick and wobble pairs, one can use the symbols "–" for both AU and GC pairs and "•" for the wobble GU pair [4], but the convention "–" for AU pairs, "=" for GC pairs, and "•" for GU wobble pairs is more explicit [5] and allows the use of "•"as a generic designation for non-Watson–Crick pairs in text. A set of black-and-white symbols to accurately specify each kind of non-Watson–Crick edge-to-edge pairing interaction was proposed based on the use of three symbols to designate the interacting edges: circles for Watson–Crick edges, squares for Hoogsteen edges, and triangles for Sugar edges (Table 1). Filled and open symbols distinguish the *cis* and *trans* base pairs. When the two interacting bases use the same edge only one symbol is necessary (e.g. *cis* Watson–Crick/Watson–Crick or *trans* Hoogsteen/Hoogsteen). When an interaction involves two different edges, it is necessary to designate which edge corresponds to which base. For example, "A•G *cis* Watson–Crick/Hoogsteen" designates a pair in which the Watson–Crick edge of the A interacts with the Hoogsteen edge of the G. To distinguish the X•Y from Y•X pairs in such cases, a composite symbol is generated by linking the edge symbols by a line as shown in Table 1. Finally a red or dotted arrow can be used to indicate points in the covalent chain at which reversals in strand orientation occur.

Examples of 2D representations of RNA motifs

To illustrate these conventions, we present in Figure 11 examples of 2D representations of RNA motifs starting with simple hairpin loops and proceeding to more complex motifs. A hairpin loop is form when an RNA strand folds back on itself.

Figure 11 shows examples of recurrent hairpin motifs, taken from the structure of the 23S ribosomal RNA of *Haloarcula marismortui* (NDB file rr0033 [6]). The first two hairpins are essentially the same motif, although the base sequences differ. The diagram makes the similarity obvious. Both hairpin loops are closed by a

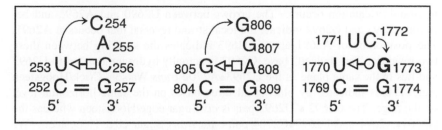

Fig. 11. Schematic representations of hairpin loops.

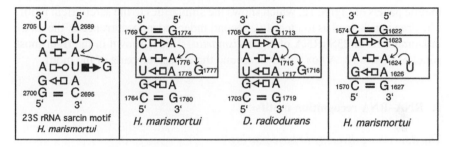

Fig. 12. Sarcin/ricin motif and related motifs. From [7].

trans Hoogsteen/Sugar edge base pair, A808•G805 in one case and C256•U253 in the other. The *trans* Hoogsteen/Sugar edge pairs are designated with open symbols (indicating the *trans* geometry) consisting of squares, placed next to A808 or C256 (for the Hoogsteen edge), linked to triangles, placed next to G805 or U253 (for the Sugar edge). The strand direction reverses direction immediately after G805 or U253. Furthermore, the corresponding bases 806–808 in the first hairpin and 254–256 in the second are stacked as indicated by placing these bases one on the other. In fact, the two hairpins are superimposable in 3D space. By contrast, the third hairpin is very different and defines a different motif. The closing base pair G1773•U1770 is *trans* Watson–Crick/Sugar edge and G1773 is in the *syn* configuration, indicated by the bold font. The strand reversal occurs between the third and fourth nucleotides of the hairpin loop (C1772–G1773). U1771 and C1772 are not stacked on each other.

The next example (Figure 12) is a motif related to the sarcin/ricin motif, a highly recurrent motif found throughout the ribosome world [8]. Almost universally, the sarcin-like motifs serve as sites for specific RNA–RNA, RNA-protein, and in a few cases RNA-drug interactions. The sarcin/ricin motif also occurs in loop E of eukaryal 5S ribosomal RNA but should not be confused with bacterial loop E. The motif is an asymmetric "internal loop" in which a local change

in strand orientation occurs. The arrows between U2690 and A2691 and between A2691 and G2692 indicate the local strand reversal that occurs at A2691.
The positioning of A2691 above U2693 indicates the stacking between these
two residues. The "bulged" base, G2692 is actually hydrogen bonded to U2693
and lies in the same plane as the U2693•A2702 *trans* Watson–Crick/Hoogsteen
pair. This is indicated by placing all three bases on the same horizontal level
on the page. The G2692 • U2693 pair is *cis* Sugar edge/Hoogsteen whereas the
G2701•A2694 and U2690•C2704 pairs are *trans* Sugar edge/Hoogsteen. We
have identified a related motif in a highly conserved stem loop in Domain IV
of 23S rRNA. The *H. marismortui* and *D. radiodurans* versions are shown in the
middle panels of Figure 12, which shows the similarities to the sarcin/ricin motif.
The drawings helped us to identify a second independent occurrence of the motif
in Domain III of 23S rRNA of *H. marismortui*. This is shown in the right-most
panel of Figure 12. A box is drawn around the conserved parts of the motif.

3. RNA–RNA recognition motifs

The flourishing diversity of base pairing

Because base pairing is so diverse and because almost several combinations of
bases can be observed in various geometries, the preceding definitions are useful
to characterize and organize base pairs. However, some pairs involve non standard H-bonding rules or intermediate geometries. Thus, so-called bifurcated (it
would be more appropriate to call them "chelated") H-bonds have been observed.
The involvement of C–H bonds in some sort of H-bonding interaction cannot be
dismissed owing to their frequent observations in high resolution crystal structures and to their surprising stabilities in long molecular dynamics simulations.
For example, the *trans* Sugar edge/Hoogsteen G•A pair covaries frequently with
a *trans* Sugar edge/Hoogsteen A•A pair in which the short distance between
N7(A) and H–C2(A) indicates the presence of a C–H. . .N H-bond. In a discon-
certing way, some pairs are mediated via one or more inserted water molecules.
More puzzling, is the example of a *trans* Hoogsteen/Watson–Crick A•C base pair
without any direct H-bond but with only water-mediated H-bond between N4(C)
and N7(A) [9].

H-bonding in the shallow groove is common and versatile

The subtle and unforeseen roles of H-bonding in the shallow groove of RNA helices are among the important contributions of the recent RNA structures. The
new pairs all involve the O2′ sugar hydroxyl group and adenine residues are the
most frequent ones found interacting with the Sugar edge sites of another base,

Watson-Crick / Sugar edge pairings

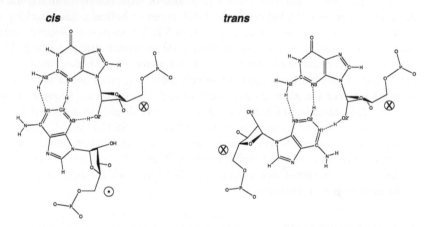

Fig. 13. Idealized drawings of Watson–Crick/Sugar edge base pairs (top) and Sugar edge/Sugar edge base pairs (bottom). The strand direction is indicated (with the convention as described in the text). Note the systematic use of the sugar hydroxyl group O2'H.

except in side-by-side pairs of the "AA-platform" motif [10]. The adenine base interacts via its Watson–Crick sites (N1 and N6), Hoogsteen sites (N6 and N7), or Sugar edge sites (N3 and C2-H) with the Sugar edge sites (N3(R), N2(G), O2(Y) and the O2'H (sugar)) of another base, which is often engaged itself in a Watson–Crick or Hoogsteen pair (Figure 13). The type of atom interacting with the O2' hydroxyl group is different in the *cis* or *trans* pairs. Thus, in a *cis*

Watson–Crick/Sugar edge pair, the N1 nitrogen of A binds the O2$'$ sugar hydroxyl group, while it is the N6 amino group in a *trans* pair. Similarly, in a *trans* Sugar edge/Sugar edge pair, the N1 binds the O2$'$ sugar hydroxyl group while, in a *cis* pair, it is the N3 nitrogen. The G•A pairs closing the GNRA tetraloops belong to the family of the *trans* Hoogsteen/Sugar edge pairs. The surprising AA-platform motif [10] in which two consecutive nucleotides stay side-by-side in the same plane involve a *cis* Hoogsteen/Sugar edge contact between the 3$'$ base and the 5$'$ base. As in the other *trans* Hoogsteen/Sugar edge pairs, the 3$'$-base of AA-platform is engaged in a Watson–Crick or Hoogsteen pairing. As a matter of fact, side-by-side platforms are not restricted to 5$'$AA3$'$ dinucleotides; 5$'$GU3$'$ platforms are observed in the sarcin loop [9] and the L11 complex [11].

Inter-strand or cross-strand stacking

In standard B-DNA types of structures, base stacking occurs mainly between bases on the same strand with a minor influence of sequence (intra-strand stacking). However, in RNA helices (or A-DNA types of helices), base stacking is strongly influenced by sequence: generally, in 5$'$R-Y3$'$ steps, one observes intra-strand stacking and, in 5$'$Y-R3$'$ steps, there is definite inter-strand stacking. This tendency is accentuated in non-Watson–Crick pairs. A well described example is that of wobble G•U pairs for which there is a pronounced inter-strand stacking between the guanines in tandem of G•U pairs with the sequence order 5$'$UG3$'$ [12]. Stretches of non-Watson–Crick pairs display pronounced purine stacks; one of the best example is again the loop E of 5S rRNAs [9]. The stabilizing effects of several layers of purine stack is seen also in bent junctions. For example, in tRNAs, the two bulging residues 59 and 60 stack on each other and on the two non-Watson–Crick *trans* pairs U8•A14/R15•Y48, stabilizing the 90° interface between the two arms.

RNA self-assembly motifs

While the Watson–Crick pairs between complementary bases are a necessity for forming the helical framework of a complex RNA, the non-Watson–Crick pairs are pivotal in RNA–RNA and RNA-protein recognition. The complementary Watson–Crick base pairs, with *cis* glycosyl bonds, form the only set of pairs which are isosteric in antiparallel helices. Thus, they promote the formation of helices with quasi-regular sugar-phosphate backbones which define the secondary structure. In single-stranded RNA molecules, stacking and base pairing drive the folding of the chain on itself through the formation of helical regions linked by non-helical elements, hairpin loops, internal bulges, and multiple junctions. RNA tertiary structure will therefore comprise those RNA–RNA interactions involving (i) two helices; (ii) two unpaired regions or (iii) one unpaired region and a double-stranded helix [13, 14].

The interactions between two helices are basically of two types: either two helices with a contiguous strand stack on each other or two distant helices position themselves so that H-bonding between sugar-phosphate backbones occur in the shallow grooves. The second type of contacts, observed as intermolecular crystal contacts [15, 16], is beautifully present in the P4–P6 structure [17]. An unpaired region belongs to either a single-stranded stretch (forming an internal loop or a bulge) or a hairpin loop closing a helix. Intramolecular interactions between two unpaired regions can be mediated by standard Watson–Crick pairing leading to the tertiary motif called pseudoknot if a single loop is involved (with possibility of co-axial stacking between the formed helices) or to loop–loop motifs (or kissing loops) otherwise [18]. Interactions between an unpaired region and a double-stranded helix can lead to various types of motifs, always involving non-Watson–Crick pairs. H-bonding of a single-stranded stretch, to sites either in the deep or the shallow groove of a double helix, leads to triples; they can always be looked at as an ensemble of two types of exclusive base pairs. Two RNA–RNA self-assembly motifs are known in which the unpaired region constitutes a terminal hairpin loop: –GNRA-tetraloops bind to the shallow groove of a RNA helix [19, 20], while –GAAA-tetraloops bind to a specific 11-nucleotide receptor (Figure 14) [17, 21]. Both motifs had been predicted on the basis of sequence analysis, coupled to molecular modelling, chemical probing, and *in vitro* selection [19, 21]. The first motif was observed as an intermolecular contact in crystals of the hammerhead ribozyme [20], while the second motif links the two main helical domains of the P4–P6 structure [17]. Internal loops, like the loop E motifs, contribute also to 3D folding motifs, since their bases are engaged in non-Watson–Crick pairings leading to compact and helix-like regions which often bind magnesium ions [7, 8, 22]. The P4–P6 structure contains a A-rich loop which is organized around two magnesium ions and presents adenines for interacting with a helix [17].

Unpaired regions of the secondary structure are structured

In RNA, secondary structure is usually defined in terms of contiguous regions of *cis* Watson–Crick pairs (including wobble GoU pairs) forming helices. Formally, the RNA folding problem is simpler than the protein folding problem [23]. Indeed, the energy content of the secondary structure is large compared to that of the tertiary structure so that the energy of the interactions maintaining the three-dimensional architecture can be considered as a perturbation on the energy of the overall system. Experimentally, this hierarchical view of RNA folding is observed in UV melting of folded RNAs where the cooperative melting of

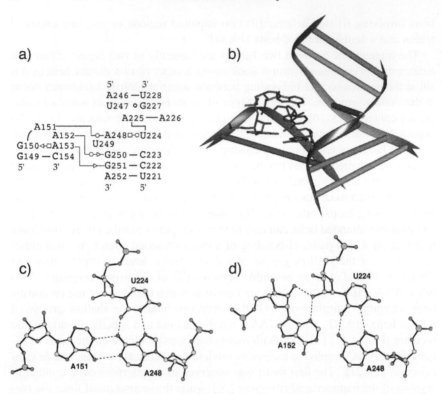

Fig. 14. a) An adenine rich tetraloop –GAAA-bind to a 11-nucleotides motif, as seen in the crystal structure of P4–P6 domain (b). The AA platform (A225–A226) is sandwiched between a *trans* Hoogsteen/Watson–Crick pair (A248•U224) and a wobble pair U247oG227. A248 forms a *trans* Watson–Crick/Watson–Crick pair with the apical base A151 of the –GAAA-tetraloop (c). In d) are illustrated direct interactions between the Sugar edge sites of A152 and the O2′H hydroxyl group of U224. Bases A151 and A152 are not in the same plane.

the tertiary structure is first observed before the broad and sequential melting of the secondary structure elements [24–27]. Besides, the melting of the tertiary structure depends strongly on divalent ion concentrations, especially magnesium ions, implying that specific ion binding sites are created during tertiary folding [24, 27–30]. On the other hand, monovalent ions influence the stability of secondary structure elements [24, 31, 32]. The separability between 2D and 3D structures is commonly observed during in vitro experiments [33]. The hierarchy in RNA folding formed the basis of a modelling approach in which preformed RNA modules were assembled into complex architectures via defined tertiary contacts [13, 19]. It is now clear that the single-stranded interhelical segments are rarely unpaired and form instead structured regions which tend to be helical-like and in which magnesium ions are bound. This is beautifully illustrated by the loop E domain of 5S rRNA. Solution data concluded that magnesium ions were necessary for the structuring of that loop [34,35]. Later, NMR evidence indicated the presence of several non-Watson–Crick pairs in the loop E of eukaryotic 5S rRNA [36] which were seen, afterwards, at high resolution by X-ray crystallography in the sarcin loop [9]. Thus, most of the unpaired regions in the secondary structures of ribosomal RNAs or large catalytic RNAs are in fact structured and organized.

4. Roles of RNA motifs in RNA-protein recognition

The A-form RNA double helix is a poor target for specific interactions

RNA secondary structure is defined by the formation of contiguous canonical Watson–Crick pairs by hydrogen bonding between the complementary bases A–U and G=C. The definition of secondary structure in RNA includes the most common non-Watson–Crick pair: the wobble GoU pair. The stacking of canonical pairs gives rise to double-stranded helices. Regular A-form helices are the basic building blocks of all RNA architectures known so far. In contrast to B-form DNA with its wide major and narrow minor groove, A-form RNA has a narrow deep and a wide shallow groove (Figure 15). The discriminatory major groove edges of the base pairs are buried in the inaccessible deep groove [37] whereas the shallow groove permits access to the rather uniform minor groove side of canonical pairs. Moreover, the polar groups of the Watson–Crick face of the bases, potential sites for hydrogen bonding, are engaged in base pair interactions. Thus, regular A-form RNA helices exhibit little potential for the specific recognition by proteins. In the present overview, we would like to emphasize the central role and importance of non-Watson–Crick pairs. In the DNA field, a mismatch (i.e. a base pair involving non-complementary bases) is functionally a potential locus for deficient or carcinogenic biological development. In the RNA world, however, non-Watson–Crick pairs are key determinants for proper native folding of the RNA and for RNA recognition by proteins or other ligands like ions or antibiotics. Therefore, to stress that fact, we will avoid term "mismatch" when describing non-Watson–Crick in RNAs.

The dual role of non-Watson–Crick pairs: groove distortion and presentation of unique hydrogen bonding sites

Pairwise combinations of hydrogen-bonded coplanar bases other than the Watson–Crick arrangement give rise to non-canonical or non-Watson–Crick pairs (re-

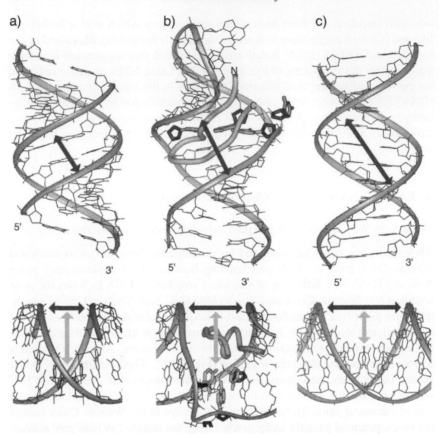

Fig. 15. Dimensions of the major/deep groove in nucleic acids duplexes shown from the side (top) and looking into the groove (bottom). a) In regular A-form RNA helices, the major groove is deep and narrow. b) Non-Watson–Crick base pairs, triples and adjacent loops distort the A-form geometry of RNA helices, leading to a expanded deep groove without reducing its characteristic depth. In the complex between BIV Tat peptide and TAR RNA [38, 39], the peptide (tube) binds in a β-turn conformation to the RNA deep groove widened by an U•A–U triple in which adenine participates in non-canonical pairing with one of the uracils (grey sticks). Adjacent to the triple, an unpaired nucleotide (dark grey sticks) facilitates the widening of the deep groove. c) In B-DNA, the major groove is much wider but less deep than in double-stranded RNA. From [40].

viewed in [2]). The GoU wobble base pair along with various types of G•A pairings are the most common non-Watson–Crick pairs in large RNA molecules such as ribosomal RNA [12]. Non-Watson–Crick interactions between nucleotides are also found in triples where a third base forms hydrogen bonds with a canonical base pair (Figure 16).

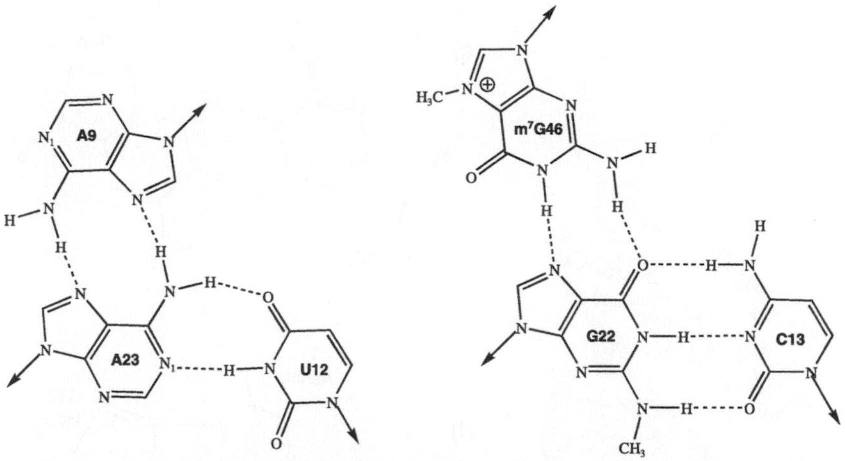

Fig. 16. Examples of triple base pairings as observed in transfert RNA.

Only a small number of non-Watson–Crick base pairs can be incorporated within stacked RNA stems without disrupting the helical structure but they do distort the regular A conformation (usually in the *cis* Watson–Crick/Watson–Crick family like the GoU wobble, a G•A pair or some Y•Y pairs). Usually, non-Watson–Crick pairs occur as a group and in a definite order, defining a RNA motif. Often, the non-Watson–Crick pairs affect especially the lateral dimension of the deep groove while its characteristic depth is maintained (Figure 15). Unpaired nucleotides are often adjacent to non-canonical pairs and increase the flexibility of the RNA backbone, thereby facilitating the widening of the deep groove. Non-Watson–Crick pairs and triples can distort the RNA deep groove to an extent allowing the accommodation of protein domains with ordered and regular secondary structure elements such as β-turns and α-helices (Figure 17).

Beyond their role of deforming the shape of the deep groove in RNA helices, non-Watson–Crick base pairs serve as specific recognition sites in hydrogen bonding interactions with proteins. In non-canonical pairs, due to the altered hydrogen bonding patterns and arrangements of bases as compared to canonical pairs, alternative and ordered sets of the polar donor and acceptor groups of the bases are available for intermolecular contacts.

Protein binding sites by combination of non-Watson–Crick pairs, triples and loops

The protein and peptide binding sites in three-dimensional structures of RNA complexes known so far suggest that single non-Watson–Crick base pairs are of-

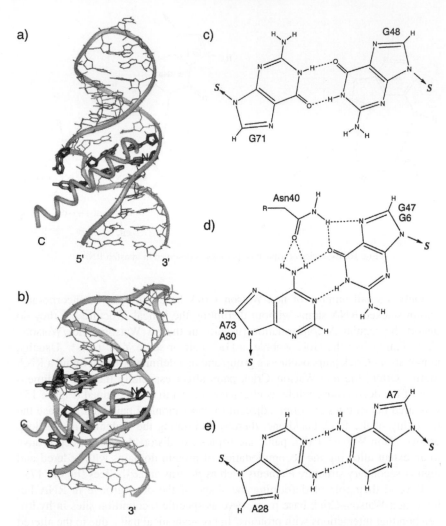

Fig. 17. Peptide-binding sites involving stacks of tandem non-canonical base pairs in RNA duplexes. a) In the complex between an HIV-1 Rev peptide and RRE RNA [41], the peptide binds as an α-helix into the deep groove widened by a G•G (c) following on a *cis* Watson–Crick/Watson–Crick G•A pair (d). b) The same peptide binds in the deep groove of an RNA aptamer [38] which contains an identical G•A pair and a symmetric A•A (e) isosteric with the corresponding G•G mismatch in the RRE RNA. A U•A–U triple identical to the triple in the BIV TAR RNA participates in the peptide binding site. d) In both complexes, the side chain of the same asparagine residue forms specific hydrogen bonds with polar groups located in the deep groove edge of the G•A pair. G47•A73 and G6•A30 correspond to the RRE RNA and the aptamer, respectively. Unpaired nucleotides in (a) and (b) are also shown in grey sticks. From [40].

ten accompanied by additional RNA motifs in order to form recognition surfaces for the substrates.

Tandem stacks of two non-Watson–Crick base pairs are found in the Rev peptide-binding sites of the HIV-1 Rev-response element (RRE) RNA [41] and a Rev-specific aptamer RNA [42] (Figure 17). In the Rev complex of RRE as well as of the aptamer, the peptide binds in α-helical conformation to the RNA deep groove which is widened by a *cis* Watson–Crick/Watson–Crick G•A pair (Figure 17d) followed by either a *trans* Watson–Crick/Watson–Crick G•G (RRE) or the isosteric A•A pair (aptamer) (Figure 17c and Figure 17e). Adjacent bulged-out nucleotides facilitate the widening of the deep groove. Consistent in both complexes, the G•A pair is recognized by the same asparagine residue in the peptide forming intermolecular hydrogen bonds with the two purines simultaneously (Figure 17d). The symmetric G•G and A•A pairs in, respectively, RRE and the aptamer are not involved in direct contacts to the peptide. The opening of the peptide binding pocket in the deep groove of RRE, however, strictly requires a homo purine pair isosteric to G•G. RRE variants obtained by *in vitro* selection displayed high affinity for the Rev protein only when a G•G or A•A mismatch could be formed between the base positions 48 and 71 [43].

In the Rev-specific aptamer RNA, a U•A–U triple contributes to the peptide binding site [42]. The arrangement of bases in the triple corresponds to a classical U•A–U trimer in which both Watson–Crick and Hoogsteen base-pairing sites of adenine are engaged simultaneously. Stereochemically identical U•A–U triples are opening up the deep groove for substrate binding in BIV TAR RNA [38, 39] in complex with a Tat peptide (Figure 17b) and a class II Rev-aptamer RNA [44] bound to a Rev peptide (Figure 18a). In the TAR complex, an isoleucine side chain of the Tat peptide in the deep groove packs against the hydrophobic C5–C6 edge of the uracil base [39, 42] which binds to the Hoogsteen face of adenine in the U•A–U triple.

The BIV TAR RNA and the class II aptamer along with the boxB RNA [45] (Figure 18b) provide examples for peptide-binding sites formed by combinations of base mismatches with adjacent loops. In the complexes between these RNAs and, respectively, Tat, Rev and N peptide, the peptide-binding pocket opens up towards a loop which folds away from the groove in order to allow the entry of the substrate. *Trans* Sugar edge/Hoogsteen G•A pairs terminate the loops and participate in hydrogen bonding to arginine residues of the peptide in the class II Rev aptamer and the boxB RNA complexes (Figure 18). The arrangement of bases in such G•A pairs projects the Hoogsteen edge of the guanine towards the deep groove readily available for contacts with amino acid side chains (Figure 18c) and Figure 18d)). Among non-Watson–Crick pairs involved in protein recognition, *trans* Sugar edge/Hoogsteen G•A pairs stand out given their wide distribution in natural RNAs.

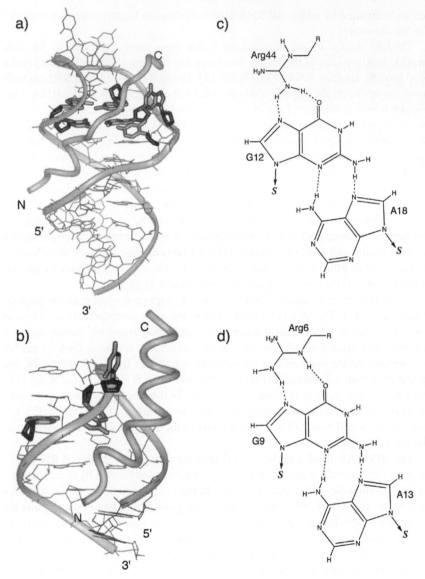

Fig. 18. Peptide-binding sites involving *trans* Sugar edge/Hoogsteen G•A pairs adjacent to a loop. In the complexes of HIV Rev peptide bound to a class II RNA aptamer [44] (a) and N peptide bound to boxB RNA [45] (b), an arginine residue forms hydrogen bonds to the Hoogsteen edge of a guanine in a *trans* Sugar edge/Hoogsteen G•A pair. The arginine binds via a single amino group (c) in the aptamer complex and, in addition, via the secondary amino group (d) in the boxB complex. Bulged-out nucleotides are also shown in grey stick representation.

GNRA tetraloops as protein recognition sites

The pentaloop in the boxB RNA adopts a GNRA tetraloop-like conformation (N is any nucleotide; R is a purine) by extrusion of one nucleotide [45] (Figure 18b) induced by binding of the N peptide. GNRA tetraloops are very frequent in large RNA folds due to both their conformational stability and their ability to participate in tertiary contacts with other RNA motifs [46]. A characteristic structural feature of GNRA loops is the terminating *trans* Sugar edge/Hoogsteen G•A pair which involves the first and last residues of the tetraloop. This *trans* Sugar edge/Hoogsteen G•A contributes to specific intermolecular hydrogen bonds to the N peptide in the boxB RNA complex and, thus, protein recognition of GNRA motifs is likely to be a general theme of RNA/protein interactions.

Recognition of non-canonical pairs in the shallow groove

The structural uniformity of the shallow groove edges of Watson–Crick base pairs renders them poor targets for specific recognition but remarquable for recognition of regular RNA helices by unpaired residues, especially adenines (see for example [46]). Non-canonical base pairs introduce asymmetries in the shallow groove of RNA duplexes which allow subtle structural discrimination in ligand binding.

Specific recognition of a base pair in the shallow groove has been observed for GoU wobble base pairs. The alanine tRNA contains a single GoU pair in the acceptor stem which serves as a major determinant for specific aminoacylation by tRNA^{Ala} synthetase [47,48]. Variant tRNAs in which the GoU pair is mutated or guanine is replaced by inosine, which lacks the 2-amino group of guanine, are not aminoacylated by the synthetase [48]. These findings indicate that tRNA^{Ala} synthetase recognizes guanine in the shallow groove by its exocyclic 2-amino group which is not involved in base pairing in GoU wobble pairs.

The shallow groove recognition of GoU wobble pairs is facilitated by the geometry of the base pair, with the uracil pushed in the deep groove creating a depression on the shallow groove surface. This site is in many cases occupied by a water molecule bridging the two bases of the wobble pair. The specific binding of ligands to the shallow groove edge of GoU pairs is thus likely to involve displacement of a bridging water molecule.

tRNA/synthetase interactions provide yet another example for mismatch recognition in the shallow groove. A sequence comparison analysis has shown that the first [32] and last [38] residues of the seven-membered tRNA anticodon covary so as to maintain characteristic bifurcated H-bonded pairs [49]. In the complex between tRNA^{Gln} and its cognate synthetase, a contact between an aspargine side chain and a uracil within a single-hydrogen-bonded U32•U38 pair has been discovered [50]. A hydrogen bond is formed between the amide group of the

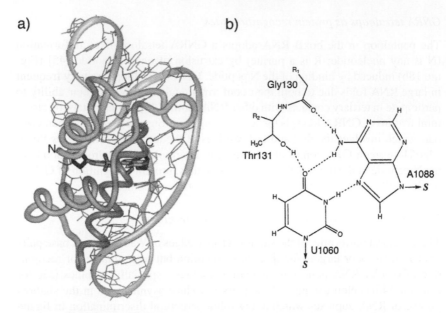

Fig. 19. a) Recognition of non-Watson–Crick pairs in the complex between the ribosomal L11 protein and a fragment of the 23S rRNA [11]. In this complex, a non-canonical *cis* Hoogsteen/Watson–Crick A•U pair is involved in protein contacts (b). This A•U base pair forms hydrogen bonds with a threonine side chain and the peptide backbone within an α-helix that binds to a shallow groove face of the RNA fold.

asparagine side chain and the O2 carbonyl atom of the uracil projecting into the shallow groove. Despite the recurrent occurrence of non-Watson–Crick pairs in tRNAs, examples for recognition of non-canonical pairs by tRNA-binding proteins are scarce. Clearly, most of them are necessary for maintaining the native architecture of tRNAs. Further, one may speculate that the numerous modifications of nucleotides may be preferred as specific identity (or anti-determinant) elements of tRNA structures.

The three-dimensional structure of a 58-nucleotide RNA fragment of 23S rRNA in complex with ribosomal L11 protein [11, 51] displays the most extensive case of shallow groove recognition in an RNA/protein complex yet. The L11 protein binds with a 15-residue α-helix to a shallow groove surface of the RNA fold (Figure 19a). Two consecutive amino acids within the α-helix, namely Gly130 and Thr131, are involved in hydrogen bonds to a non-canonical A•U pair in the RNA (Figure 19b). The *cis* Hoogsteen/Watson–Crick A•U is formed by a long-range tertiary interaction between an adenosine and a uridine which ties together the RNA fold.

5. Conclusions

Canonical Watson–Crick pairs in RNA can be considered as the most basic unit for building three-dimensional frameworks essentially made of rather regular helices interrupted at defined positions by unique interaction or recognition motifs promoting RNA–RNA or RNA-protein contacts. Because of their protean diversity, the simplest motifs generating irregularities and asymmetries suitable for specific interactions are non-canonical base pairs. The three-dimensional structures of RNA/peptide and RNA/protein complexes reveal non-Watson–Crick base pairs as key elements of RNA recognition. At the interface between the worlds of RNA and protein molecules, the deviations and distortions from the Watson–Crick geometry do indeed rule the geometry of base pairs involved in specific protein binding.

Further, it is not clear how, in the RNA world, the appropriate functional equilibrium between Watson–Crick and non-Watson–Crick pairs is determined. Divalent ions could be an important factor for the maintenance of the correct distribution of base pairs. While monovalent ions stabilize helices (and thus secondary structure), divalent ions do not influence much the stability of helices but on the contrary stabilize tertiary folding [52]. The recent crystal structures display a wealth of new structural information and insight on divalent ion binding to RNA, especially the whole important magnesium ion [53]. Although magnesium ions bind frequently to (and often link) the anionic phosphate oxygen atoms and the Hoogsteen sites of guanines, the non-Watson–Crick pairs, because of their effects on the groove sizes and on the sugar-phosphate backbone path, mould often ion binding cavities. Thus, the balance between Watson–Crick and non-Watson–Crick pairs could be strongly dependent on the concentration of divalent ions.

References

[1] Leontis, N.B. and Westhof, E. (2001) Geometric nomenclature and classification of RNA base pairs. *Rna*, **7**, 499–512.

[2] Leontis, N.B., Stombaugh, J. and Westhof, E. (2002) The non-Watson–Crick base pairs and their associated isostericity matrices. *Nucleic Acids Res*, **30**, 3497–3531.

[3] Westhof, E. (1992) Westhof's rule. *Nature*, **358**, 459–460.

[4] Cech, T.R., Damberger, S.H. and Gutell, R.R. (1994) Representation of the secondary and tertiary structure of group I introns. *Nat Struct Biol*, **1**, 273–280.

[5] Michel, F., Jacquier, A. and Dujon, B. (1982) Comparison of fungal mitochondrial introns reveals extensive homologies in RNA secondary structure. *Biochimie*, **64**, 867–881.

[6] Ban, N., Nissen, P., Hansen, J., Moore, P.B. and Steitz, T.A. (2000) The complete atomic structure of the large ribosomal subunit at 2.4 A resolution [see comments]. *Science*, **289**, 905–920.

[7] Leontis, N.B., Stombaugh, J. and Westhof, E. (2002) Motif prediction in ribosomal RNAs Lessons and prospects for automated motif prediction in homologous RNA molecules. *Biochimie*, **84**, 961–973.

[8] Leontis, N.B. and Westhof, E. (1998) A common motif organizes the structure of multi-helix loops in 16 S and 23 S ribosomal RNAs. *J Mol Biol*, **283**, 571–583.

[9] Correll, C.C., Wool, I.G. and Munishkin, A. (1999) The two faces of the Escherichia coli 23 S rRNA sarcin/ricin domain: the structure at 1.11 A resolution. *J Mol Biol*, **292**, 275–287.

[10] Cate, J.H., Gooding, A.R., Podell, E., Zhou, K., Golden, B.L., Szewczak, A.A., Kundrot, C.E., Cech, T.R. and Doudna, J.A. (1996) RNA tertiary structure mediation by adenosine platforms. *Science*, **273**, 1696–1699.

[11] Wimberly, B.T., Guymon, R., McCutcheon, J.P., White, S.W. and Ramakrishnan, V. (1999) A detailed view of a ribosomal active site: the structure of the L11-RNA complex. *Cell*, **97**, 491–502.

[12] Masquida, B. and Westhof, E. (2000) On the wobble GoU and related pairs. *Rna*, **6**, 9–15.

[13] Westhof, E. and Michel, F. (1994), *RNA-Proteins interactions: Frontiers in molecular biology*. IRL Press at Oxford University Press, pp. 25–51.

[14] Batey, R.T., Rambo, R.P. and Doudna, J.A. (1999) Tertiary Motifs in RNA Structure and Folding. *Angew Chem Int Ed Engl*, **38**, 2326–2343.

[15] Masquida, B. and Westhof, E. (1999) In Neidle, S. (ed.), *Oxford handbook of nucleic acid structures*. Oxford University Press, Oxford, UK, pp. 533–565.

[16] Jaeger, L., Michel, F. and Westhof, E. (1994) Involvement of a GNRA tetraloop in long-range RNA tertiary interactions. *J Mol Biol*, **236**, 1271–1276.

[17] Cate, J.H., Gooding, A.R., Podell, E., Zhou, K., Golden, B.L., Kundrot, C.E., Cech, T.R. and Doudna, J.A. (1996) Crystal structure of a group I ribozyme domain: principles of RNA packing. *Science*, **273**, 1678–1685.

[18] Westhof, E. and Jaeger, L. (1992) RNA Pseudoknots: Structural and functional aspects. *Curr. Op. Struct. Biol.*, **2**, 327–333.

[19] Michel, F. and Westhof, E. (1990) Modelling of the three-dimensional architecture of group I catalytic introns based on comparative sequence analysis. *J Mol Biol*, **216**, 585–610.

[20] Pley, H.W., Flaherty, K.M. and McKay, D.B. (1994) Model for an RNA tertiary interaction from the structure of an intermolecular complex between a GAAA tetraloop and a nRNA helix. *Nature.*, **372**, 111–113.

[21] Costa, M. and Michel, F. (1995) Frequent use of the same tertiary motif by self-folding RNAs. *Embo J*, **14**, 1276–1285.

[22] Leontis, N.B. and Westhof, E. (2003) Analysis of RNA motifs. *Curr Opin Struct Biol*, **13**, 300–308.

[23] Brion, P. and Westhof, E. (1997) Hierarchy and dynamics of RNA folding. *Annu Rev Biophys Biomol Struct*, **26**, 113–137.

[24] Brion, P., Michel, F., Schroeder, R. and Westhof, E. (1999) Analysis of the cooperative thermal unfolding of the td intron of bacteriophage T4. *Nucleic Acids Res*, **27**, 2494–2502.

[25] Banerjee, A.R., Jaeger, J.A. and Turner, D.H. (1993) Thermal unfolding of a group I ribozyme: the low-temperature transition is primarily disruption of tertiary structure. *Biochemistry*, **32**, 153–163.

[26] Jaeger, L., Westhof, E. and Michel, F. (1993) Monitoring of the cooperative unfolding of the *sunY* group I intron of the Bacteriophage T4. *J. Mol. Biol.*, **234**, 331–346.

[27] Tanner, M. and Cech, T. (1996) Activity and thermostability of the small self-splicing group I intron in the pre-tRNA(Ile) of the purple bacterium Azoarcus. *Rna*, **2**, 74–83.

[28] Cate, J.H. and Doudna, J.A. (1996) Metal-binding sites in the major groove of a large ribozyme domain. *Structure*, **4**, 1221–1229.

[29] Cate, J.H., Hanna, R.L. and Doudna, J.A. (1997) A magnesium ion core at the heart of a ribozyme domain. *Nat Struct Biol*, **4**, 553–558.

[30] Misra, V.K. and Draper, D.E. (1998) On the role of magnesium ions in RNA stability. *Biopolymers*, **48**, 113–135.

[31] Tinoco, I., Jr. and Bustamante, C. (1999) How RNA folds. *J Mol Biol*, **293**, 271–281.

[32] Costa, M., Fontaine, J.M., Loiseaux-de Goer, S. and Michel, F. (1997) A group II self-splicing intron from the brown alga Pylaiella littoralis is active at unusually low magnesium concentrations and forms populations of molecules with a uniform conformation. *J Mol Biol*, **274**, 353–364.

[33] Silverman, S.K., Zheng, M., Wu, M., Tinoco, I., Jr. and Cech, T.R. (1999) Quantifying the energetic interplay of RNA tertiary and secondary structure interactions. *Rna*, **5**, 1665–1674.

[34] Leontis, N.B., Ghosh, P. and Moore, P.B. (1986) Effect of magnesium ion on the structure of the 5S RNA from Escherichia coli. An imino proton magnetic resonance study of the helix I, IV, and V regions of the molecule. *Biochemistry*, **25**, 7386–7392.

[35] Romby, P., Westhof, E., Toukifimpa, R., Mache, R., Ebel, J.P., Ehresmann, C. and Ehresmann, B. (1988) Higher order structure of chloroplastic 5S ribosomal RNA from spinach. *Biochemistry*, **27**, 4721–4730.

[36] Wimberly, B., Varani, G. and Tinoco, I., Jr. (1993) The conformation of loop E of eukaryotic 5S ribosomal RNA. *Biochemistry*, **32**, 1078–1087.

[37] Weeks, K.M. and Crothers, D.M. (1993) Major groove accessibility of RNA. *Science*, **261**, 1574–1577.

[38] Ye, X., Kumar, R.A. and Patel, D.J. (1995) Molecular recognition in the bovine immunodeficiency virus Tat peptide-TAR RNA complex. *Chem Biol*, **2**, 827–840.

[39] Puglisi, J.D., Chen, L., Blanchard, S. and Frankel, A.D. (1995) Solution structure of a bovine immunodeficiency virus Tat-TAR peptide-RNA complex. *Science*, **270**, 1200–1203.

[40] Hermann, T. and Westhof, E. (1999) Non-Watson–Crick base pairs in RNA-protein recognition. *Chem Biol*, **6**, R335–343.

[41] Battiste, J.L., Mao, H., Rao, N.S., Tan, R., Muhandiram, D.R., Kay, L.E., Frankel, A.D. and Williamson, J.R. (1996) Alpha helix-RNA major groove recognition in an HIV-1 rev peptide-RRE RNA complex. *Science*, **273**, 1547–1551.

[42] Ye, X., Gorin, A., Ellington, A.D. and Patel, D.J. (1996) Deep penetration of an alpha-helix into a widened RNA major groove in the HIV-1 rev peptide-RNA aptamer complex. *Nat Struct Biol*, **3**, 1026–1033.

[43] Giver, L., Bartel, D.P., Zapp, M.L., Green, M.R. and Ellington, A.D. (1993) Selection and design of high-affinity RNA ligands for HIV-1 Rev. *Gene*, **137**, 19–24.

[44] Ye, X., Gorin, A., Frederick, R., Hu, W., Majumdar, A., Xu, W., McLendon, G., Ellington, A. and Patel, D.J. (1999) RNA architecture dictates the conformations of a bound peptide. *Chem Biol*, **6**, 657–669.

[45] Legault, P., Li, J., Mogridge, J., Kay, L.E. and Greenblatt, J. (1998) NMR structure of the bacteriophage lambda N peptide/boxB RNA complex: recognition of a GNRA fold by an arginine-rich motif. *Cell*, **93**, 289–299.

[46] Nissen, P., Ippolito, J.A., Ban, N., Moore, P.B. and Steitz, T.A. (2001) RNA tertiary interactions in the large ribosomal subunit: the A-minor motif. *Proc Natl Acad Sci U S A*, **98**, 4899–4903.

[47] Musier-Forsyth, K., Usman, N., Scaringe, S., Doudna, J., Green, R. and Schimmel, P. (1991) Specificity for aminoacylation of an RNA helix: an unpaired, exocyclic amino group in the minor groove. *Science*, **253**, 784–786.

[48] Frugier, M. and Schimmel, P. (1997) Subtle atomic group discrimination in the RNA minor groove. *Proc Natl Acad Sci U S A*, **94**, 11291–11294.

[49] Auffinger, P. and Westhof, E. (1999) Singly and bifurcated hydrogen-bonded base-pairs in tRNA anticodon hairpins and ribozymes. *J Mol Biol*, **292**, 467–483.

[50] Rath, V.L., Silvian, L.F., Beijer, B., Sproat, B.S. and Steitz, T.A. (1998) How glutaminyl-tRNA synthetase selects glutamine. *Structure*, **6**, 439–449.

[51] Conn, G.L., Draper, D.E., Lattman, E.E. and Gittis, A.G. (1999) Crystal structure of a conserved ribosomal protein-RNA complex. *Science*, **284**, 1171–1174.

[52] Serra, M.J., Baird, J.D., Dale, T., Fey, B.L., Retatagos, K. and Westhof, E. (2002) Effects of magnesium ions on the stabilization of RNA oligomers of defined structures. *Rna*, **8**, 307–323.

[53] Klein, D.J., Moore, P.B. and Steitz, T.A. (2004) The contribution of metal ions to the structural stability of the large ribosomal subunit. *Rna*, **10**, 1366–1379.

Course 3

REGULATION OF TRANSCRIPTION BY RNA POLYMERASE II

James T. Kadonaga

Section of Molecular Biology
University of California, San Diego
9500 Gilman Drive
La Jolla, CA 92093-0347, USA

D. Chatenay, S. Cocco, R. Monasson, D. Thieffry and J. Dalibard, eds.
Les Houches, Session LXXXII, 2004
Multiple aspects of DNA and RNA: from Biophysics to Bioinformatics

73

Contents

Contents

1. Introduction

The growth, development, and maintenance of our cells are dependent upon the proper regulation of gene expression. Each of our tens of thousands of genes has its own unique program of activity that controls *where*, *when*, and *how much* each gene is to be expressed. The resulting pattern of gene expression specifies the identity and function of cells. Abnormalities in the regulation of gene expression result in cell death or in misregulated cell growth, such as the uncontrolled proliferation of cells in cancer.

The study of gene regulation is a formidable informational and mechanistic challenge. How are the spatial, temporal, and quantitative expression parameters of each of the tens of thousands of genes controlled? To approach this problem, it is necessary, at the least, to understand the fundamental aspects of the transcription process.

This chapter will focus on the regulation of gene expression in eukaryotes at the level of transcription by RNA polymerase II – that is, the synthesis of mRNA, which is destined to be translated into proteins. There is a nearly overwhelming number of factors that are involved in the regulation of eukaryotic gene expression (Fig. 1). In spite of this complexity, it is possible to outline the general categories of factors and their functions. Thus, I will first describe the cis-acting DNA elements and trans-acting protein factors and coregulators that are involved in RNA polymerase II transcription. In addition, I will provide an overview of the role of chromatin and chromatin-modifying factors in the regulation of gene activity.

For background reading, there are many review articles that describe different aspects of transcriptional regulation and chromatin dynamics. A sampling of review articles and treatises are as follows: eukaryotic transcription (White 2001; Levine and Tjian 2003); RNA polymerase II core promoter (Smale and Kadonaga, 2003); basal/general transcription factors for RNA polymerase II (Orphanides *et al.* 1996); sequence-specific DNA-binding regulatory factors (Kadonaga 2004); transcriptional coactivators (Taatjes *et al.* 2004); transcriptional elongation (Sims *et al.* 2004); insulator/boundary elements (West *et al.* 2002); enhancers (Blackwood and Kadonaga 1998; Bulger and Groudine 1999); chromatin (Wolffe 1998; Kornberg and Lorch 1999); chromatin remodeling factors (Becker and Hörz 2002; Lusser and Kadonaga 2003); histone modifying

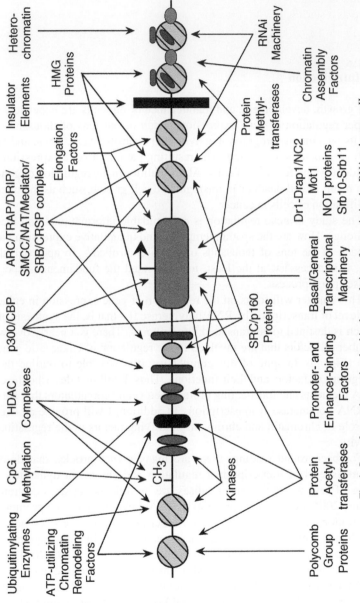

Fig. 1. Some factors that are involved in the regulation of transcription by RNA polymerase II.

enzymes (Strahl and Allis 2000; Berger 2002); chromatin assembly (Haushalter and Kadonaga 2003; Lusser and Kadonaga 2004). These articles would also serve as a starting point for a more thorough analysis of the literature.

2. DNA regulatory elements

A variety of cis-acting DNA elements are involved in the control of gene activity (Fig. 2). The cis-acting sequences in the DNA template contain the genetic component of the information that directs the transcription of each gene. It has been a key challenge to decipher the gene expression code that is embedded in the primary DNA sequence information. Some of our knowledge of this code is as follows.

First, the core promoter, which is typically about 40 bp in length, encompasses and specifies the transcription start site. There are different types of core promoters, which vary with respect to the presence or the absence of specific sequence motifs. These core promoter motifs include the TATA box, initiator (Inr), TFIIB recognition element (BRE), motif ten element (MTE), and downstream core promoter element (DPE) (for review, see: Smale and Kadonaga 2003) (Fig. 3). The key factor that mediates the recognition of the core promoter motifs is termed

DNA Regulatory Elements (cis elements) for Transcription of Protein-coding Genes by RNA Polymerase II

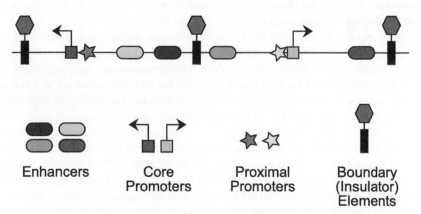

| Enhancers | Core Promoters | Proximal Promoters | Boundary (Insulator) Elements |

Fig. 2. Cis-acting DNA regulatory elements for transcription by RNA polymerase II.

Core Promoter Elements

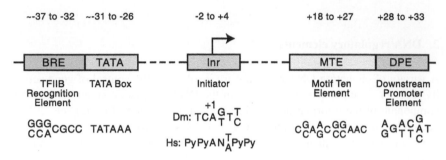

Fig. 3. Core promoter motifs.

transcription factor IID (TFIID; discussed later). It is also important to note that some transcriptional enhancers function with DPE-dependent core promoters but not TATA-dependent core promoters, and vice versa. Thus, the core promoter is not only important for the basic transcription process, but it is also a regulatory element.

The region that is immediately upstream of the core promoter, from about −30 to −250 relative to the transcription start site, is typically termed the proximal promoter. The proximal promoter is usually bound by sequence-specific DNA-binding transcription factors. Some particular sequence-specific transcription factors, such as Sp1 and CBF/NF-Y, are commonly found in the proximal promoter. In some instances, the proximal promoter appears to function as a conduit between distant enhancer elements and the core promoter.

Transcriptional enhancers are located at variable distances, as far as 80 kbp or longer, from the RNA start site (for reviews, see: Blackwood and Kadonaga 1998; Bulger and Groudine 1999). Moreover, enhancers act from upstream as well as downstream of the start site. Enhancers are typically about 0.5 to 1 kbp in length, and they contain clusters of recognition sites for the binding of sequence-specific transcription factors. Although enhancers were discovered over 20 years ago, their mechanism of action remains a mystery. One model posits looping of the DNA between the enhancer and proximal and/or core promoter. A second model postulates that there is scanning of DNA-bound factors between the enhancer and the proximal/core promoter. Then, a third model suggests that a short protein-DNA loop is formed at the enhancer that subsequently scans to the proximal/core promoter. Most genes are regulated by their enhancers, and thus, it is important to understand how enhancers work. In addition, there are repressive

DNA elements, which are sometimes termed silencers, which function to inhibit transcription. Like enhancers, silencers contain recognition sites for sequence-specific DNA-binding factors.

Why do enhancers function from such large distances both upstream and downstream of the transcription start site? One possibility is that this property enables multiple enhancers with specific functions to regulate transcription from a single core promoter. For example, if a gene needs to be expressed in the liver and the brain, this goal can be achieved by the combination of a brain-specific enhancer and a liver-specific enhancer. Moreover, the location of these enhancers relative to the core promoter need not be precise, because there is considerable flexibility in the distance from which enhancers can activate transcription from the core promoter.

Another class of cis-acting DNA elements are insulator (also known as boundary) elements (for a review, see: West *et al.* 2002). As suggested by their name, transcriptional insulators function to demarcate functional regions of the genome. In the strictest sense, a true insulator possesses neither positive nor negative transcriptional activity, but rather functions as a neutral boundary across which neither transcriptional activators nor transcriptional repressors (including repressive heterochromatin effects) can exert their influence. Like other transcriptional elements, insulators contain recognition sites for sequence-specific DNA-binding proteins. It is not yet known how transcriptional insulators function. One model postulates that insulators are sites of attachment to a 'fixed' moiety, such as the nuclear matrix, and that the DNA (or perhaps, more accurately, chromatin) in each loop constitutes a genetic domain. We are still at a relatively early stage in our understanding of insulators, however. It is possible, for instance, that many 'insulators' are not purely transcriptionally neutral.

3. Basal/general transcription factors

RNA polymerase II is a multisubunit enzyme that is able to synthesize RNA from a DNA template, but by itself, the polymerase is not able to initiate transcription accurately from a core promoter. To perform this task, the polymerase requires additional auxiliary factors that are commonly known as the 'general' or 'basal' factors (for a review, see: Orphanides *et al.* 1996). The basal/general factors that mediate transcription initiation from TATA-dependent core promoters are designated as TFIIA, TFIIB, TFIID, TFIIE, TFIIF, and TFIIH ('TFII' = Transcription Factor for RNA polymerase II).

A short description of the function of each of the basal/general factors is given in Fig. 4, in which the factors are listed roughly in the order in which they are thought to be assembled into a complex (termed the preinitiation complex) prior

Basal ('General') Transcription Factors for RNA Polymerase II

- TFIID – consists of TBP (TATA-box binding protein) + TAFs (TBP-associated factors). Binds to core promoter motifs. TAFs interact with activator proteins. The first step in basal transcription is probably binding of TFIID to the core promoter.

- TFIIA – three (or two) small subunits. Increases affinity of TBP for DNA in vitro. Not needed for transcription in vitro. Could be an anti-inhibitor.

- TFIIB – one subunit of 35 kDa. Binds to TBP and the BRE.

- RNA Polymerase II – consists of two large subunits (IIa and IIb) as well as about ten smaller subunits. Unique feature of largest (IIa) subunit is the C-terminal domain (CTD), which is an imperfectly-repeated heptapeptide motif, YSPTSPS.

- TFIIF – also known as RAP30/74. Binds to RNA polymerase II. Two subunits of 30 and 74 kDa. Functions in transcription initiation and elongation.

- TFIIE – two polypeptides of 34 and 56 kDa. Required for assembly of TFIIH into the transcription preinitiation complex (PIC).

- TFIIH – ten polypeptides. Core TFIIH has seven subunits, which include 5'–>3' and 3'–>5' DNA helicases, and is also involved in nucleotide excision repair. Also has a three subunit Cdk7/MO15 + Cyclin H + MAT1 kinase complex that phosphorylates Ser5 of the CTD during transcription initiation.

Fig. 4. The basal/general transcription factors for TATA-dependent promoters.

to initiation of transcription. [Alternatively, it has also been postulated that these factors exist as a large supercomplex (termed the RNA polymerase holoenzyme) in which most of the basal factors associate with the promoter in a single step.] One notable factor is TFIID, which is a multisubunit protein that consists of TBP (TATA box binding protein) and about 11 or so TAFs (TBP-associated factors). TFIID is the key factor that is responsible for sequence-specific recognition of the core promoter. The TBP and TAF subunits of TFIID make distinct contacts with the TATA, Inr, MTE, and DPE motifs. In some promoters, TFIIB interacts with the BRE element that is located immediately upstream of a subset of TATA box motifs. Another interesting factor is TFIIH, which contains DNA helicase activity that mediates the unwinding of the DNA to allow entry of the polymerase. TFIIH functions in both transcription and DNA repair (nucleotide excision repair). Moreover, mutations in TFIIH subunits in humans are responsible for diseases such as xeroderma pigmentosum, Cockayne syndrome, and trichothiodystrophy. In addition to the initiation of transcription, it is important to consider transcription elongation (for a review, see: Sims *et al.* 2004) and termination as well as the posttranscriptional processing of the RNA, such as pre-mRNA splicing.

4. Sequence-specific DNA-binding factors

The recognition of specific sequence motifs in promoters, enhancers, silencers, and insulators is mediated by thousands of different sequence-specific DNA-binding factors. These sequence-specific factors are thus the key factors that are responsible for the interpretation of the information encoded in the primary DNA sequence. There are probably several thousand genes that encode sequence-specific transcription factors. Moreover, due to alternative pre-mRNA splicing (which results in multiple mRNA species from a single gene), there may be several tens of thousands (or more!) different sequence-specific DNA-binding proteins that are involved in the regulation of transcription. Basically, it appears that the regulation of gene expression is a complex process that requires (perhaps somewhat inelegantly) many different regulatory factors.

Some key properties of the sequence-specific factors are listed in Fig. 5. One notable feature of these factors is that they are modular in structure. That is, they contain DNA binding modules, transcriptional activation/repression modules, multimerization modules, and regulatory modules. There are multiple versions of each type of module. For example, some DNA-binding modules include

Sequence-specific DNA-binding Transcription Factors (RNA Pol II)

- Modular Structure
  - Sequence-specific DNA-binding Modules
  - Transcriptional Activation/Repression Modules
  - Regulatory Modules (inter- or intramolecular)
  - Multimerization Modules (homo- and heterotypic interactions)
- Regulate Transcription via Recruitment of Coactivators and Corepressors
- Chromatin Is an Integral Component in the Function of Sequence-specific Factors
- Sequence-specific Factors Can Be Regulated by Post-translational Modifications
- Sequence-specific Factors Are Often Members of Multiprotein Families
- Recognition Sites for Sequence-specific Factors Tend to Be Located in Clusters
- Sequence-specific Factors Typically Bind to DNA with Relatively Low Specificity
- Sequence-specific Factors Can Affect Transcription Initiation and/or Elongation
- A Subset of Factors Are Commonly Found in Proximal Promoter Regions
- Sequence-specific Factors Bind to Boundary/Insulator Elements
- Some Sequence-specific Factors Can Bend DNA

Fig. 5. Properties of sequence-specific DNA-binding transcription factors.

Nuclear Receptors Are an Interesting Family of Sequence-specific DNA-binding Transcription Factors

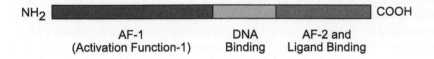

NH$_2$ COOH

AF-1   DNA   AF-2 and
(Activation Function-1)   Binding   Ligand Binding

- Sequence-specific DNA-binding proteins

- Upon binding of their cognate ligands (agonists), they activate transcription.

- Thus, nuclear receptors function as both the receptor for the signals (agonists) as well as sequence-specific DNA-binding transcriptional activators.

- Inactivated by antagonists, which are ligands that resemble the agonists, but block activation functions.

- Examples include estrogen receptor, androgen receptor, glucocorticoid receptor, vitamin D receptor, thyroid hormone receptor.

Fig. 6. Nuclear receptors.

zinc fingers, homeodomains, HMG domains, basic leucine zippers, and so on. The structure and properties of sequence-specific factors are nicely exemplified by the nuclear receptors, which are a superfamily of DNA-binding transcription factors that additionally function as receptors for a variety of molecules that range from estrogens and testosterones to vitamin D, retinoic acid, and thyroid hormone. Nuclear receptors generally share a common structure that is depicted in Fig. 6, where it can be seen that they possess DNA-binding, activation, and regulatory motifs. Nuclear receptors also form homodimers and/or heterodimers with other nuclear receptor family members. For a more extensive discussion of the properties of these factors, see Kadonaga (2004).

How do the sequence-specific factors work? It appears that there are direct and indirect means of communication between the sequence-specific factors and the basal/general factors that carry out the transcription process. There is evidence for direct interactions between activators and the TAF subunits of TFIID. Then, there are a variety of coactivator proteins, such as Mediator (also known as SRB complex, TRAP, DRIP, SMCC, NAT, ARC, CRSP), p300/CBP, and SRC/p160, that appear to function as a physical bridge between the DNA-binding factors and the basal/general machinery. A simple first-order model involving direct macromolecular interactions is thus as follows. First, sequence-specific factors recognize and bind to enhancers and proximal promoters, and are converted, when necessary, to an activated form. [Note that sequence-specific transcription factors can be bound to DNA in an inactive form that can be converted to an active form. For example, a nuclear receptor bound to DNA in the absence of a ligand is not active (and is sometimes repressive), and then becomes activated upon binding of an agonist.] Second, the sequence-specific factors recruit coactivators (and TFIID, which is a special DNA-binding basal/general factor that interacts with sequence-specific activators) via direct protein-protein interactions. Third, the coactivators recruit the basal/general transcriptional machinery to the core promoter via direct protein-protein interactions. Thus, the basic activation process appears to involve a series of protein-DNA and protein-protein interactions. Moreover, as discussed in the next section, chromatin is also an important component with which coregulators indirectly affect the activation process.

5. Chromatin and transcription

Chromatin is an integral component of transcriptional regulation (Wolffe 1998; Kornberg and Lorch 1999). The fundamental repeating unit of chromatin is the nucleosome, which consists of a core histone octamer (which contains two copies each of the core histones H2A, H2B, H3, and H4), one molecule of linker histone (such as histone H1), and about 180 bp to 200 bp of DNA. It is well-recognized that a primary function of chromatin is the packaging of DNA. In humans, for instance, approximately two meters of DNA is packaged into a nucleus that has an average diameter of about 10 microns. It is also important to consider, however, that chromatin has existed for several hundred million years, and thus, DNA-utilizing processes in the nucleus have evolved to function optimally in chromatin rather than with plain ('naked') DNA. Moreover, it is interesting to note that histones are highly conserved in eukaryotes and are even present in archaebacteria (but not in eubacteria, such as *Escherichia coli*).

Chromatin represses the basal transcription process. It appears that chromatin is not a general repressor of all transcription factors, but rather a selective repressor of the basal/general transcriptional machinery. In this manner, a gene in the absence of activators would be inactive. Then, the sequence-specific factors – many of which have been found to bind efficiently to chromatin – function to counteract the chromatin-mediated repression. The sequence-specific factors appear to achieve this task by several means. First, the protein contacts that

J.T. Kadonaga

Chromatin Remodeling Factors

• Pure nucleosomes are effectively immobile on a biological time scale under 'physiological' conditions.

• The mobilization and/or disruption of nucleosomes is catalyzed by ATP-utilizing proteins termed 'chromatin remodeling factors'.

• All known chromatin remodeling factors possess a subunit that is a member of the SNF2-like family of DNA-stimulated ATPases.

• Many but not all chromatin remodeling factors are multiprotein complexes.

• Chromatin remodeling factors appear to function by different mechanisms. Some factors disrupt nucleosomes, whereas other factors translocate nucleosomes along the DNA.

Fig. 7. Chromatin remodeling factors.

form between sequence-specific factors, the coactivators, and the basal transcription factors probably facilitate the activation process in the context of chromatin. Second, the sequence-specific factors recruit ATP-dependent chromatin remodeling enzymes to disrupt or to mobilize nucleosomes (for reviews, see: Becker and Hörz 2002; Lusser and Kadonaga 2003) (Fig. 7). Third, there are a variety of enzymes that covalently modify histones (Strahl and Allis 2000; Berger 2002) (Fig. 8). This modification of histones is thought to create signals that are recognized by other factors that affect gene activity – this concept is known as the histone code hypothesis (Strahl and Allis 2000). One example of a histone modification acting as a signal or code is the binding of heterochromatin protein-1 (HP-1) to histone H3 that is methylated at lysine 9, but not to histone H3 that is not methylated at lysine 9. It is additionally possible that some histone modifications, such as acetylation, alter the physical properties of chromatin, such as the affinity of the histones to DNA, and thus affect the extent to which chromatin represses transcription.

Hence, the model for the activation of gene transcription should include not only the binding of sequence-specific factors to DNA and the recruitment of coactivators and basal/general factors, but also the remodeling of nucleosomes, the covalent modification of histones, and the factors that are affected by the specific histone modifications.

Core Histones Are Subjected to a Variety of Covalent Modifications

- Acetylation of Lysine residues

- Methylation of Lysine and Arginine residues

- Phosphorylation of Serine residues

- Ubiquitylation of Lysine residues

- Poly(ADP)ribosylation at Glutamate (Aspartate) residues

General Trends

- Histone acetylation generally correlates with gene activation.

- Methylation of histone H3 Lysine 4 correlates with activation.

- Methylation of histone H3 Lysine 9 correlates with repression.

Fig. 8. Covalent modification of histones.

6. Conclusions and speculations

The control of gene transcription in eukaryotes is a complex yet solvable process. In theory, there should be a gene transcription code in the primary DNA sequence that provides much, but certainly not all, of the information regarding when, where, and to what extent each gene will be activated. The sequence-specific DNA-binding factors 'read' the specific DNA motifs in core promoters, proximal promoters, enhancers, silencers, and insulators. In a sense, the DNA sequence represents the 'genetic' component of the regulatory code. There is also, however, an epigenetic component to the control of gene expression that is not strictly dependent on the precise primary DNA sequence. Such phenomena include CpG methylation of DNA, which generally correlates with transcriptional repression such as that seen in the inactive X chromosome. The CpG methylation state is maintained through many cell divisions and is thus passed from parental cells to daughter cells. It is possible that some forms of histone modifications may be similarly transmitted through multiple cell divisions. Also, another form of transcriptional control involves small RNA molecules (for review, see: Meister and Tuschl 2004). Thus, we still have much to learn about the factors and mechanisms that control gene transcription (and even more to learn about all of the other stages, such as RNA processing, translation, and posttranslational processing, at which gene expression is regulated).

For the immediate future, it seems that significant effort should be devoted toward understanding the basic mechanisms of gene expression. As our knowledge of these processes is expanded and refined, computational approaches (performed within the framework of knowledge) involving large sets of data generated through bioinformatics will continue to reveal key insights into gene expression codes in different organisms.

It is reasonable to ask – is every gene regulated by an entirely unique mechanism, or are there common themes and mechanisms that apply to larger groups of genes? Or, more generally, it can be asked how many generally applicable themes and concepts can be found in biology relative to those in physics or chemistry? In my own very naïve and limited view, general concepts, theories, and laws seem to be dictated by energy, and in biology, the energetic efficiency of processes is of secondary importance relative to the maintenance of life. In other words, the cell tolerates many inefficient processes because it cannot go into a dormant state, re-engineer the basic underpinnings of the inefficient processes, and then come back as a new, improved cell. The consequence of this limitation is that the cell has many processes that are not strictly dictated by energetics, and thus, the 'simplest' and most efficient (and probably generalizable) mechanisms are not necessarily employed. Nevertheless, in spite of the potential absence of a logical design for a gene expression code, we have, as outlined in this chapter, been able to establish a reasonable foundation of knowledge of the factors and mechanisms that control transcription by RNA polymerase II.

In conclusion, the mechanisms of gene expression are complex and perhaps not optimally efficient, but the process is of immense biological importance. In the future, we will need to gain a better understanding of the energetics of these processes as well as to devise quantitative theories and models that fit the large amounts of experimental data that are being generated. In these and other ways, the field of gene expression would be significantly strengthened by the contributions of physicists and mathematicians.

Acknowledgments

I am grateful to Timur Yusufzai, Tammy Juven-Gershon, Chin Yan Lim, and Alexandra Lusser for critical reading of the manuscript. Research in my laboratory is supported by grants from the National Institutes of Health (GM41249, GM58272).

References

[1]   P.B. Becker and W. Hörz, *Annu. Rev. Biochem.* **71** (2002) 247–273.

[2]   S.L. Berger, *Curr. Opin. Genet. Dev.* **12** (2002) 142–148.

[3]   E.M. Blackwood and J.T. Kadonaga, *Science* **281** (1998) 60–63.

[4]   M. Bulger and M. Groudine, *Genes Dev.* **13** (1999) 2465–2477.

[5]   K.A. Haushalter and J.T. Kadonaga, *Nature Rev. Mol. Cell Biol.* **4** (2003) 613–620.

[6]   J.T. Kadonaga, *Cell* **116** (2004) 247–257.

[7]   R.D. Kornberg and Y. Lorch, *Cell* **98** (1999) 285–294.

[8]   M. Levine and R. Tjian, *Nature* **424** (2003) 147–151.

[9]   A. Lusser and J.T. Kadonaga, *BioEssays* **25** (2003) 1192–1200.

[10] A. Lusser and J.T. Kadonaga, *Nature Methods* **1** (2004) 19–26.

[11] G. Meister and T. Tuschl, *Nature* **431** (2004) 343–349.

[12] G. Orphanides, T. Lagrange and D. Reinberg, *Genes Dev.* **10** (1996) 2657–2683.

[13] R.J. Sims, III, R. Belotserkovskaya and D. Reinberg, *Genes Dev.* **18** (2004) 2437–2468.

[14] S.T. Smale and J.T. Kadonaga, *Annu. Rev. Biochem.* **72** (2003) 449–479.

[15] B.D. Strahl and C.D. Allis, *Nature* **403** (2000) 41–45.

[16] D.J. Taatjes, M.T. Marr and R. Tjian, *Nature Rev. Mol. Cell Biol.* **5** (2004) 403–410.

[17] A.G. West, M. Gaszner and G. Felsenfeld, *Genes Dev.* **16**, (2002) 271–288.

[18] R.J. White, *Gene transcription: mechanisms and control* (Blackwell Science, Ltd., Oxford, UK, 2001).

[19] A.P. Wolffe, *Chromatin: structure and function*, 3rd Ed. (Academic Press, San Diego, CA, 1998).

Course 4

BASIC CONCEPTS OF STATISTICAL PHYSICS OF POLYMERS

A.R. Khokhlov and E.Yu. Kramarenko

Physics Department, Moscow State University, Moscow, 119992, Russia

D. Chatenay, S. Cocco, R. Monasson, D. Thieffry and J. Dalibard, eds.
Les Houches, Session LXXXII, 2004
Multiple aspects of DNA and RNA: from Biophysics to Bioinformatics

Contents

1. Introduction to polymer physics

1.1. Fundamentals of physical viewpoint in polymer science

Polymer chains of different chemical structure have, of course, different properties. However, there are many common properties characteristic for large classes of polymer systems. For example, all rubbers (cross-linked polymer networks, see below) exhibit the property of *high elasticity*, all polymer melts are *viscoelastic*, all polyelectrolyte gels absorb a large amount of water, etc. Such properties can be described on a molecular level taking into account only general polymer nature of constituent molecules, rather than the details of their chemical structure. It it these properties that are studied by polymer physics.

What are the main factors governing the general physical behavior of polymer systems? *Three* of them should be mentioned on the first place.

First of all, polymers are long molecular chains. In Fig. 1 three most common polymer chains with carbone backbone are shown. One can see that small atomic groups (monomer units) are connected in linear chains by covalent chemical bonds. *Chain structure* of constituent molecules is the first most fundamental feature of polymer systems. In particular, this means that monomer units do not have the freedom of independent translational motion, and therefore polymers do not possess the entropy associated with this motion (the so-called translational entropy). This is sometimes expressed as follows: polymer systems are *poor in entropy*.

Second, number of monomer units in the chain N, is large: $N \gg 1$ (otherwise we have oligomer, not polymer). For macromolecules synthesized in the

$$-CH_2-CH_2-CH_2-CH_2- \quad \text{poly(ethylene)}$$

$$-CH_2-CH-CH_2-CH- \quad \text{poly(styrene)}$$

$$-CH_2-CH-CH_2-CH- \quad \text{poly(vinyl chloride)}$$
$$ClCl$$

Fig. 1. Common polymer chains.

95

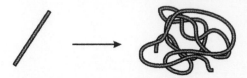

Fig. 2. Polymer chains are generally flexible, they normally take the conformation of the coil, not of the rigid rod.

chemical laboratory normally $N = 10^2 \div 10^4$. For biological macromolecules the values of N can be much larger, for example, the longest polymer chains are those of DNA molecules: $N \sim 10^9 \div 10^{10}$. Such large objects can be seen by normal optical microscope (if DNA is labeled with fluorescence dyes), since the linear size of DNA coil turn out to be larger than the wavelength of light.

Third, polymer chains are generally *flexible* (see Fig. 2), they normally take the conformation of the entangled coil, rather than that of a rigid rod. We will discuss in detail the notion of polymer chain flexibility in section 1.2.

Chain structure, large number of monomer *units* in the chain and chain *flexibility* are three main factors responsible for specific properties of polymer systems.

1.2. Flexibility of a polymer chain. Flexibility mechanisms

Let us outline main flexibility mechanisms.

1.2.1. Rotational-isomeric flexibility mechanism

Let us consider a simplest polyethylene chain (Fig. 1) and let us ask ourselves for which conformation do we have the *absolute energetic minimum?*

Such conformation corresponds to a straight line and is shown schematically in Fig. 3. For this conformation all the monomer units are in the so-called *trans-position*. This would be an equilibrium conformation at $T = 0$.

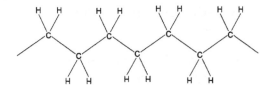

Fig. 3. The rectilinear conformation of polyethylene chain.

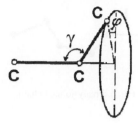

Fig. 4. Definition of the valency angle γ and angle of internal rotation φ for carbon backbone.

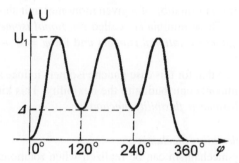

Fig. 5. The typical dependence of energy vs. the internal rotation angle φ.

At $T \neq 0$ due to the thermal motion the deviations from the minimum-energy conformation are possible. According to the *Boltzmann law* the probability of realization of the conformation with the excess energy U over the minimum-energy conformation is

$$P(U) \sim \exp\left(-\frac{U}{kT}\right) \tag{1.1}$$

What are the possible conformational deviations from the structure shown in Fig. 3? For carbon backbone the *valency angle* γ (see Fig. 4) should be normally considered as fixed (for different chains $50° < \gamma < 80°$). However the rotation with fixed γ by changing the angle of internal rotation φ (see Fig. 4), is possible. Any value $\varphi \neq 0$ gives rise to the deviations from the rectilinear conformation, i.e. to the chain flexibility.

The typical dependence of the energy of internal rotation on φ is shown in figure 5.

There are several minima separated by energetic barriers. The height of the barriers U_1 is of order of 3 Kcal/mol, which is much higher than kT, while the

Fig. 6. Freely-jointed chain.

value Δ of the energetic difference between the minima (see Fig. 5) is normally less then 1 Kcal/mol, i.e. of order kT. Therefore, taking into account eq. 1.1, one can say that the conformation of a given monomer unit should correspond to one of the minima. These minima are called *rotational isomers* (for $\varphi = 120°$ and $\varphi = 240°$ — *gauche rotational isomers* and for $\varphi = 0°$ — *trans rotational isomers*).

It is easy to realize that for this case gauche isomers induce sharp bends of the chain and give dominant contribution to the flexibility. This kind of flexibility is called a *rotational-isomeric flexibility mechanism*.

1.2.2. Persistent flexibility mechanism

Another flexibility mechanism can be realized when rotational isomers are not allowed, e.g. in α-helical polypeptides or DNA double helix. The conformations of these chains are stabilized by hydrogen bonds and internal rotation is impossible. In this case *small thermal vibrations* around the equilibrium conformation play the most important role. Via the accumulation over large distances along the chain, these vibrations give rize to the deviations from the rectilinear conformation, i.e. to the chain flexibility. This is a *persistent flexibility mechanism*, it is analogous to the flexibility of a homogeneous elastic filament.

1.2.3. Freely-jointed flexibility mechanism

Another mechanism of flexibility is realized in the so-called *freely-jointed model* of a polymer chain. In this model the flexibility is located in the freely-rotating junction points. This mechanism is normally not characteristic for real chains, but it is frequently used for model theoretical calculations (see Fig. 6).

1.3. Types of polymer molecules

Polymer chains shown in Fig. 1 are the simplest ones. In this section we will describe some of the typical more complicated structures.

Frist of all, the polymer shown in Fig. 1 are *homopolymers:* all their monomer units are identical. The opposite case is that of *copolymers:* copolymer chain contains monomer units of different types. Most important example of copolymers

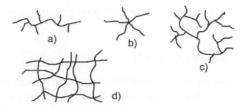

Fig. 7. Types of branched polymer molecules.

are main biological macromolecules: proteins are polymer chains composed of 20 types of monomer units (corresponding to 20 types of aminoacid residues), DNA chains contain 4 types of monomer units (corresponding to 4 types of nucleotides). Sequence of monomer units in a copolymer chain is called the *primary structure* of the chain.

One can say that primary structure of a protein corresponds to some text in 20-letter alphabet. These texts were selected in the course of *molecular evolution*, they contain the biologically important *information* (for example, all the genetic information is written in the 4-letter text of DNA primary structure). On the contrary, copolymers which are synthesized in the chemical laboratory normally correspond to *random* sequences of monomer units, or sequences with some simple short-range correlations, so the primary structure of such copolymers does not contain any valuable information.

Second, linear polymer chains shown in Fig. 1 do not contain *branching points*. On the other hand, branching points can easily emerge in the course of polymer synthesis in the presence of multifunctional units (having more than two free valencies) leading to *branched macromolecules*. Typical branched macromolecules are shown in Fig. 7. If the process of joining of multifunctional units to the growing chain is not specifically controlled, one normally ends up with *randomly* branched macromolecules (Fig. 7c), however there are methods to obtain more sophisticated *comb-like* (Fig. 7a) and *star-like* (Fig. 7b) macromolecules. If the randomly branched chain grows further, a *polymer network* of the type shown in Fig. 7d finally appears. This is a giant molecule of macroscopic size (ca. 1 cm) where all the monomer units are connected in the molecular network by covalent chemical bonds.

Finally, it is worth-while to mention here *ring macromolecules* which can be obtained by a chemical reaction between the chain ends with the formation of a covalent bond. The importance of ring macromolecules can be illustrated by the fact that in many cases the native DNA macromolecules in living cells correspond to closed ring form. For ring macromolecules very important are *topological restrictions*: polymer chains can not "go through" each other without the chain disintegration. Therefore, the type of *knot* which ring macromolecule had at the moment of its formation is conserved forever.

1.4. Physical states of polymer materials

Suppose now that we have a polymer material composed of polymer chains (one polymer component only). What are the possible *physical states* for such a material? According to the traditional classification of physical states of low-molecular substances, they can exist in one of the three states, namely that of a *gas, liquid,* or *crystalline solid*.

However, for polymers such classification is not informative. Indeed, it is very difficult to convert polymers to a gaseous state (this would correspond to very exotic conditions of extremely high vacuum and absence of gravity).

Perfect crystals are very rarely formed with polymers because of the reason illustrated in Fig. 8. Suppose that we have an entangled polymer melt at high temperatures (Fig. 8b) which can in principle crystallize into perfect crystal (shown in Fig. 8c). Then, upon cooling we will have random formation of the nuclei of crystalline phase (Fig. 8a) which will effectively "freeze" all the macromolecules entering in at least one crystalline nucleus. Therefore, the large-scale rearrangements of polymer chains resulting in the perfect structure of Fig. 8c become impossible, and the system stays in the partially crystalline state where crystalline nuclei are separated by amorphous layers.

Thus, polymer materials are generally *liquids* (in the sense that they are neither gases nor perfect crystals). Because of this, the following classification of physical states of these liquids is used instead of traditional one. Pure polymer materials can exist in one of the four states.

1. *Partially crystalline* state. This state is shown in Fig. 8a, and it was described above.

2. *Viscoelastic* state (or *polymer melt*). This state appears at high enough temperatures when entangled polymer chains can move with respect to each other;

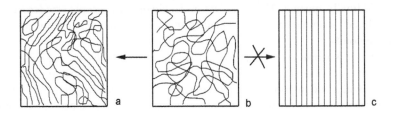

Fig. 8. Physical states of polymer materials.

it is schematically shown in Fig. 8b. One of the main specific polymer properties in this state is that of *viscoelasticity*: for rapid enough external actions such melt behaves as elastic solid, while for slow actions it flows, i.e. behaves as viscous fluid.

3. *Highly elastic* state. If in the polymer melt (Fig. 8b) polymer chains are cross-linked and form a network (Fig. 7d), the large scale motion of one chain with respect to each other is forbidden, while at smaller scales the mutual motion is still free. This state of polymer material is called a highly elastic one, because of the property of *high elasticity* characteristic for this state (i.e. ability to undergo large reversible deformations at moderate applied stresses). In the ordinary life highly elastic polymers are called *rubbers*.

4. *Glassy* state. If upon cooling crystallization is impossible (e.g. we have a random copolymer or an atactic polymer), polymer material becomes glassy. Most of the *plastics*, in particular, organic glasses of poly(styrene) or poly(methyl methacrylate) correspond to this state. Glassy state is very characteristic for polymer materials at low temperatures, because each partial crystallization normally requires special conditions (such as chain stereoregularity).

1.5. Polymer solutions

In the previous section we described the physical states of one-component polymer systems. Multicomponent polymer systems are also of great practical and theoretical interest. Among these systems the special role belongs to *polymer solutions*, i.e. two-component systems consisting of polymer chains mixed with low-molecular solvent.

Fig. 9 illustrates possible regimes of behaviour of polymer solutions, depending on the polymer concentration. In *dilute solutions* (Fig. 9a) polymer coils are separated from each other, and by studying the solution properties in this regime we can obtain information about properties of *isolated macromolecular coils* in solution. In a sense, this regime is equivalent to the ideal gas regime for low-molecular substances.

When the polymer concentration increases and the spheres surrounding macromolecular coils start to overlap (Fig. 9b), the solution undergoes *crossover* from dilute to semidilute regime. Upon further increase of polymer concentration we convert to a regime of *semidilute solution* (Fig. 9c) where the chains are already entangled with each other, but the volume fraction occupied by polymer in the solution is still small. The existence of this regime is a specific polymer property (not available for low-molecular solutions) connected with the fact that *volume fraction* occupied by a polymer chain inside a coil is *very small*.

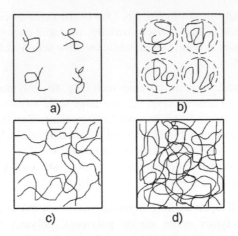

Fig. 9. Concentration regimes of polymer solutions.

When polymer concentration increases further, and polymer volume fraction in the solution reaches 20–30%. The solution should be called *concentrated*, rather than semidilute (Fig. 9d). Finally, in the absence of solvent (polymer volume fraction equal to unity) we return to a one-component polymer system which can exist in one of the four states outlined in the previous section.

2. Single ideal polymer chain

2.1. Definition of ideal polymer chain

From the consideration of the previous section it is clear that systematic description of physical properties of polymers should start with the study of conformations of a *single polymer chain* which can be observed in *dilute solution* (Fig. 9a). In addition to this, in this chapter we study the most fundamental case of single *ideal* polymer chain

By definition, in the *ideal polymer chain* we take into account only the interactions of close neigbours along the chain. The interactions of monomer units which are far from each other along the chain are neglected. Polymer chains behave as ideal ones in the so-called θ-conditions (see below).

2.2. Size of ideal freely-jointed chain. Entangled coil

Let us consider ideal N-segment freely-jointed chain with each segment of length l (see Fig. 10).

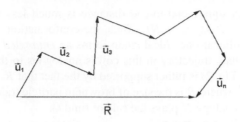

Fig. 10. The model of freely-jointed chain.

The size of such chain can be characterized by end-to-end vector $\vec{R}$ (see Fig. 10). However, this vector will change rapidly in the course of the thermal motion. Important characteristic would be the average size R of a polymer coil. This average can not be characterized by $\langle \vec{R} \rangle$ since all the segments orientations are equal, therefore $\langle \vec{R} \rangle = 0$. That is why the size of coil is usually characterized by mean-square end-to-end distance $R \sim \sqrt{\langle R^2 \rangle}$. Let us calculate this value for our model. The end-to-end vector is the sum of the segments vectors (see Fig. 10):

$$\vec{R} = \sum_{i=1}^{N} \vec{u}_i \tag{2.1}$$

Thus the square of the end-to-end distance is:

$$R^2 = \left(\sum_{i=1}^{N} \vec{u}_i \right) \left(\sum_{j=1}^{N} \vec{u}_j \right) = \sum_{i=1}^{N} \sum_{j=1}^{N} \vec{u}_i \vec{u}_j, \tag{2.2}$$

and the average of this value

$$\langle R^2 \rangle = \sum_{i=1}^{N} \sum_{j=1}^{N} \langle \vec{u}_i \vec{u}_j \rangle = \sum_{i=1}^{N} \langle \vec{u}_i^2 \rangle + \sum_{i=1}^{N} \sum_{j=1, j \neq i}^{N} \langle \vec{u}_i \vec{u}_j \rangle \tag{2.3}$$

In the last equality in eq. 2.3 we have separated the terms with $i = j$ from all the other terms. Taking into account that $\langle \vec{u}_i^2 \rangle = l^2$ and $\langle \vec{u}_i \vec{u}_j \rangle_{i \neq j} = 0$ (because the orientations of different chain segments in the freely-jointed chain model are not correlated), the final result is

$$R \sim \sqrt{\langle R^2 \rangle} = N^{1/2} l \tag{2.4}$$

Note that the mean square end-to-end distance is much less than the contour length of the chain: $R \ll L = Nl$. Thus, the conformation of an ideal chain is far from the rectilinear one. Ideal chain forms an *entangled coil* (Fig. 2, 9a). The fact that the chain trajectory in this coil is analogous to the trajectory of a *brownian particle* (Fig. 2) is futher supported by the fact that $R \sim N^{1/2}$ (cf. with the result $R \sim t^{1/2}$, where R is the size of brownian particle trajectory during the time t; thus in this analogy N plays the role of time t).

Eq. 2.4 can be also rewritten as

$$\langle R^2 \rangle = Ll \tag{2.5}$$

since $L = Nl$.

2.3. *Size of ideal chain with fixed valency angle*

The conclusion $R \sim N^{1/2}$ is actually valid for ideal chain with any flexibility mechanism (not only for a freely-jointed chain model). E.g., let us consider the *model with fixed valency angle* γ between the segments of length b and free internal rotation ($U(\varphi) = 0$) (see Fig. 4). As follows from section 1.2, this model is close to real chain with rotational-isomeric flexibility mechanism.

As for the freely-jointed model, using the same notation (see Fig. 4) we can write

$$< R^2 > = \sum_{i=1}^{N} < \vec{u}_i^2 > + \sum_{i=1}^{N} \sum_{j=1, j \neq i}^{N} < \vec{u}_i \vec{u}_j > \tag{2.6}$$

as before $< u_i^2 > = b^2$, but now the value of $< \vec{u}_i \vec{u}_j >$ for $i \neq j$ is not generally equal to zero: $< \vec{u}_i \vec{u}_j > = b^2 < \cos \vartheta_{ij} >$, where ϑ_{ij} is the angle between segments i and j.

$$< R^2 > = Nb^2 + 2b^2 \sum_{1 \leq i < j \leq N} \sum < \cos \vartheta_{ij} > \tag{2.7}$$

To calculate $< \cos \vartheta_{ij} >$, let us consider first the simplest case, then i and j segments are neighbours. In this case it is clear $< \cos \vartheta_{i,i+1} > = \cos \gamma$.

For the average angle between segments i and $i + 2$ let us decompose the vector $\vec{u}_{i+2}$ into components parallel and perpendicular to the vector $\vec{u}_{i+1}$ (see Fig. 10). When the rotation of the segment $i + 2$ with respect to $i + 1$ is taken into account, it is clear that the average value of perpendicular projection is zero, and of the parallel one is $\cos \gamma$. By allowing now the rotation of the segment $i + 1$ with respect to the segment i, the average value of the projection of $\vec{u}_{i+2,i}$ on the direction of the vector $\vec{u}_i$ will give another factor $\cos \gamma$. Thus, we have $< \cos \vartheta_{i,i+2} > = (\cos \gamma)^2$.

By continuing this line of arguments we obtain

$$< \cos \vartheta_{i,i+k} > = (\cos \gamma)^k \tag{2.8}$$

Thus from eqs. 2.7, 2.8 we get

$$
\begin{aligned}
\langle R^2 \rangle &= Nb^2 + 2b^2 \sum_{i=1}^{N} \sum_{k=1}^{N-i} (\cos \gamma)^k = Nb^2 + 2b^2 \sum_{i=1}^{N} \frac{\cos \gamma}{1 - \cos \gamma} \\
&= Nb^2 + 2Nb^2 \frac{\cos \gamma}{1 - \cos \gamma} = Nb^2 \frac{1 + \cos \gamma}{1 - \cos \gamma}
\end{aligned} \tag{2.9}
$$

In the second of equations 2.9 we used the formula $\sum_{k=1}^{\infty} a^k = \frac{a}{1-a}$ and assumed that the value of $N - i$ is large, so that we can replace it with infinity.

Conclusions: 1) For the model with fixed valency angle

$$R \sim \sqrt{< R^2 >} = N^{1/2} b \sqrt{\frac{1 + \cos \gamma}{1 - \cos \gamma}} \tag{2.10}$$

We see that the average size of the chain is still proportional to $N^{1/2}$, so in this model the chain is also in the *entangled coil* conformation. In fact this is a *general property of ideal polymer chains* independently of the model.

2) At $\gamma < 90°$ the value of R is larger than for freely-jointed chain, while at $\gamma > 90°$ the relationship is reverse.

2.4. Kuhn segment length of a polymer chain

We have seen above that for ideal chain always $< R^2 > \sim N \sim L$ (at large values of contour length L). Therefore, the ratio $< R^2 > /L$ should be independent of L and should give a measure of chain flexibility. By definition, the *Kuhn segment length* of a polymer chain is introduced as

$$l = \frac{< R^2 >}{L} \quad \text{(at large } L\text{)} \tag{2.11}$$

I.e. the equality $< R^2 > = Ll$ is exact by definition.

The physical meaning of l follows from comparison of this equality with eq. 2.4 valid for freely-jointed chain. Such comparison shows that if we try to chose a *freely-jointed equivalent* to a given chain with the same values of $< R^2 >$ and L, the segment length for this equivalent chain should be equal to l. I.e. l is the length of the *equivalent segment*, or *approximately straight* part of the chain. Thus, it is a quantitative characteristic of chain flexibility.

A. Khokhlov and E.Yu. Kramarenko

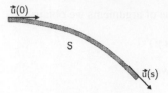

Fig. 11. Persistent polymer chain.

2.5. Persistent length of a polymer chain

Let us return to the result 2.8 and rewrite it in the following form

$$\langle \cos \vartheta_{i,i+k} \rangle = (\cos \gamma)^k = \exp(k \ln \cos \gamma) = \exp(-k|\ln \cos \gamma|)$$
$$= \exp\left(-\frac{kb}{b/|\ln \cos \gamma|}\right) = \exp(-s/\tilde{l}) \tag{2.12}$$

where $s = kb$ is the contour distance between two monomer units along the chain and $\tilde{l} \equiv b/|\ln \cos \gamma|$. Thus we conclude that the orientational correlations exponentially decay along the chain with some characteristic decay length $\tilde{l}$.

This result can be reformulated as follows (see Fig. 11) Let $\vec{u}(s)$ be the tangential unit vector along the chain as a function of the contour distance and let us represent equation (2.12) as

$$< \cos \vartheta_{\vec{u}(0),\vec{u}(s)} > \sim \exp(-s/\tilde{l}) \tag{2.13}$$

where $\vartheta_{\vec{u}(0),\vec{u}(s)}$ is the angle between unit vectors $\vec{u}(0)$ and $\vec{u}(s)$ (see Fig. 11).

This formula was derived for the model with fixed valency angle γ, however it is valid for any model: orientational correlations always decay exponentially along the chain. The characteristic length of this decay, $\tilde{l}$, is called a *persistent length* of the chain.

The physical meaning of this characteristics follows from eq. (2.13). At $s \ll \tilde{l}$ we have $< \cos \vartheta > \approx 1$, so the chain is approximately rectilinear, while at $s \gg \tilde{l}$ we obtain $< \cos \vartheta > \approx 0$, i.e. the memory of chain orientation is lost.

It is worth-while to emphasize here the advantages and disadvantages of using l and $\tilde{l}$ as quantitative characteristics of chain flexibility. The advantage of l is that it can be directly experimentally measured (the values of $< R^2 >$ and L can be determined from the light scattering experiments — see below). The advantage of $\tilde{l}$ is that it has a direct microscopic meaning (see Fig. 11). Depending on what is more important in the specific problem, one may use $\tilde{l}$ or l to characterize the chain flexibility.

One can show that always $l \sim \tilde{l}$.

For example let us examine this relationship for the model with fixed valency angle. Since

$$< R^2 >= Lb\frac{1+\cos\gamma}{1-\cos\gamma} \qquad (2.14)$$

We have

$$l = b\frac{1+\cos\gamma}{1-\cos\gamma} \qquad (2.15)$$

On the other hand,

$$\tilde{l} = \frac{b}{|\ln\cos\gamma|} \qquad (2.16)$$

Thus

$$\frac{l}{\tilde{l}} = |\ln\cos\gamma|\frac{1+\cos\gamma}{1-\cos\gamma} \qquad (2.17)$$

We see that the ratio $l/\tilde{l}$ is always close to 2. For the limit $\gamma \to 0$ we have the exact equality $l = 2\tilde{l}$. This limit corresponds to the chain with *persistent flexibility*. Indeed, let $\gamma \to 0$, $N \to 0$, $b \to 0$ in such a way that

$$Nb = L = \text{const}, \quad \text{and}$$
$$l = b\frac{1+\cos\gamma}{1-\cos\gamma} \approx \frac{2b}{1-1+\gamma^2/2} = \frac{4b}{\gamma^2} = \text{const} \qquad (2.18)$$

Then we arrive at the limit of a filament of length L with homogeneously distributed flexibility (*persistent chain*).

For persistent chain the equality $l = 2\tilde{l}$ is *exact*. The explanation of this relationship is that the orientational correlations spread in both directions along the chain, thus the average length of the rectilinear segment should be twice the persistent length.

2.6. *Stiff and flexible chains*

Now we have quantitative parameters characterizing the chain stiffness: Kuhn segment length l and persistent length $\tilde{l}$ (which in most cases is approximately

$l/2$). The value of l is normally larger than the *contour length per monomer unit* l_0. The ratios l/l_0 for most common polymers are shown below.

poly(ethylene oxide)	2.5
poly(propylene)	3
poly(methyl methacrylate)	4
poly(vinyl chloride)	4
poly(styrene)	5
poly(acrylamide)	6.5
cellulose diacetate	26
poly(para-benzamide)	200
DNA (in double helix)	300
poly(benzyl glutamate) (in α-helix)	500

From macroscopic viewpoint a polymer chain can be always represented as some filament which is characterized by two lengths: *Kuhn segment length l* and *characteristic* chain *diameter d*. Depending on the relationship between these two lengths, we can now introduce the notion of stiff and flexible chains. *Stiff* chains are those for which $l \gg d$, while for *flexible* chains $l \sim d$. The examples of stiff chains are DNA, helical polypeptides, aromatic polyamides etc. Examples of flexible chains are polyethylene, polystyrene, etc. — most carbon backbone polymers.

2.7. Gaussian distribution for the end-to-end vector for ideal chain

Up to now we have considered mainly average size of a polymer coil $R \sim \sqrt{\langle R^2 \rangle}$. However, the vector $\vec{R}$ fluctuates due to the thermal motion, therefore in addition to the average values it is of interest to introduce $P_N(\vec{R})$: probability distribution for the end-to-end vector $\vec{R}$ of N-unit chain.

Let us at first consider this function for the freely-jointed chain (see Fig. 10). Since each step (segment) gives *independent* contribution to $\vec{R}$, by analogy with the trajectory of Brownian particle, the Gaussian distribution should be valid for $\vec{R}$:

$$P_N(\vec{R}) = \left(\frac{3}{2\pi N l^2} \right)^{3/2} \exp\left(-\frac{3R^2}{2Nl^2} \right) \tag{2.19}$$

Therefore, the ideal coil is sometimes called a *Gaussian coil*.

Since $P_N(\vec{R})$ is a probability distribution, the normalization condition

$$\int P_N(\vec{R}) d^3R = 1 \tag{2.20}$$

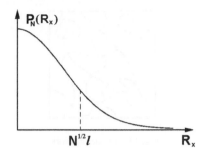

Fig. 12. Probability distribution function for the x-projection of the end-to-end vector of the Gaussian chain.

is valid. The coefficient before exp is just chosen in order to satisfy this condition. Also, since $R^2 = R_x^2 + R_y^2 + R_z^2$, we have $P_N(\vec{R}) = P_N(R_x)P_N(R_y)P_N(R_z)$ with

$$P_N(R_x) = \left(\frac{3}{2\pi N l^2}\right)^{1/2} \exp\left(-\frac{3R_x^2}{2N l^2}\right) \tag{2.21}$$

The one-dimensional distribution function (2.21) is shown in figure 12. From this plot we see that the function P_N decays at the distances $\sim N^{1/2}l$, however at $R < N^{1/2}l$ the values of P_N only weakly depend on R.

Thus we conclude that the value of $\vec{R}$ undergoes strong fluctuations: any value of $R \leq N^{1/2}l$ can be realized with more or less equal probability.

For other models, different from freely-jointed chain, Gaussian distribution should be still valid, since orientational correlations decay exponentially. Indeed, we can rewrite eq. 2.19 in the form

$$P_N(\vec{R}) = \left(\frac{3}{2\pi \langle R^2 \rangle}\right)^{3/2} \exp\left(-\frac{3R^2}{2\langle R^2 \rangle}\right) \tag{2.22}$$

which is independent of any specific model, $\langle R^2 \rangle = Ll$.

3. High elasticity of polymer networks

3.1. The property of high elasticity

It is to be reminded that polymer network consists of long polymer chains which are *crosslinked* with each other and form a continuous *molecular framework* (see figure 13).

A. Khokhlov and E.Yu. Kramarenko

Fig. 13. Polymer network.

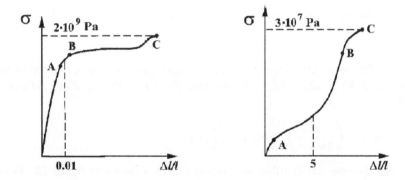

Fig. 14. Typical stress-strain curves for steel (left) and for rubber (right).

All polymer networks (which are not in the glassy of partially crystalline states) exhibit the property of *high elasticity*, i.e. *the ability to undergo large reversible deformations at relatively small applied stress*. High elasticity is the most specific property of polymer materials and it is connected with the fundamental features of ideal chains considered above. In everyday life highly elastic polymer materials are called *rubbers*. To illustrate the property of high elasticity in more detail, let us compare the typical stress-strain curves for steel and highly-elastic rubber (see figure 14). Different points in this picture have the following meaning: point A is the upper limit for stress-strain linearity; point B is the upper limit for reversibility of deformations, while point C is the fracture point.

From the comparision of two stress-strain curves one can make the following conclusions:

1. Characteristic values for the deformation $\Delta l / l$ are much larger for rubber (cf. the values $\Delta l / l \sim 5$ for rubber and $\Delta l / l \sim 0.01$ for steel).

2. Characteristic values for the strain σ are much larger for steel ($\sim 10^9$ for steel and $\sim 10^7$ for rubber).

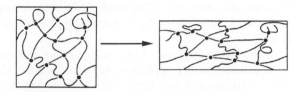

Fig. 15. Molecular picture of high-elastic deformation.

3. The two previous conclusions lead to the result that the characteristic values for Young moduli (defining the initial slope of the stress-strain curve) are enormously larger for steel ($E \approx 2 \times 10^{11} Pa$) than for rubber ($E \leq 10^6 Pa$).

4. For steel linearity and reversibility are lost practically simultaneously, while for rubbers there is a very *wide region of nonlinear reversible deformations*.

5. For steel there is a wide region of plastic deformations (between points B and C) which is practically absent for rubbers.

As it was already mentioned, the property of high elastisity can be understood on the molecular level. Molecular picture of high-elastic deformation is shown in figure 15.

One can see that the elasticity of rubber is composed from the elastic responses of the chains crosslinked in the network sample. Therefore, we start below with the description of elasticity of a single polymer chain.

3.2. Elasticity of a single ideal chain

Let us consider a single polymer chain stretched at the end by the external force f (Fig. 16). What is the origin for chain elasticity? It is well known that for ordinary crystalline solids (like steel) the elasticity response appears because external stress changes the equilibrium interatomic distances and increases the *internal energy* of the crystal (*energetic elasticity*). Since the energy of ideal polymer chain is equal to zero, the elastic response can appear only by purely

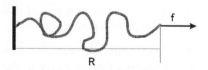

Fig. 16. Stretching of a polymer chain. The external force cf induces the average end-to-end distance $\vec{R}$.

entropic reasons (*entropic elasticity*). This kind of elasticity emerges because under stretching the long chain adopts the less probable conformation and thus its entropy decreases.

Let us calculate the elastic response of the ideal polymer chain qualitatively. We assume that the external force $\vec{f}$ applied to the end of polymer chain induces the average end-to-end distance $\vec{R}$ (see figure 16). According to Boltzmann, the entropy is

$$S(\vec{R}) = k \ln W_N(\vec{R}) \tag{3.1}$$

where k is the Boltzmann constant and $W_N(\vec{R})$ is the number of chain conformations compatible with the end–to–end distance $\vec{R}$. The function $W_N(\vec{R})$ is proportional to the probability distribution $P_N(\vec{R})$ (2.22), so we can write $W_N(\vec{R}) = \text{const}\, P_N(\vec{R})$ where const stands for the total number of conformations of the chain of N segments, this value is independent of R. Thus

$$S(\vec{R}) = k \ln P_N(\vec{R}) + \text{const} \tag{3.2}$$

or, after inserting here eq. (2.22)

$$S(\vec{R}) = -\frac{3kR^2}{2Ll} + \text{const} \tag{3.3}$$

where const always defines some R-independent item. For the system kept at constant temperature T the important thermodynamic potential is not the entropy itself but the free energy which consists of energetic and entropic part, $F = E - TS$. For our case the internal energy E is zero. Thus,

$$F = -TS = \frac{3kTR^2}{2Ll} + \text{const.} \tag{3.4}$$

If we assume that under the action of external force $\vec{f}$ the end-to-end distance changed from $\vec{R}$ to $\vec{R} + d\vec{R}$, the work done by the external force is $\vec{f} d\vec{R}$, and it is equal to the corresponding increase of the free energy. Thus, $\vec{f} d\vec{R} = dF$ and

$$\vec{f} = \frac{dF}{d\vec{R}} = \frac{3kT}{Ll}\vec{R} \tag{3.5}$$

The equation 3.5 gives the dependence of the applied force on the induced "deformation" $\vec{R}$, i.e. it describes the elastic response of a single polymer chain. From eq. 3.5 the following conclusions can be made.

1. The chain is elongated in the direction of $\vec{f}$ and $\vec{f} \sim \vec{R}$ (kind of a Hooke law). However, it should be emphasized that, contrary to the Hooke law, we can not introduce here the relative deformation $\Delta l / l_0$, since there is no parameter playing the role of the size of the undeformed sample l_0.

2. "Elastic modulus" in this Hooke law is equal to $\frac{3kT}{Ll}$. It should be noted that this modulus

   (a) is proportional to $1/L$, i.e. is very small for large values of L. This means that *long polymer chains are very susceptible to external actions*;

   (b) is proportional to kT which is the indication to *entropic nature* of elasticity: the higher is the temperature the more is the elastic response.

Some limitations for the validity of eq. 3.5 should be mentioned here. In the derivation we used the formula 2.22, which means that the probability $P_N(\vec{R})$ should be Gaussian. This is the case only for not too strongly elongated chains; for very stretched chains other expressions for $P_N(\vec{R})$ should be used.

3.3. Elasticity of a polymer network (rubber)

Now let us turn to the derivation of the elastic response of a macroscopic network sample. To this end we consider densely packed system of crosslinked chains (freely jointed chains of contour length L and Kuhn segment length l) (see figure 15). Let us introduce the coordinate axes and assume that the relative deformation of the sample along the axes x, y, z is $\lambda_x, \lambda_y, \lambda_z$, i.e. the sample dimensions along the axes are $a_x = \lambda_x a_{x0}, a_y = \lambda_y a_{y0}, a_z = \lambda_z a_{z0}$ (a_{x0}, a_{y0} and a_{z0} are the dimensions of undeformed sample). In principle, the chains in the network sample are strongly entangled with each other, and it is not clear whether we can use the formulae of the section 3.2 derived for a single ideal chain. However, here the so-called *Flory theorem* appears to be of great help. According to this theorem, the statistical properties of polymer chain in the dense system are equivalent to those for single ideal chains. We will comment more on the Flory theorem later. At the moment we will just use it to apply single-chain formulae for the chains of the densely packed network. Another assumption which we will use in derivation is that the crosslink points are deformed affinely together with the network sample (*affinity assumption*). this means that if in the initial state the end–to–end vector had the coordinates $\{R_{0x}, R_{0y}, R_{0z}\}$, in the deformed state its coordinates become $\{R_x = \lambda_x R_{0x}, R_y = \lambda_y R_{0y}, R_z = \lambda_z R_{0z}\}$. Actually this assumption is not very essential for the validity of the final results, but we will use it for the simplicity of the derivation. Accordingly to eq. 3.5, the change of the free energy of the chain between two crosslink points upon extension is

$$\begin{aligned} \Delta F &= \frac{3kT}{2Ll}(\vec{R}^2 - \vec{R}_0^2) = \frac{3kT}{2Ll}\{(R_x^2 - R_{0x}^2) + (R_y^2 - R_{0y}^2) + (R_z^2 - R_{0z}^2)\} \\ &= \frac{3kT}{2Ll}\{R_{0x}^2(\lambda_x^2 - 1) + R_{oy}^2(\lambda_y^2 - 1) + R_{0z}^2(\lambda_z^2 - 1)\} \end{aligned} \quad (3.6)$$

For the whole sample $\Delta\mathcal{F} = \nu V < \Delta F >$, where ν is the number of network chains per unit volume and V is the volume of the sample. Thus, we have

$$\Delta\mathcal{F} = \frac{3kT}{2Ll}\nu V\{(\lambda_x^2 - 1)\langle R_{0x}^2 \rangle + (\lambda_y^2 - 1)\langle R_{0y}^2 \rangle + (\lambda_z^2 - 1)\langle R_{0z}^2 \rangle\}. \quad (3.7)$$

On the other hand, since the axes x, y and z are equivalent,

$$\langle R_{0x}^2 \rangle = \langle R_{0y}^2 \rangle = \langle R_{0z}^2 \rangle = \frac{\langle R_0^2 \rangle}{3} = \frac{Ll}{3} \quad (3.8)$$

Therefore,

$$\Delta\mathcal{F} = \frac{1}{2}kT\nu V\{\lambda_x^2 + \lambda_y^2 + \lambda_z^2 - 3\} \quad (3.9)$$

The result 3.9 appears to be astonishingly simple, the dependences on L and l dropped out from the final formulae. This indicates to the *universality* of the theory, i.e. its independence of the specific polymer chain model and possible polydispersity of the chains. The only important assumption which we used above was about Gaussian character of the chains. Let us apply the general formula 3.9 for the case of uniaxial extension ($\lambda_x = \lambda > 1$) or compression ($\lambda_x = \lambda < 1$) along the axis x. The values of deformation in the perpendicular directions λ_y and λ_z ($\lambda_y = \lambda_z$) can then be obtained from the *incompressibility condition*. Indeed, at a characteristic values of stress which are normally applied to rubbers ($\sim 10^5 \div 10^6$ Pa) the interatomic distances in the dense network sample practically do not change ($\sim 1\%$ of change require the stress of $\sim 10^7$ Pa). Therefore, the volume of the sample is kept fixed under the deformation. Since $V = \lambda_x a_{x0}\lambda_y a_{y0}\lambda_z a_{z0} = \lambda_x\lambda_y\lambda_z V_0$ we have from the incompressibility condition ($V = V_0$): $\lambda_x\lambda_y\lambda_z = 1$. We have $\lambda_x = \lambda$ and $\lambda_y = \lambda_z$, therefore

$$\lambda\lambda_y^2 = 1 \quad \text{or} \quad \lambda_y = \lambda_z = \frac{1}{\sqrt{\lambda}} \quad (3.10)$$

Substituting the values in eq. 3.9, we obtain for the uniaxial extention-compression:

$$\Delta\mathcal{F} = \frac{kT}{2}\nu V\{\lambda^2 + \frac{2}{\lambda} - 3\} \quad (3.11)$$

The applied stress σ is equal to the uniaxial for $\partial(\Delta\mathcal{F})/\partial a_x$ divided by the sample crossection $a_{y0}a_{z0}$, thus

$$\sigma = \frac{1}{a_{0y}a_{0z}}\frac{\partial(\Delta\mathcal{F})}{\partial a_x} = \frac{1}{a_{0x}a_{0y}a_{0z}}\frac{\partial(\Delta\mathcal{F})}{\partial\lambda} = \frac{1}{V}\frac{\partial(\Delta\mathcal{F})}{\partial\lambda} \quad (3.12)$$

Therefore, we have

$$\sigma = kTv \left(\lambda - \frac{1}{\lambda^2} \right) \tag{3.13}$$

Eq. 3.13 is one of the main results of the classical theory of high elastisity. The dependence 3.13 is shown in figure 17 by the bold line.

In connection with the result 3.13 the following points should be mentioned.

1. Modulus of elasticity is $E = 3kTv$. For loosely crosslinked networks it is very small. Indeed, from the dense-packing condition we have $vNv = 1$, where N is the number of monomer units in one network chain and v is the volume of a monomer unit, thus $v \sim 1/Nv$ and

$$E \sim \frac{kT}{Nv} \tag{3.14}$$

The large parameter $N \gg 1$ in the denominator makes the values of E very small. Thus is just *the origin of the high elasticity of rubbers*.

2. The final formula predicts not only modulus, but also *nonlinear elasticity*. It is therefore one of the rare occasions when the nonlinear response of some material can be calculated *exactly*.

3. Analogous formulae can be obtained for other kinds of deformation (shear, twist, etc.).

4. The final formula is universal, i.e. independent of specific chain model. The reason for this is connected with the fact that entropic elasticity is caused by *large-scale properties* of polymer coils, rather than by short-scale details of the chain structure.

5. Main assumptions which were used in the above derivation are: (i) the chains were assumed to obey *Gaussian* statistics; (ii) restrictions imposed by other chains to the conformation of a given chain were neglected.

6. If $\sigma = \text{const} > 0$ and the temperature T increases, according to eq. 3.13, the value of λ should decrease, i.e. the *rubber shrinks upon heating* (contrary to gases) and vice versa. Also: *at adiabatic extension the rubber is heated* (contrary to gases) because the work done by the external force is transformed into internal energy of the sample. These facts are the direct consequences of *entropic* character of elasticity.

7. Comparision of the results of classical theory of high elasticity vs. experimental data is schematically illustrated in figure 17. Normally, at $0.4 < \lambda < 1.2$ the agreement is very good, at $1.2 < \lambda < 5$ the theory slightly overestimates

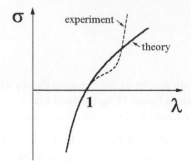

Fig. 17. Comparison of the results of classical.

stress at a given strain. According to the modern theoretical development, the reason for this is connected with the *mutual steric restrictions* of strongly entangled chains. Finally, at $\lambda > 5$ the theory significantly underestimates stress at a given strain. The reason for this is a *finite extensibility* of the chains: at high values of λ the chains are close to the limit of full extension, and their statistics is no longer Gaussian.

4. Viscoelasticity of entangled polymer fluids

4.1. Main properties of entangled polymer fluids

Entangled polymer fluids are polymer melts and concentrated or semidilute polymer solutions. In these systems polymer coils strongly overlap with each other, and polymer chains are entangled.

Experimentally we know several specific properties of entangled polymer fluids which distinguish them from normal fluids:

1. Entangled polymer fluids normally have a very *high viscosity*.

2. Such fluids keep for a long time the *memory* about the history of flow.

3. Such fluids exhibit the property of *viscoelasticity*: at *fast* (high frequency) external action the response of a fluid is elastic, while for *slow* (low frequency) external action the response is viscous (i.e. flow starts). In other words, for fast external action entangled polymer fluid behaves as elastic solid, while for slow action its behavior corresponds to viscous fluid.

The main of those properties is that of viscoelasticity. It is observed for all entangled polymer fluids. In this chapter we consider this property in more detail, and give its molecular explanation. But before that in the next section we give a brief reminder of the basic physical definitions and relations connected with the viscosity of fluids.

4.2. *Viscosity of fluids*

The simplest setup for the viscosity determination is shown in Figure 18. The fluid is confined in the slit of width d between the two plates. The lower plate is immobile, while the upper plate moves with some constant velocity v under the action of the force f. The surface areas of both plates is S. Newton has established that for these conditions

$$f = \eta \frac{Sv}{d} \tag{4.1}$$

(Newton-Stokes law). The proportionality coefficient η is called the *viscosity of the fluid*.

In the differential form the Newton-Stokes law can be written as follows

$$\sigma = \frac{f}{S} = \eta \frac{dv}{dz} \tag{4.2}$$

where σ is the imposed stress and z is the coordinate perpendicular to the plates (see Figure 18; in writing $v/d = dv/dz$ we used the fact that the velocity of the fluid changes linearly with z).

Alternative way of writing eq. 4.2 for the simple shear flow shown in Fig. 18 is to introduce a shear angle γ in the following way:

$$\sigma = \eta \frac{dv}{dz} = \eta \frac{d(dx/dt)}{dz} = \eta \frac{d^2 x}{dt \, dz} = \eta \frac{d(dx/dz)}{dt} = \eta \frac{d\gamma}{dt}. \tag{4.3}$$

I.e. for fluids the stress is proportional not to deformation γ (as for solids), but to the time derivative of the deformation.

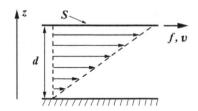

Fig. 18. Simplest setup for viscosity determination.

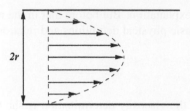

Fig. 19. Flow in capillary viscosimeter.

Viscosity is measured in *poise* (1 poise=1$g/(cm \cdot sec)$); the viscosity of water at 20° is $\sim 10^{-2}$ poise, while viscosity of polymer melts can be of order $10^{10} \div 10^{12}$ poise, or even higher.

Viscosity is measured by viscosimeters. Most common are *capillary viscosimeters*; in these viscosimeters the flow of a fluid through a thin capillary of the radius r is studied

The method of measurements is then based on the *Poiseille equation*:

$$Q = \frac{\pi r^4 \Delta Pt}{8\eta l} \tag{4.4}$$

where Q is the mass of the fluid flown through the capillary during the time t, ΔP is the pressure difference at the ends of the capillary of length l and radius r.

4.3. The property of viscoelasticity

The property of viscoelasticity is a characteristic specific property of polymer fluids. Let us describe this property in more precise terms. Assume that a step-wise shear stress starting at $t = 0$ is applied to a fluid (Figure 20(a)).

The reaction of the *ordinary fluid* to such a stress would be a normal flow (after some initial equilibriation period), i.e. the shear angle γ will vary with time as $\gamma = \frac{\sigma t}{\eta}$ where η is the viscosity of the fluid (see Figure 20(b)).

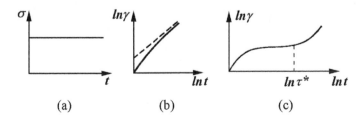

Fig. 20. Reaction of an ordinary fluid (b) and polymer fluid (c) to a step-wise external stress (a).

On the contrary, the typical reaction of the *polymer fluid* is shown in Figure 20(c). At $t \ll \tau^*$ the value of γ is practically constant $\gamma \approx \frac{\sigma}{E}$ and only at $t \gg \tau^*$ the flow starts $\gamma \sim \sigma \frac{t}{\eta}$.

In other words for entangled polymer fluid at $t \ll \tau^*$ we have the *elastic* response $\gamma \approx \frac{\sigma}{E}$ where E is the effective Young modulus, while at $t \gg \tau^*$ we have $\gamma \sim \frac{\sigma t}{\eta}$ i.e. the response is *viscous*. This is just the property of *viscoelasticity*.

The relations $\gamma \sim \frac{\sigma}{E}$ and $\gamma \sim \frac{\sigma t}{\eta}$ should match each other at $t \sim \tau^*$. Thus, we have

$$\frac{1}{E} \sim \frac{\tau^*}{\eta} \qquad \text{or} \qquad \eta \sim \tau^* E, \tag{4.5}$$

i.e. the viscosity of a polymer fluid is just a product of the crossover time τ^* and the Young modulus E corresponding to the plateau in Figure 20(c).

The described property of *viscoelasticity* is a *general property* of all entangled polymer fluids, as long as they are not crystalline, glassy or crosslinked. Therefore, as for the property of high elasticity, the general *molecular explanation* should be possible which is based on the fact of the *chain structure* of polymer molecules, without the explicit reference to the specific chemical nature of monomer units.

Such explanation was developed by de Gennes, Doi and Edwards (1971–1979); it is called the theory of *reptations*.

4.4. Theory of reptations

Let us consider one chain entangled with many others (Figure 21) and let us *"freeze"* for a moment the conformations of other chains. This gives rise to a certain *"tube"*: the given chain cannot escape in the directions perpendicular to the tube axis. Therefore the only allowed type of motion is the *snake-like diffusion along the tube axis* (Figure 22). This type of motion is called *reptations*.

If other chains are *"defrozen"* the competing mechanizm appears: *"tube renewal"*, but it can be shown that *reptations always give a dominant contribution*.

Fig. 21. One chain in entangled polymer fluid: a tube constraint.

Fig. 22. Reptation – type of motion.

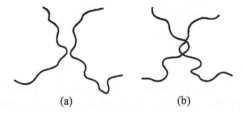

(a)   (b)

Fig. 23. (a) Two neighboring chains not forming a'quasi-cross-link'; (b) Two neighboring chains forming a "quasi-cross-link".

The neighboring chains forming the "*walls*" of the tube create *restrictions* for the motion of a given chain and are in this sense analogous to *cross-links*. But these are "*quasi-cross-links*" with *finite lifetime*: they "*relax*" after some time τ^* which is needed for the chain to *leave the initial tube*. After the time interval τ^* *all the neighbouring chains* for a given chain *are new*.

From this viewpoint, the following *molecular interpretation* can now be given to the *property of visoelasticity*: at $t < \tau^*$ the polymer fluid behaves as network with "*quasi-cross-links*", and the response is *elastic*; while at $t > \tau^*$ "quasi-cross-links" *relax*, and the response is *viscous*. Thus for the time τ^* in Fig. 20 the following molecular interpretation can be given: this is the time required for reptating chain to leave an initial tube.

What is the elastic modulus E for the network of "quasi-cross-links"? Accoding to the classical theory of high elasticity (see eq. 3.14) $E \sim kT\nu$, where ν is the number of elastic chains per unit volume, $\nu \sim 1/Na^3$, N being the number of units between two cross-links. It is normally assumed that $N \sim N_e$, where N_e is the *number of units between two "quasi-cross-links"*. The value of N_e is assumed to be some constant for a given polymer depending on its ability to form entanglements. Most often the constant N_e is in the interval from 50 to 500; this is a phenomenological parameter reflecting that *not each contact acts like cross-link*. The latter fact is illustrated in Figure 23.

Thus,

$$E \sim kT/N_e a^3 \tag{4.6}$$

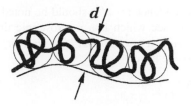

Fig. 24. Chain in a tube.

It should be noted that, of course, the model of reptations and the motion of tube constraint are valid only for $N \gg N_e$.

On the basis of eq. (4.6) and preceeding discussion the following *picture of a tube* can be drawn (see figure 24).

The chain is a sequence of *subcoils*, each containing N_e links and having the size $d \sim N_e^{1/2}a$. At the scales smaller than d the chain is not entangled and does not feel the tube constraint. The *width* of the tube is $d \sim N_e^{1/2}a$. The *length* of the tube is

$$\Lambda \sim \frac{N}{N_e}d \sim \frac{Na}{N_e^{1/2}} \tag{4.7}$$

which is much smaller than the contour length of the tube $L = Na$. The *diffusion coefficient* corresponding to reptations along the tube is $D_t = kT/\mu$, where μ is the corresponding friction coefficient. In dense system the friction of each monomer unit is independent, therefore $\mu = N\mu_0$ (μ_0 being the friction coefficient for one monomer unit). So

$$D_t = kT/\mu_0 N \tag{4.8}$$

On the other hand, from the molecular interpretation of the time τ^* it should be $\Lambda^2 \sim D_t\tau^*$, thus

$$\tau^* \sim \frac{\Lambda^2}{D_t} \sim \frac{N^2a^2\mu_0 N}{N_e kT} \sim \frac{N^3}{N_e} \cdot \frac{\mu_0 a^2}{kT} \sim \frac{N^3}{N_e}\tau_0, \tag{4.9}$$

where $\tau_0 \sim \mu_0 a^2/kT \sim 10^{-12}$ sec is the characteristic microscopic time.

Therefore, we will obtain for the time τ^*

$$\tau^* \sim \frac{N^3}{N_e}\tau_0 \tag{4.10}$$

The very strong N - dependence ($\sim N^3$) should be noted. For $N \sim 10^5$, $N_e \sim 10^2$, this gives $\tau^* \sim 10$ sec which is a macroscopic time scale. This is the fundamental *reason for slow relaxation and high viscosity in polymer fluids.*

Thus, using eq. (4.5), we have

$$\eta \sim E\tau^* \sim \frac{kT}{N_e a^3} \frac{N^3}{N_e} \tau_0 \sim \frac{N^3}{N_e^2} \eta_0 \qquad (4.11)$$

where $\eta_0 \sim kT\tau_0/a^3$ is a characteristic viscosity for low molecular fluid ($\eta_0 \sim 1$ poise).

Therefore, the reptation theory gives for the viscosity of entangled polymer melt the law $\eta \sim N^3$. This relationship is not far from the best experimental fit $\eta \sim N^{3.4}$.

The model of reptations was successfully applied not only to the determination of viscosity, but to many other problems of dynamic behavior of concentrated polymer solutions and melts. This is the first *molecular theory of dynamics of entangled polymer fluids.*

4.5. The method of gel-electrophoresis in application to DNA molecules

Macromolecules containing charged monomer units are called *polyelectrolytes.* Charged monomer units appear as a result of dissociation of neutral monomer unit with the formation of the charged monomer unit and counter ion.

Most impotant polyelectrolytes are biological macromolecules: DNA and proteins.

As an application of the reptation theory, let us consider the *gel-electrophoresis* of polyelectrolytes, having in mind mainly DNA molecules. This method is used in practice for the *separation* of DNA fragments of different length and composition (for *DNA sequencing*).

Gel is a swollen polymer network. In the process of gel-electrophoresis the negatively charged DNA molecules (total charge Q) move through the gel in external electric field $\vec{E}$ (Figure 25).

The drift velocity $\vec{v}$ depends on the length of the chains; in this way, the separation of DNA molecules of different length is achieved.

In the absence of the gel this would not be the case. Indeed, the force $\vec{F}$ acting on the free DNA chain in the solution is $\vec{F} = Q\vec{E}$. On the other hand, $\vec{F} = \mu\vec{v}$, where μ is the friction coefficient of DNA chain. Thus, $v = QE/\mu$. But $Q \sim L$ and $\mu \sim L$ (different parts of DNA molecule exhibit independent friction), therefore v is *independent* of L and DNA molecules cannot be separated in the solution without the gel.

In the gel DNA molecules are in the effective *"tubes"* and move via *reptations.* Let us divide the molecule in small fragments and count only the forces acting

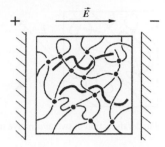

Fig. 25. DNA molecule moving in the gel under the influence of electric field.

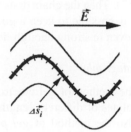

Fig. 26. DNA chain in effective tube in the course of foretic motion.

along the tube. The forces in the direction perpendicular to the tube axis are counterbalanced by the reaction of the tube.

Then we have (Figure 26):

$$F_t = \sum_i \frac{Q \Delta \vec{S}_i}{L} \vec{E} = \frac{Q}{L} \left(\sum_i \Delta \vec{S}_i \right) \vec{E} = \frac{Q \vec{R} \vec{E}}{L} \qquad (4.12)$$

The drift velocity along the tube is $v_t = F_t / \mu$, where $\mu = \mu_0 N$, μ_0 being the friction coefficient for one monomer unit. Thus,

$$v_t = \frac{QRE}{L \mu_0 N}. \qquad (4.13)$$

We have $Q \sim L$, $R \sim N^{1/2}$ (in weak field), $N \sim L$, therefore v_t is proportional to $1/L^{1/2}$ (or $1/N^{1/2}$). This means that *in the gel shorter chains move faster*, and the *separation of DNA fragments of different lengths is possible.*

However, in the stronger fields DNA molecules *stretch*, R becomes proportional to N, and the resolution of gel-electrophoresis vanishes. To deal with this problem the following method was proposed. The direction of the field was

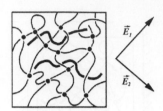

Fig. 27. DNA molecules in the gel under the action of two alternative mutually perpendicular electric fields.

turned at 90° after each time interval τ^* (τ^* being the time of reneval of initial tube via reptations) (Figure 27). Then the chain does not have time to stretch during one cycle. In this way it is possible to keep a good resolution of the method of DNA gel-electrophoresis even in strong enough field.

4.6. Gel permeation chromatography

Since we have considered the method of gel electrophoresis which is used for the separation of polyelectrolytes of different lengths, it is worth-while to consider briefly at this stage also the method of *gel permeation chromatography* which is also based on the idea of the separation of chains of differnt lengths in the process of their motion through microporous (gel) medium-*chromatographic column*. The differences of this method from gel-electrophoresis can be summarized as follows.

1. The driving force for the motion is *pressure gradient* (due to the pumping of polymer solution through chromatographic column), not electric field. Therefore, the method can be applied *to all polymers*, not only to charged ones.

2. The gel medium is normally *solid microporous material*, rather than swollen soft gel. Contrary to gel-electrophoresis, the *size of the largest pores* is usually much *higher than the coil size* (although the pore sizes exhibit very wide distribution).

3. Contrary to gel-electrophoresis, in the normal *exclusion regime* of gel permeation chromatography (no specific interactions of polymers with the column) *longer chains move faster*. Explanation: shorter chains can penetrate even in small pores of microporous system, while long chains move only through largest pores. Therefore, the *"effective way" for the long chain is shorter*.
There is another regime, called *adsorption chromatography*, when polymer chains are attracted to the walls of the microporous system of the column and "stick" to them. In this case, the "sticking energy" for *longer chains* is higher, and they move *slower*.

References

[1] A.Yu. Grosberg, A.R. Khokhlov. *Giant Molecules: Here, There, and Everywhere...* Academic Press, New York, 1997.

[2] A.Yu. Grosberg, A.R. Khokhlov. *Statistical Physics of Macromolecules,* New York: AIP Press, 1994.

[3] P.J. Flory. *Principles of Polymer Chemistry,* Cornell University Press, Ithaca, New York, 1953.

[4] P.G. de Gennes. *Scaling Concepts in Polymer Physics,* Ithaca, NY: Cornell University Press, 1979.

[5] M. Doi, S.F. Edwards, *The Theory of Polymer Dynamics,* Oxford: Oxford University Press, 1986.

[6] M. Doi. *Introduction to Polymer Physics* Clarendon Press, Oxford, 1996.

[7] M.V. Volkenstein. *Molecular Biophysics,* New York: Academic Press, 1977; *Physics and Biology,* New York: Academic Press, 1982; *General Biophysics,* New York: Academic Press, 1983; *Physics Approaches to Biological Evolution,* Berlin and New York: Springer-Verlag, 1994.

[8] H. Morawetz. *Polymers. The Origins and Growth of Science,* New York: Willey, 1985.

[9] M.D. Frank-Kamenetskii. *Unraveling DNA,* New York: VCH, 1993.

[10] M. Rubinstein, R. Colby. *Polymer Physics* Oxford University Press, 2003.

Course 5

THE PHYSICS OF DNA ELECTROPHORESIS

T.A.J. Duke

Cavendish Laboratory, Madingley Road, Cambridge CB3 0HE, UK

D. Chatenay, S. Cocco, R. Monasson, D. Thieffry and J. Dalibard, eds.
Les Houches, Session LXXXII, 2004
Multiple aspects of DNA and RNA: from Biophysics to Bioinformatics

Contents

Contents

Abstract

The ability to sort DNA molecules according to size is crucial to modern genetics and cell biology. The current method of DNA sequencing relies on high-resolution separation of mixtures of single-stranded molecules containing every size up to several hundred nucleotides, and the physical mapping of genomes requires the separation of double-stranded DNA containing hundreds of thousands of base pairs. Gel electrophoresis is the traditional method used to separate DNA, but more recently novel solid-state devices have been invented to accomplish the task. Using microfabrication techniques, miniature obstacle courses can be engraved on a silicon chip. DNA molecules, propelled through the device by an electric field move with a complex dynamics. By carefully designing the pattern of obstacles, the migration speed of the DNA molecules can be tailored to provide excellent separation within a specified range of molecular sizes. In this course, I will describe the physics underlying a wide variety of separation techniques, and emphasize the productiveness of a close interplay between theory and experiment in meeting the technical challenges of modern biology.

1. Importance of DNA sorting in biology and how physics can help

The entire genomic information of a living organism is contained in a handful of chromosomes, each of which is a single molecule of DNA. Modern genetics is concerned with understanding how this information is organized and with deciphering the message encoded in the sequence of nucleotides that form the links of the DNA chain. Inevitably, this involves cutting the enormously long chromosomes into pieces that are small enough to examine in detail. It is no surprise, then, that an efficient method of sorting DNA fragments is of paramount importance to biological research. This need has been emphasized by the advent of genomics, which aims to study the similarities and differences between the genomes of different organisms. Genomics involves two different tasks (Figure 1a). The first is to create a library containing hundreds of thousands of DNA fragments, covering the entire genome. When the chromosome of origin of each fragment has been identified, together with the location at which the fragment belongs, this collection constitutes a 'physical map' of the genome. Once accomplished, researchers wishing to study a particular gene may simply access the library to obtain a clone, rather then laboriously isolating the gene themselves.

131

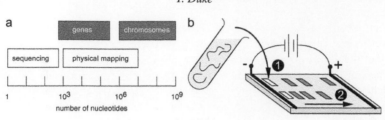

Fig. 1. a: Range of sizes of important elements in the organization of the genome and the typical sizes of DNA fragments analysed in sequencing and in mapping. In sequencing, the locations at which a particular nucleotide appears in a DNA fragment are deduced by a precise measurement of the spectrum of lengths of a set of single-stranded chains which are synthesized using the DNA fragment as a template, starting at one end of the molecule and finishing up at one of the various positions where the nucleotide occurs. Mapping makes use of enzymes that recognize and cut the chromosome at specific sequence motifs which occur at random intervals along the DNA, to create a set of restriction fragments which typically contain many thousands of nucleotides. b: A typical electrophoresis procedure. (1) The mixture of DNA fragments is introduced into the electrophoresis matrix, which is traditionally a gel but may alternatively be a microfabricated chamber. (2) An electric field is applied and the DNA molecules migrate towards the anode. Molecules move at a speed that depends on their size so that, after an interval of time, fragments of different length resolve into spatially segregated bands. The sizes of the molecules in the mixture may be inferred from the spatial distribution of the bands, or alternatively by examining the temporal distribution of molecules arriving at the anode.

The second task is to establish the precise nucleotide sequence of each entry in the library. The value of this information may be appreciated when one knows that genetic diseases can be caused by the modification or absence of just two or three nucleotides in a gene. These two tasks require the ability to analyse DNA molecules with sizes that differ by many orders of magnitude.

The task of separating DNA molecules according to size is not an easy one because a DNA mixture is a tangled mass of long, flexible polymers. Imagine that, given a pot of boiling spaghetti, you were asked to separate all the broken pieces from the whole ones. While you might manage to pick them out one by one, a reliable automatic procedure is less evident. All the more so, considering that big DNA fragments correspond to spaghetti over a kilometre long! Centrifugation is one possible approach, but even in an ultra-centrifuge the force acting on DNA molecules proves too small to provide efficient fractionation. So electrophoresis, in which the DNA is driven by an electric field, is the preferred method (Figure 1b). Straightforward electrophoresis in solution does not work, however, because molecules of different length all migrate at the same speed [1, 2]. A variety of tricks is required to achieve good separations. Passing the DNA through a gelified medium, for example, works well in many instances. Indeed, gel electrophoresis has been phenomenally successful and is now so ubiquitous that it is impossible to leaf through a molecular biology journal without seeing images of gels. It played a vital role in the successful completion of the Human Genome

Project, where it is was used both to sequence DNA and to make physical maps. However, in many respects gel electrophoresis is unsatisfactory and one wonders whether there may not be a more effective alternative.

The advent of new technologies in the physical sciences permits novel approaches to old problems. In recent years, the technique of microlithography has been developed to engineer microscopic features on the surface of silicon chips. This technology is being adapted to create specially-designed electrophoretic chambers – miniature obstacle courses engraved on a silicon wafer. In part, this work is motivated by the immediate need to accomplish better, more efficient, fractionation of DNA molecules. In part, by the more grandiose conception that microfabrication techniques promise a new level of convenience to biological researchers. One day, all the individual steps of an experiment – isolation, manipulation and analysis – could be accomplished without test-tubes. A miniture laboratory on a single chip. Electrophoresis is, in many respects, an ideal case where physics can help the biological sciences. On the practical side, technologies provide the capacity to engineer precisely controlled electrophoretic environments and observe the dynamics of individual molecules. On the theoretical side, DNA isolated from the cell behaves as a purely physical system and reliably obeys the laws of classical dynamics and statistical mechanics. So theory and experiment can advance hand in hand to create new solutions to the challenges posed by DNA sorting.

2. Physical description of DNA

The enormous length of DNA molecules confers several advantages when it comes to a physical description. First, their detailed chemical structure (even their celebrated double-helical nature) can be neglected. They can simply be treated as long chains and a minimum number of physical parameters characterizing length, flexibility and charge density suffice to describe them. Second, the comportment of individual molecules can be adequately described by statistical methods.

Measuring along the polymer backbone, a DNA molecule has an overall length L proportional to the number of nucleotides that it contains. This can be very long: even the smallest chromosome, containing a few million nucleotides has $L \approx 1$ mm – a macroscopic size! However, in solution the polymer has a much smaller overall dimension. This is because the DNA is bent this way and that as it gets buffeted by the water molecules and consequently becomes randomly coiled. Were you to follow along the backbone, you would lose track of the original direction in which you were headed after a short distance. This length scale evidently depends on the inherent flexibility of the DNA and on the magnitude of the Brownian forces. It is called the 'Kuhn length' b. For double-helical DNA $b \approx 100$ nm, equivalent to 300 nucleotides, while the more flexible denaturized single strand has $b \approx 5$ nm. Since the molecule looks rigid on scales shorter than the Kuhn length, but is randomly coiled on larger scales, it may be conveniently pictured as a random walk made up of $N = L/b$ straight segments of length b. The overall size of the coil R may be estimated from the statistics of the random walk:

$$\langle R^2 \rangle = Nb^2 = Lb. \tag{2.1}$$

For a fragment containing a milllion nucleotides, $R \approx 5\,\mu$m, big enough to resolve with a light microscope.

The arguments above apply to the case where no external forces are acting on the polymer. During electrophoresis, however, the electric field driving the motion might also deform the chain. What typical magnitude of force would need be applied to stretch the DNA? Suppose that one could grasp a molecule by the ends and pull them apart. As the end-to-end separation R increases, the number w of corresponding random configurations of the DNA decreases, resulting in an entropy loss. Equilibrium is reached when the spatial gradient of $kT \log w$ is equal to the applied force F. The statistics of the random walk (the probability distribution of the end- to-end vector is approximately Gaussian with variance given by Eq. 2.1) then leads to

$$F \approx \frac{kTR}{3Lb}, \quad R \ll L. \tag{2.2}$$

At small extensions, the DNA acts as a Hookean spring, with spring constant $kT/3Lb$. The longer the DNA, the easier it is to deform. Eq. 2.2, also indicates that the typical force required to stretch the DNA to almost its full length is

$$F_{stretch} \sim kT/b. \tag{2.3}$$

For double-stranded DNA, $F_{stretch} \approx 0.1$ pN. As we shall see, this value is not so high and it is relatively easy to stretch out DNA molecules.

3. Electrophoretic force

DNA is an acid, so in solution it is negatively charged. In fact, there is one negative charge per nucleotide, but this attracts positive counterions in the buffer solution which form a so-called 'double-layer' sheathing the DNA polymer [1]. The ions in the inner layer, which has a thickness of a few angstroms, are tightly bound to the molecule, while those in the outer layer are mobile. The concentration of ions in the outer later declines exponentially over the Debye length, which is typically a few nanometers in the buffer solutions used in electrophoresis. This screening of electrostatic interactions on a scale shorter than the Kuhn length ensures that the DNA remains randomly coiled in solution. It also has important consequences for the dynamical behaviour when an electric field is applied. Because the field acts on both the negatively charged DNA and the positively charged counterions, its effect is to generate a shear flow in the mobile outer layer of counterions, and consequently to establish relative motion between the surface of the DNA molecule and the distant, neutral fluid. This flow screens the hydrodynamic interaction between different parts of the DNA molecule, so that the friction is essentially local and may be characterized by a uniform friction coefficient ζ_0 per Kuhn segment. The relative velocity between the molecule and the fluid depends on the residual charge of the DNA plus the bound counterions, which in turn depends on the composition of the buffer. But the net effect is that the DNA interacts with the external field as if each Kuhn segment has an effective charge q_0. For double-helical DNA, $q_0 \approx 50e$, and for single-stranded DNA $q_0 \approx e$.

In free solution, then, both the total force acting on a DNA molecule and the total friction opposing its motion are proportional to its length. As a result, the electrophoretic mobility $\mu \equiv v/E$ is independent of size, $\mu = \mu_0 = q_0/\zeta_0$, and molecules cannot be fractionated. The simple expedient of using a gel matrix to support the electrophoretic solution, however, can result in excellent separations [3]. This is particularly true of sequencing electrophoresis, in which single-stranded DNA molecules containing several hundred nucleotides can be separated in a polyacrylamide gel; experimentally, it is found that the mobility declines roughly inversely with the length of the molecule [4]. How does the gel network help? One might think that it is just acting as a sieve, blocking the big molecules and letting the little ones through. But it is not so simple, because DNA is flexible and even very long molecules can slither through small holes.

4. DNA sequencing: gel electrophoresis of single-stranded DNA

A long polymer, which when randomly coiled is too big to fit in a single open space in a gel, must thread its way from one pore to another. The description of its dynamics is based on Edward's hypothesis [5] that the polymer is effectively confined to the contorted tube formed by this sequence of pores (Fig. 2). Buffeted by Brownian impulses but prevented from moving laterally by the gel fibres, the molecule can only wriggle along the tube axis. So globally, the polymer diffuses backwards and forwards along its own contour or 'reptates' like a snake slithering through thick grass [5, 6].

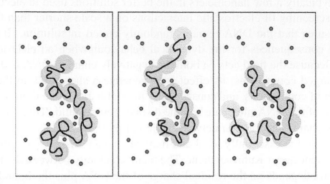

Fig. 2. DNA in a gel is effectively confined to a tube (shaded) by the surrounding gel fibers (drawn in cross-section as circles). As the molecule diffuses longitudinally along the tube, new sections of the tube are created at one end, and old sections are deleted at the other end.

4.1. Reptative dynamics

In the simplest 'primitive path' formulation of reptation, the molecule is represented by a freely-jointed chain of N segments, each of linear dimension equal to the average size of the gel pores. If the polymer is flexible on the scale of the pores (the usual case in gel electrophoresis), each primitive chain segment corresponds to a coil of the polymer containing $p = (a/b)^2$ Kuhn lengths. Thus the parameters of the primitive path picture may be expressed in terms of the molecular parameters introduced above: the number of primitive chain segments is $N = L/pb$ and the charge and fiction per segment are $q = pq_0$ and $\zeta = p\zeta_0$ respectively. Within the tube the DNA moves as if it were a chain of beads connected by springs, according to the description of Rouse [5]. So the curvilinear diffusion coefficient D_s of the primitive chain along the tube may be identified as the Rouse diffusion coefficient of the molecule;

$$D_s = kT/N\zeta. \tag{4.1}$$

Owing to the random path of the tube, the translational diffusion coefficient D of the molecule in real space is smaller than the curvilinear diffusion by a factor N so that

$$D \sim N^{-2}. \tag{4.2}$$

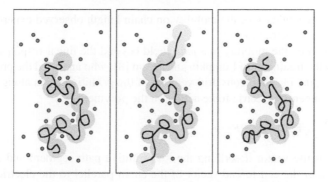

Fig. 3. The diffusion of 'stored length' of DNA along the tube, which ultimately leads to the reptative motion of the centre of mass, also causes the tube length to fluctutate.

The primitive path dynamics describes the *average*, coherent motion of the polymer along the tube. The other mode of motion associated with reptation concerns the longitudinal *fluctuations* of the molecule within the tube [5] (see Fig. 3). In the field-free situation, these fluctuations are governed by Rouse dynamics; during a short time interval t, the motion of a given part of the chain is correlated only with that of the adjacent section of the molecule whose Rouse relaxation time is equal to t. One consequence is that the tube length is not fixed, but undergoes equilibrium fluctuations of magnitude ΔN which varies as

$$\Delta N \sim (t/\tau)^{1/4}, \quad t < N^2\tau, \tag{4.3}$$
$$N^{1/2}, \quad t > N^2\tau, \tag{4.4}$$

where

$$\tau = \frac{a^2\zeta}{kT} \tag{4.5}$$

is the Rouse relaxation time of a primitive path segment.

4.2. Biased reptation

The application of the reptation model to gel electrophoresis, where in addition to Brownian impulses the polymer also experiences a force due to the external field, was first suggested by Lerman and Frisch [7]. They pointed out that in the *zero-field limit*, the mobility must be related to the centre-of-mass diffusion coefficient via the Einstein relation. Since the total driving force on the DNA is proportional to its length, this implies that $\mu \sim N^{-1}$. Thus reptation accounts for the inverse dependence of the mobility on chain length observed experimentally in sequencing electrophoresis.

To determine the mobility in a *finite* field beyond the linear response regime, we follow the treatment of Lumpkin and Zimm [8], who analysed the problem by considering the instantaneous drift velocity of the primitive chain along the tube. The total tangential electric force acting on the polymer is

$$F_s = \sum q\mathbf{E}.\mathbf{R_i} = qER_{\parallel}/a, \tag{4.6}$$

where $\mathbf{R_i}$ is the vector describing the ith primitive path segment and $R_{\parallel}$ is the component of the end-to-end vector of the chain parallel to the electric field $\mathbf{E}$. From the Einstein relation, Eq. 4.1 implies that the effective friction coefficient of the primitive chain in the tube is $\zeta_s = N\zeta$ so that its instantaneous drift velocity along the axis is

$$v_s = F_s/\zeta_s = \epsilon \frac{R_{\parallel}}{Na} \frac{a}{\tau}. \tag{4.7}$$

where the parameter

$$\epsilon = qEa/kT \tag{4.8}$$

is a dimensionless measure of the field strength.

The centre-of-mass velocity v of the DNA in the field direction is related to the instantaneous curvilinear velocity v_s along the tube by

$$v = v_s \frac{R_{\parallel}}{Na}, \tag{4.9}$$

so that the electrophoretic mobility is

$$\mu \equiv \langle v \rangle / E = \mu_0 \frac{\langle R_{\parallel} \rangle}{(Na)^2}, \tag{4.10}$$

with $\mu_0 = q_0/\zeta_0$, as before. Thus the mobility depends on the average tube conformation and the problem is to determine what this is. The question of the perturbation of the equilibrium configuration due to the external field is a delicate one. During the reptation process, perturbation of the tube can originate only at its ends. It was initially argued [9, 10] that, as the chain slides along the tube, the end segment of the primitive path acts like a dipole which the field (in competition with the randomizing effect of thermal motion) tends to orient. The argument was unreliable, however, because by using the primitive path picture it failed to take account of the detailed, short-timescale motion of the chain ends which is governed by longitudinal fluctuations of the chain in the tube.

The problem of the perturbation of the conformation, which is really a dynamical one, may be tackled using a quasi-equilibrium approach. To do so, however, careful consideration must be taken of the lengthscale on which the chain has the opportunity to equilibrate before it gets restricted in the tube [11]. This is determined by a competition between fluctuations and drift: fluctuations liberate the end section of the chain from the tube while the advance of the molecule through the gel seeks to imprison it. After a short time interval t, the number of tube segments that have been altered by longitudinal fluctuations is given by

$$n_{fluc} \sim (t/\tau)^{1/4}. \tag{4.11}$$

Meanwhile, the primitive path has drifted through a number of pores equal to

$$n_{drift} \sim \frac{v_s}{a} t. \tag{4.12}$$

Equality of these two quantities

$$n = n_{fluc} = n_{drift}, \tag{4.13}$$

determines the size n of the terminal section of the chain that has time to equilibrate before it gets constricted within the tube. One may imagine the terminal section of the primitive path to rapidly sample alternative routes through the gel, influenced by both the electric field and the randomizing effect of thermal agitation. The longer the terminal section, the greater the total force acting on it and the more it tends to get oriented. As the molecule advances, this orientation is transferred to the entire tube so that if θ is the angle that a tube segment makes with the field direction, one obtains

$$\langle \cos \theta \rangle \sim n\epsilon. \tag{4.14}$$

Equations 4.11–4.13 together with Eq. 4.7 provide a set of equations for the orientation that must be solved self-consistently. In the long chain limit where the Gaussian component of the chain conformation is negligible so that $\langle R_{\parallel} \rangle = Na \langle \cos \theta \rangle$, the solution is

$$\langle \cos \theta \rangle \sim \epsilon^{1/2}. \tag{4.15}$$

Thus long chains are oriented with an end-to-end distance that varies with the square-root of the field strength. Equation 4.10 then implies that their mobility varies linearly with the field strength and is independent of molecular size. For shorter chains, $N < N^*$, it is the orientational component of the conformation that is negligible compared to the Gaussian part so that $\langle R_{\parallel}^2 \rangle \sim Na^2$. Then, from Eq. 4.10, the mobility is inversely proportional to the chain length and the linear response regime is recovered. To summarize, the scaling behaviour predicted by the theory is

$$\mu/\mu_0 \sim \epsilon^0 N^{-1}, \quad N \ll N^*, \tag{4.16}$$
$$\epsilon^1 N^0, \quad N \gg N^*, \tag{4.17}$$
$$N^* \sim \epsilon^{-1}. \tag{4.18}$$

Translating from the primitive path parameters back to the molecular parameters reveals the full dependence of the mobility on the field strength E, molecular contour length L and pore size a

$$\mu/\mu_0 \sim (L/b)^{-1}(a/b)^2, \quad L \ll L^*, \tag{4.19}$$
$$(E/E_0)(a/b)^3, \quad L \gg L^*, \tag{4.20}$$
$$L^*/b \sim (E/E_0)^{-1}(a/b)^{-1}. \tag{4.21}$$

where $E_0 = kT/q_0 b$. The limit of separation L^* may be increased by reducing the field strength or by using gels with smaller pores, but since the variation is only linear in each of these variables a substantial improvement is difficult to obtain. Note that for sequencing electrophoresis, $E_0 \approx 40\,000$ V/cm, so that using a polyacrylamide gel with pores that are several nanometers in size and applied fields of the order of 100 V/cm, it is possible to separate DNA molecules containing hundreds of nucleotides.

4.3. Repton model

While the above scaling arguments are plausible, they are not entirely rigorous, so it is encouraging that they are supported by studies of a lattice model of reptation, known as the 'repton model' [12]. This model captures, in a simplified manner, the two main aspects of reptative dynamics: confinement of the chain within a tube; and internal breathing modes of the chain within the tube, which lead to tube length fluctuations. As indicated in Fig. 4, the chain is divided into a number of sections which slightly exceeds the average number of primitive path segments. The end-point of each section is then replaced by a 'repton', which is projected onto the closest site on a one-dimensional lattice, directed along the field. This procedure yields an ordered chain of reptons, in which successive reptons occupy either adjacent lattice sites, or the same lattice site.

The dynamics of the model consists of permitting any repton which has one neighbour located at the same site, and the other neighbour at an adjacent site, to hop stochastically to the adjacent site. This represents the diffusion of what de Gennes termed a 'length defect' along the tube [6]. In the absence of a field, hops occur with equal rates in either direction. When a field is present, hops up-

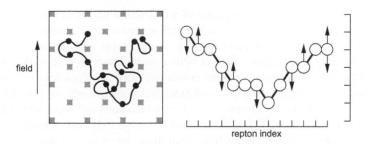

Fig. 4. Lattice model of biased reptation. The DNA molecule is represented as a chain of reptons on a 1-dimensional lattice. Individual reptons can hop to adjacent lattice sites, as indicated by the arrows.

and downfield occur at rates w_+ and w_- which satisfy the principle of detailed balance:

$$\frac{w_-}{w_+} = \exp(-\epsilon). \tag{4.22}$$

Special treatment is required for the end reptons. If an end repton occupies the same site as its neighbour, it may hop to either adjacent site with probabilities biased by the field:

$$p_\pm = \frac{1}{[1 + \exp(\mp\epsilon)]}. \tag{4.23}$$

This represents a section of the chain coming out of the old tube, and exploring a new gel pore. An end repton that occupies a different site to its neighbour, on the other hand, can only hop to join its neighbour. This represents a section of the chain retracting into the tube.

Because the repton model can be mapped onto an Ising-like spin model, many powerful analytical techniques may be brought to bear on it to study the dynamics [13, 14]. Thus a number of results have been rigorously proved. Efforts to derive the scaling of the mobility of long polymers, Eq. 4.17, have so far been unsuccessful, however. Nevertheless, extensive numerical simulation [15] of the repton model indicates that the mobility does indeed scale as $\mu \sim \epsilon$ when $N \gg 1/\epsilon$.

4.4. Strategies for DNA sequencing

Interesting though the physics may be, for practical purposes the orientation is a nuisance, for it limits the length of fragment that can be sequenced. Presently, about 500 nucleotides is the typical extent of sequence that can be determined at

one go. This means that the sequence of a restriction fragment has to be laboriously reconstructed from about one hundred separate pieces. Various strategies
have been proposed to augment the size of fragments that can be sequenced.
One possibility is to replace the gel with a more temporary network, like the
one formed by concentrated polymer solutions; since the obstacles to motion are
continually changing, the DNA is not held back long enough for orientation to
build up [16]. The best performance to date, which achieved sequencing of 1000
nucleotides in an hour [17], was achieved by using long, thin capillaries filled
with a viscous polymer solution. The small diameter of the capillary permits the
use of high fields without too much heating, and the long run length provides
good resolution. The sequencing mixture containing the fluorescently labelled
oligonucleotides is loaded at one end, and the molecules are detected as they
reach the far end of the capillary. From Eq. 4.19, the passage time increases linearly with the molecular length, so oligonucleotides of different length emerge
evenly spaced in time.

5. Gel electrophoresis of long double-stranded DNA molecules

Sorting restriction fragments requires the ability to fractionate long, double-stranded DNA molecules. A natural question arises in comparison with the task of
sequencing single-stranded DNA. Is it possible to scale up all dimensions – the
contour length, the Kuhn length and the pore size – and maintain the same dynamical behaviour? Does biased reptation describe the motion of a 50, 000 base-
pair fragment in an agarose gel with a typical pore size of 500 nm, for example?
Alas, no! At large scales the dynamics becomes much more complicated. The
tube hypothesis is no longer valid and the reptation model is inappropriate. This
is because when the gel fibres are widely spaced, the entropy loss associated with
a loop of DNA squeezing between a pair of obstacles can be outweighed by a gain
in electrostatic energy, even at very low field strengths. Consequently there is no
free energy barrier to prevent the DNA leaking out of the sides of the tube, and
forming a 'hernia' (see Fig. 5).

5.1. Complex dynamics in constant fields

The gel-electrophoretic dynamics of long double-stranded DNA molecules was
first described by Deutsch [18] using numerical simulations. The DNA executes
an episodic type of motion that, when viewed under a microscope, appears almost animate [19] (see Fig. 6). Typically, a molecule is oriented along the field
and migrates in that direction. But frequently a loop of the molecule, instead of
obediently following the route taken by the head, passes a different way around

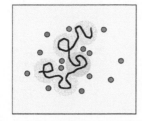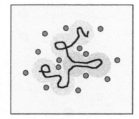

Fig. 5. Initiation of a hernia, as a loop of DNA slips between two gel fibres. The entropy loss associated with a hernia spanning n pores is $\Delta S \sim nkT$, but the gain in free energy is $\Delta U \sim n^2 \epsilon kT$, so the free energy barrier to for hernia formation is $\Delta F \sim kT/\epsilon$. The barrier is negligible when $\epsilon > 1$, which is typically the case for electrophoresis of double-stranded DNA in agarose.

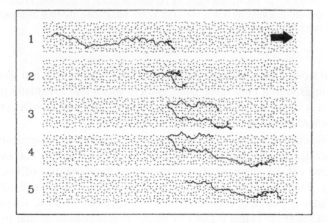

Fig. 6. Typical cyclic motion of a 100 kilobase DNA molecule in an agarose gel.

a gel fibre and sets off in pursuit. Loop and head advance simultaneously, until the former unravels as the tail of the molecule passes along it. This leaves the DNA hooked over an obstacle in a long U shape. In this configuration, the DNA can be very highly extended. As mentioned previously, the force required to stretch double-stranded DNA, given by Eq. 2.3, is about 0.1 pN. In a typical field $E = 1\,\mathrm{V/cm}$, this corresponds to the electrophoretic force acting on a 50, 000 nucleotide piece of DNA. Molecules longer than this, when hooked on an obstacle, are almost fully extended by the electric field pulling on both arms. At this point, the migration of the DNA can be almost completely arrested. This is particularly the case if the arms of the U are of nearly equal length, for then the net force acting to move the molecule, which is proportional to the end-to-end vector $R_{\parallel}$, is very small. However, the slightly greater force tugging on the

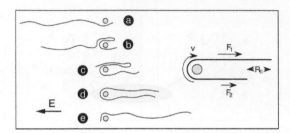

Fig. 7. Long DNA molecules undergo the episodic motion sketched in a-e. Configuration d, in which the DNA is hooked over an obstacle and stretched into a long U-shape can last for a long duration. A simplified model for the disengagement of the molecule is illustrated. The total force influencing the DNA to slide around the obstacle is equal to the difference in the forces acting on the two arms $F = F_1 - F_2 = qER_\parallel/b$. The friction opposing the motion is $N\zeta$, proportional to the DNA length. The instantaneous sliding velocity is thus $v = \mu_0 ER_\parallel/L$. Integrating, the total disengagement time is $t_{hook} \sim L\exp(L/R_\parallel)/\mu_0 E$.

longer arm slowly hauls the molecule around the obstacle and, just like a heavy rope slipping round a pulley, the DNA gathers speed and eventually pulls free. Whereupon it immediately starts to contract elastically and rapidly returns to a situation where the whole cycle can repeat itself.

Does this episodic motion result in a useful separation of fragments according to size? Experimentally there is no evidence that it does. The reasons are two-fold and can be appreciated by considering the dynamics of hooking and unhooking. First, hooking is a random process and a very wide distribution of hooking times is possible, depending on the asymmetry of the U shape formed. It is easy to see (Fig. 7) that if DNA of length L gets hooked so that its ends are initially a distance $R_\parallel$ apart, the time it takes to slide off the obstacle varies as $\exp(L/R_\parallel)$. This exponential distribution of times leads to a big variance in the migration velocity of individual molecules, causing poor electrophoretic resolution. Second, experimental observations suggest that the probability distribution of the parameter $R_\parallel/L$ characterizing hook asymmetry is independent of molecular length. In this case, the mean time that DNA requires to disengage from an obstacle varies linearly with its length. But since the molecules usually hook again as soon as they pull free, the mean distance travelled between hooking events also varies linearly with molecular size. As a result, the mean velocity is length- independent. Hooking slows down all molecules by the same amount.

5.2. *Pulsed-field gel electrophoresis: separation of restriction fragments*

While standard gel electrophoresis fails to fractionate double-stranded DNA molecules containing more than about 10 thousand nucleotides, it has been known for some time how to separate fragments containing up to 10 million nucleotides. This remarkable improvement can be achieved by pulsing the electric field [20–22]. The technique works by playing on the transient behaviour of the molecules following a sudden field switch. Unlike the steady-state mobility, this is length-dependent: small molecules can readjust rapidly to changes in the field, but long ones are too sluggish to do so.

The episodic motion of the DNA discussed above gives a clear insight into how pulsed-field electrophoresis works at the molecular level. Consider, for example a repetitive pulsing pattern in which the field is applied initially for time T in the forward direction, then time $T/2$ in the reverse direction, and so on. Small molecules will quickly readjust at each switch, then migrate steadily in the direction of the field and since the forward pulse lasts longer than the reverse, they will advance through the gel. But what about longer molecules whose average hooking time is close to T? At the end of the forward pulse, they are typically hung over an obstacle in an extended U. When the field reverses, the arms of the U retract, driven by both the field and elastic recoil. Subsequently, when the field reverts to its former direction, the arms extend again, but there is not sufficient time for the DNA to disengage from the obstacle before the field switches once more. Consequently, the molecule remains hung on the same obstacle for many successive field cycles and makes no progress.

The technique called 'field-inversion gel electrophoresis' [21] uses the above pulsing pattern and has proved successful at resolving megabase molecules. The original pulsed field method, discovered by Schwartz and Cantor [20], uses a different protocol: the field is alternately applied for equal times T along two different directions. Figure 8 shows the variation of mobility with pulse time that is observed in simulations of crossed-field electrophoresis [23]. In this case, the angle between the fields is 120° but very similar patterns of the relative mobility are obtained for a wide range of obtuse angles. The mobility curves dip at a value of the pulse time that differs as the molecular size changes. In fact, the minimum mobility occurs at a switch time that is roughly proportional to the chain length. Consequently, a clear resolution of molecules up to a given size N_{max} may be achieved by using a pulse time that varies approximately linearly with N_{max}. This behaviour is well documented experimentally, where it has long been used as a rule of thumb for calculating the pulse regime to obtain molecular separation in a given range. If the field angle is orthogonal or acute, on the other hand, the mobility of all molecules is practically independent of the pulse time and no separation is obtained.

The marked difference between pulsed-field electrophoresis using acute and obtuse angles, which is well known in practice [24], reflects a contrast in the mechanism of reorientation. This is most readily examined by observing a chain's motion under a pulse regime that substantially depresses its mobility at 120° but

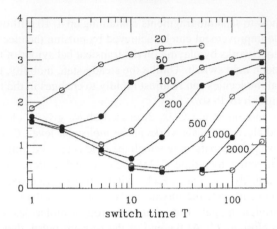

Fig. 8. Pulsed-field separation of DNA molecules. Average migration speed along the bisector of the two fields, as a function of migration time. Curves labelled by length of DNA molecule (in kbp).

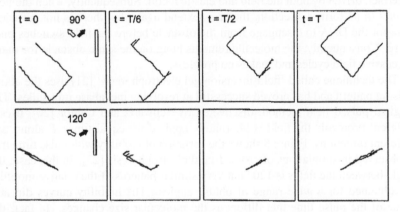

Fig. 9. Reorientation of DNA during a single pulse in crossed-field electrophoresis, with field angles of 90° and 120°.

not at 90°. Figure 9 shows the typical behaviour during a single pulse for the case where the switch time is of the same order of magnitude as the molecular reorientation time. The chain typically alternates between U-shaped configurations, aligned along the two field directions. In the case of an orthogonal angle, the U is broad-based. Following the field switch, both chain ends turn to move in the new field direction and a hernia starts to grow from the place where the tube is bent (it is here that an abrupt drop in the field gradient measured along the tube axis causes a pooling of the chain which can then seep out between gel fibres). Since there is an approximately equal force pulling on both the leading end and the hernia, the molecule remains extended as it changes direction. At the end of the pulse, it is once more in a broad U but has *advanced* along the gel. For this reason, pulsing has little effect on the mobility. In the case of an obtuse angle, on the other hand, the U is typically very narrow. When the field is changed, both ends of the chain move off in the new field direction, but also a hernia sprouts from the base of the U. Since the hernia is further downfield, is has an advantage over the ends and once the chain becomes taut, tugs them back into the arms of the U. Due to the obtuse angle, there is a component of the field that then drives the rapid retraction of the chain down the arms, leading to bunching at the base where more hernias are immediately created. These grow, sucking up the rest of the molecule, and eventually unravel to leave the molecule once again in a narrow U. During the reorientation process, the molecule makes *little progress* along the gel; it retracts backwards at first, then starts to move forward but gets held up as the hernias resolve into a U. This causes a very large reduction in its mobility.

The way that obtuse-angle crossed-field electrophoresis discriminates between molecules of different size is evident from these observations. Short molecules reorient rapidly and spend most time migrating steadily along the current field direction. They consequently have a high mobility, close in value to the resultant continuous field mobility resolved along the diagonal. Larger fragments spend as a significant proportion of the time reorienting, during which their progress is hindered. The longer the molecule, the more time it loses changing direction and the lower its mobility. It is this feature that leads to the good separation according to size. Still larger molecules whose reorientation time is longer than the pulse time, respond too slowly to entirely change direction and so are not differentiated by the pulsed field.

Many varieties of pulsed-field techniques exist, but all of them work by a similar mechanism, which may be regarded as a sort of resonance. Molecules of a characteristic length have a typical hooking time that is comparable with the pulse time T. The pulsed field drives the internal modes of these chains, making them stretch and recoil, rather than forcing their movement through the gel. Consequently, they may be separated from shorter molecules which do migrate.

5.3. Difficulty of separating very large molecules

While pulsed-field electrophoresis has enabled the separation of large restriction fragments, it suffers from a number of drawbacks that prevent it from becoming a standard laboratory procedure. First, there is no general recipe for choosing the pulsing parameters that provide fractionation in a particular range. Consequently, automation of the technique is problematic. Second, the complicated dynamics leads to a diverse range of reorientation times for molecules of the same size. This causes band broadening, limiting the resolution of the technique. Third, it has been found that very long molecules can get irreversibly trapped in the gel, in a manner that depends on the field strength [25,26]. One possible cause is that the large force pulling on the arms of a molecule in a U-shaped configuration is sufficient to locally melt the double-helix near the base of the U. If one of the melted strands also happens to be nicked, there is the possibility than when the two strands anneal again, they will enclose a gel fibre. Because the force increases linearly with DNA length, the critical field at which irreversible trapping occurs decreases inverse-linearly with DNA size. In practice, this means that fields little greater than 1 V/cm can be used to separate megabase DNA molecules and the separation is consequently very slow, often taking days.

6. Obstacle courses on microchips

The advent of optical and electron-beam lithography, permitting precise engineering on the microscopic scale, has led to the development of a whole range of nanotechnologies. These techniques may be adapted to make miniature solid-state electrophoresis chambers [27]. Two-dimensional obstacle courses of almost any pattern can be created on a silicon wafer in the following way. The surface of the chip is etched away to a depth of $0.1–1$ μm, through a mask which protects selected regions so that they remain raised. The structure is then sealed with a cover slip and the gap between silicon and glass filled with saline solution, into which a mixture of DNA molecules can be injected. The unetched regions provide obstacles to the free movement of DNA molecules migrating in the fluid. Figure 10 shows the most straightforward example of such a device – a regular lattice of cylindrical columns. Over the past decade, a variety of techniques have been developed to manufacture such arrays in a range of different materials [28]. Perhaps the most convenient method is to make a negative 'master' using lithography, and then use this to imprint the chosen design in a polymer elastomer such as PDMS.

The advantages of these solid-state devices over a conventional gel can be readily appreciated. The structure is completely regular and well-controlled and the geometry can be chosen at will. A gel, by contrast, is random and ill- characterized. Also, the scale may be varied to cover a range that is not feasible using gels. A gel with pores 1 μm in size, for example, would be too fragile to handle. Furthermore, the motion of DNA molecules in the device can easily be examined. The DNA can be stained with a fluorescent dye and observed under a light microscope. Since the arrays are 2- dimensional, the DNA can be kept in focus over a wide field of view. Videomicroscopy allows ready comparison between experiment and theory, greatly aiding the design of devices to perform specific tasks.

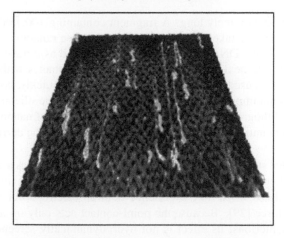

Fig. 10. Fluorescently-stained DNA molecules moving through an array of obstacles, engraved on a silicon chip. The posts have diameter 1 μm, height 0.15 μm and centre-to-centre spacing 2 μm. The longest DNA molecules in the picture contain approximately 100 000 nucleotides.

6.1. Collision of a DNA molecule with an obstacle

When long DNA molecules are observed migrating in the regular array of obstacles shown in Fig. 10, they display the same type of episodic behaviour that we have previously described in constant-field gel electrophoesis. The molecules get hooked on an obstacle, form an extended U, pull free and relax; and the cycle then repeats (see Fig. 10). Thus the array fails to separate molecules according to size for the same reason that a gel does not work: the typical cycle time is proportional to the length of the molecule, but so is the distance travelled per cycle. This suggests an obvious solution: Impose a particular frequency of hooking events. This may be done by changing the geometry of the array. For example, suppose that the majority of the obstacles are removed from the regular lattice, leaving only single rows of posts separated by long open spaces. In such a device, the DNA molecules hook only when they encounter a line of obstacles, so the distance travelled between hooking events is constant, independent of the molecular size. The average hooking time, however, remains proportional to the DNA length, so each encounter with a row delays a longer molecule more than a shorter one. The desired fractionation is achieved.

This simple example demonstrates the versatility of microfabricated arrays. Variability of the geometry permits straightforward solutions that would be impossible to implement using gels. In fact, this particular design is expected to work much better as a sequencing tool than as a method of separating long restriction fragments. The reason is that the timescale for relaxation of big DNA molecules can be extremely long. A fragment containing 100, 000 nucleotides, for example, typically takes a few seconds to relax to a random coil after it has been stretched. If the DNA does not have time to recoil as it traverses from one row of posts to the next, it might slip through the obstacles without hooking. The tiny fragments used in DNA sequencing relax very quickly, however, so the timescale does not impose a limitation. By miniaturizing the silicon arrays, using electron beam lithography to create obstacles on the scale of nanometers, it may be possible to manufacture a sequencing device that rivals the current gel-based methods.

Experimental studies of long double-stranded DNA molecules interacting with a single row of posts have nonetheless proved useful for understanding how polyelectrolytes behave when they experience both an electrophoretic force and a point-contact force [29]. Because the point-contact acts only on the DNA, and not on the counterions, its effect is not hydrodynamically screened. As a result, the argument given in the caption of Fig. 7 for the extensional force on the DNA molecules is an oversimplification. In fact, a DNA molecule hooked over an obstacle gets stretched as though it were in a hydrodynamic flow of velocity $v = \mu_0 E$ [30]. One immediate consequence is that the degree of stretching depends on the etch-depth of the array. In shallow arrays, hydrodynamic interactions are screened at short length scales, and the molecular extension is consequently greater for a given field strength. The time that a molecule takes to relax, once it has pulled free of the post, is also longer in a shallow array, for the same reason. As we shall see, some separation techniques in microarrays require the DNA to be stretched, while others require it to remain coiled, so the ability to modify the degree of extension by changing the etch-depth is a useful feature.

6.2. Efficient pulsed-field fractionation in silicon arrays

While a simple lattice of obstacles fails to sort long molecules using a continuous field, we might expect it to work well in pulsed-field conditions. Here, the regularity of the array provides a real advantage over a gel, because if the field is carefully applied along the axes of the array, the molecules can move along the channels between rows of obstacles without bumping into the posts.

Figure 11 shows a simulation of DNA moving in a regular hexagonal array under crossed-field conditions. The field alternates with switch time T between two directions aligned along axes of the array. Since the field pulls first one way, then the other, the repeated pulsing maintains the molecules in a highly extended state. This tautness prevents loops from growing in the middle of the molecule so that the motion is always led by an end. Since the molecules remain linear, the dynamics is remarkably simple. The total force F pushing a DNA molecule along its contour is proportional to the projection of the end-to-end vector in

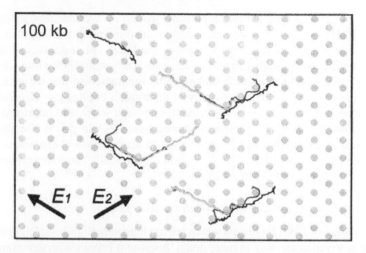

Fig. 11. Simulation of the motion of DNA in a hexagonal array of obstacles, under pulsed-field conditions. Molecular conformations are shown at the beginning of each pulse (black), and at times $t = T/6$ (dark grey) and $t = T/2$ (light grey), where T is the pulse time.

the field direction. So immediately following a field switch through an *obtuse* angle, *F changes sign*. The molecule sets off in the new direction led by what was previously its back end. This head-tail switch causes the DNA to retrace part of the route that it took during the previous pulse. If one looks along the bisector of the fields, which is the direction of net migration, the molecules first move backwards before they advance; longer molecules backtrack further and this gives rise to fractionation.

DNA fragments which are too long to realign completely before the end of a pulse fail to advance through the array, so only those molecules whose reorientation time is shorter than the pulse time can be fractionated. The total time that it takes a molecule to reorient is proportional to its length and inversely proportional to the field strength. This implies that the upper limit of the separation range increases linearly with both the pulse time and the field strength. The recipe for setting the field parameters to obtain the desired range of fractionation could hardly be simpler. Molecules smaller than the limiting size first backtrack and then move forward as they reorient, so that they make no progress during this period and advance only during the remaining fraction of the cycle. Consequently, their average migration speed falls linearly with increasing length. The backtracking mechanism leads to a clean, linear fractionation of the DNA [31].

This way of switching the field through an obtuse angle to fractionate long DNA seems almost too simple. But often the simplest methods are the best. In fact, the backtracking mechanism was previously proposed by Southern [32] to explain how pulsed-field gel electrophoresis works. As we saw, though, the dynamics of DNA in gels is much more complicated owing to the way that loops of the molecule can easily squeeze between the fibres. Microfabricated arrays, however, offer the possibility to realize the desired dynamical behaviour – by sufficiently spacing the obstacles, hooking can be suppressed so that the molecules remain linear. The resulting uniformity of the motion implies less dispersion in the migration velocity leading to a reduction in band width and improved resolution. Also, the problem of trapping of megabase DNA in gels is alleviated by using an array, so higher fields can be used to achieve faster separations. The improvement is straggering. A fractionation that typically takes several hours in a gel can be completed in just 10 seconds [33].

6.3. Continuous separation in asymmetric pulsed fields

The dynamics of DNA fragments which are too long to realign completely during a pulse is worth examining in more detail, because it reveals a *broken symmetry* in the system. Suppose that such a molecule has somehow managed to get completely aligned along one of the field directions. When the field is switched, the molecule backtracks and moves off along the new direction, but does not have enough time to reorient completely. When the field reverts, the molecule slides back into its original channel and, because it now has a head-start, it *is* able to completely realign before the end of the pulse, and can even drift a little way along the channel (see Fig. 12). With repeated pulsing, the molecule will con-

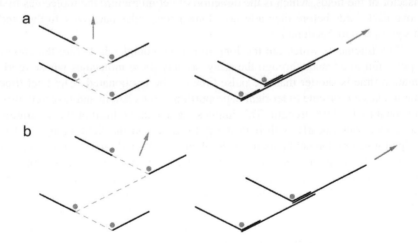

Fig. 12. Motion of short and long molecules in pulsed-field electrophoresis with equal pulse times (a); and with unequal pulse times (b). Arrows indicate the net direction of migration.

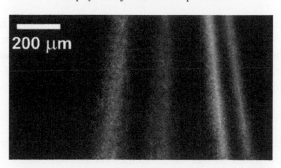

Fig. 13. Continuous separation of a DNA molecules by asymmetric pulse-field electrophoresis in a 'DNA prism'. The four streams of molecules contain DNA of sizes 61kb, 114kb, 155kb and 209kb.

tinue to advance in this direction. Clearly this argument applies equally well whichever field direction the molecule is initially aligned along. So there is a broken symmetry: very long molecules will eventually get aligned along one or other of the field directions, and will subsequently keep migrating in that direction.

At first sight, this behaviour seems to be a nuisance, since it will lead to a smeared band of long molecules along the two field directions, which could interfere with the separated bands of smaller molecules along the direction of the bisector. On reflection, however, this behaviour provides a wonderful opportunity. Consider what happens if the symmetry is *deliberately* broken by using pulses of different durations in the two field directions. Long molecules which cannot reorient during the shorter of the two pulses will move along the principal axis of the array (the direction in which the pulse is longest). The very shortest molecules, on the other hand, will be able to respond quickly to all the field switches and so will migrate in the direction of the time-averaged field. And molecules in between these two extremes will behave in an intermediate way, and migrate at an angle that depends on their size. Thus molecules of different length take different trajectories across the device, much as light of different wavelengths takes different paths across a prism [34] (see Fig. 13).

The great advantage of this method is that is allows molecules to be sorted *continuously*. A mixture of species can be injected through a narrow conduit on one side of the device. Molecules of different size then follow distinct trajectories across the array and the sorted components can be continuously collected at different locations along the opposite side. The technique is therefore especially suitable for automated applications in which molecular separation needs to be integrated with subsequent analytical steps.

Continuous sorting using asymmetric pulsed fields relies on keeping the DNA molecules almost fully stretched. It is therefore advantageous to use high electric fields, and it also helps to use a shallow device which minimizes hydrodynamic interactions, to enhance the extensional electrophoretic forces on the molecule and increase the relaxation time. By observing these principles, Huang *et al.* [34] were able to continuously sort viral chromosomes in a matter of minutes – an impressive achievement.

6.4. Asymmetric sieves for sorting DNA

The first devices that were invented to sort DNA continuously were suitable for operation at low fields, so that the molecules remain reasonably compact coils. The idea was to use an asymmetric array of obstacles to deflect the molecules away from the field direction, thereby effecting a separation in the direction transverse to the field [35, 36]. Quite generally, this can be done by choosing a periodic array of obstacles, each of which is asymmetric with respect to reflection in the field direction. The combination of the spatial asymmetry and the broken time-reversal symmetry (imposed by the flow) causes the Brownian motion of the molecules to be rectified. Since the effect depends on the thermal motion, molecules with different diffusion coefficients are deflected by different amounts, and consequently a mixture of molecules is sorted according to size.

A particular realization of such a device, shown in Fig. 14, consists of a periodic pattern of rectangular obstacles. An electric field is applied at 45° to the

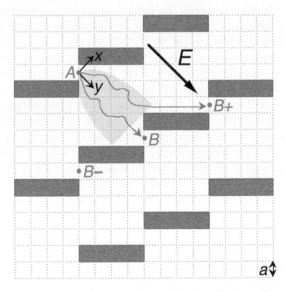

Fig. 14. Asymmetric sieve for continuous sorting. Molecules which pass through gap A diffuse laterally as they drift along the diagonal, and therefore tend to occupy the parabolic, shaded region.

principal axes of the array. In this direction, there is a clear 'line of sight' through narrow gaps between adjacent obstacles. Molecules migrating electrophoretically naturally tend to follow this pathway along the diagonal, passing through successive gaps (eg. from A to B). However, as they drift, they also diffuse. The probability distribution of a molecule, initially localised in gap A, spreads in the direction transverse to the field as the molecule advances. Consequently, there is a probability that the molecule passes around the top left corner of the obstacle located immediately above of B, in which case the field will carry it to gap B+. There is also a possibility Ñ- more remote, since it requires diffusion through a greater distance in a shorter time Ñ- of the molecule diffusing around the top left corner of the obstacle just below B, in which case it will proceed to gap B-. Owing to the different probabilities of passage through the three gaps B, B+ and B-, the molecule's line of motion will, on average, be deflected away from the field direction. Most importantly, since the deflection probabilities depend on the diffusion coefficient, the mean trajectory depends on the molecular size. Smaller, more rapidly diffusing species are deflected through a larger angle.

Effective separation in such a device requires careful matching of the timescales for diffusion and drift, and therefore depends on the Peclet number

$$\text{Pe} = \frac{va}{D}, \tag{6.1}$$

where v is the electrophoretic drift velocity of the DNA, D is the diffusion coefficient, and a is the shorter dimension of the rectangular obstacles. Simple theoretical arguments indicate that good separations are achieved in the range $\text{Pe} = 1\text{--}10$.

In practice, things are more complicated. The obstacles are insulating and deform the electric field lines, so it is a gross simplification to suppose that the electrophoretic drift is in a uniform direction. When the distortion of the field lines is taken into account, the conclusion is quite different. Indeed, one quickly realizes that the ions in solution must on average, travel straight through the device, in the direction of the electric current passing through the fluid. The trajectory of point-like ions cannot depend on their diffusion coefficient.

Nevertheless, experiments using viral DNA molecules of different size show unambiguously that the asymmetric sieves are capable of sorting molecules [37]. What is causing the separation? As shown in Fig. 15, the excluded volume interaction between the DNA molecules and the obstacles plays a decisive role [38]. Large molecules get pushed by the obstacles to the centre of gap A, and therefore tend to follow the field lines which are go fairly straight through the device to gap B. Smaller molecules can pass closer to the obstacles, and can therefore follow the distorted field lines which carry them in the transverse direction. It is thus

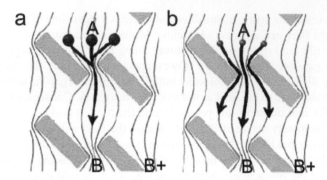

Fig. 15. Insulating obstacles distort the field lines. (a) Large molecules tend to follow only the field lines that pass through the centre of the gap. (b) Smaller molecules can follow more field lines.

much easier for the smaller molecules to diffuse onto other field lines which will carry them to gap B+.

Thus this technique works surprisingly well, but because it relies on diffusion it is quite slow. For DNA molecules containing several hundred thousand base-pairs, it is appropriate to use obstacles with size $a \sim 1\mu$m, and the right range of Peclet number is obtained with flow velocities $v \sim 1\mu$m/s.

6.5. *Rapid continuous separation in a divided laminar flow*

Far more rapid separations can be achieved in asymmetric sieves by using a different technique, in which the obstacles divide a laminar flow through the device [39]. This method is designed to work at high Peclet number, Pe $\gg 1$, and is therefore suitable for molecules which are not easily deformed by the flow, such as bacterial artificial chromosomes.

The principle of the method is illustrated in Fig. 16. The sieve consists of a periodic sequence of rows of obstacles, transverse to the flow. Each row is shifted laterally with respect to the previous row by one-third of the obstacle spacing, so that the sequence repeats every three rows. Fluid passing through a gap between adjacent obstacles in one row will bifurcate when it encounters the obstacle in the next row; two-thirds of the flow will pass to the left of the obstacle, and one-third to the right. If we divide the flow through a gap into three lanes, numbered 1,2,3 as shown in Fig. 16, we see that the flow which passes through lane 1 of the gap in the first row will pass through lane 3 of the gap in the next row, then lane 2 of the row after that. After three rows, the lanes rejoin in their original configuration.

Now consider a small particle that is able to approach the obstacles quite closely, and which can therefore take any of the three lanes through a gap. If

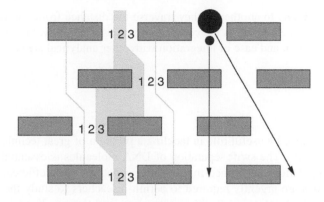

Fig. 16. Principle of separation in which asymmetrically disposed obstacles divide the laminar flow. Small molecules travel downwards, in the direction of the mean flow, but large molecules move at an angle to the flow, as indicated.

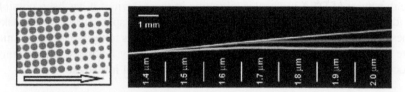

Fig. 17. Flow device in which the asymmetric disposition of obtacles is achieved by tilting the axes of a regular lattice at a slight angle to the flow. By grading the size of the gaps between the obstacles over the range $1.4\,\mu m - 2.0\,\mu m$, particles of sizes $0.8\,\mu m$, $0.9\,\mu m$ and $1.0\,\mu m$ can be made to depart from the main flow direction at different locations.

the Peclet number is high, so that diffusion is limited, the particle will continue to follow the same streamline as it passes many rows of obstacles. It will thus follow the sequence of lanes 1-3-2, and will travel straight through the device, on average. A larger particle, on the other hand, can only fit through the gap if its centre-of-mass is in lane 2. When it arrives at the next row of obstacles, its centre-of-mass will be in lane 1, but the particle will get bumped into lane 2 by the obstacle to its left. Consequently, it will follow the sequence of lanes 2-2-2, and will travel at an angle to the flow. The device thus provides a very sharp discrimination of particles that are bigger or smaller than a certain critical size, which is set by the size of the gap between obstacles. By manufacturing an array in which the gap size is graded with distance (Fig. 17), particles with a range of different sizes can be directed to different locations [39].

The most appealing feature of this technique is that the resolution improves as the flow speed increases, because particles then have less opportunity to diffuse from one flow line to another as they traverse the distance from one row to the next. Thus the method offers the combined features of rapidity, high resolution, ready automation, and ease of integration with other analytical steps in a 'lab on a chip'.

7. Summary

Physics has played a useful role in tackling a problem of great technical importance in biology. The swift separation of DNA molecules is essential in many of the manipulations performed in molecular genetics. More efficient fractionation methods are urgently required to permit reseachers to study the genomes of pathogens and commercially important crops, and to extend genomics to the study of individual cell lines. The technologies of micro- and nanofabrication offer a novel approach to DNA fractionation. Miniature electrophoretic chambers, of versatile design, may be etched from silicon chips. By investigating the dynamics of DNA molecules migrating in these devices it has been possible to identify particular designs that are ideally- suited to specific separation tasks. These devices represent the first step on a road that leads to the creation of miniature laboratories on a chip which, by automating many laborious experimental procedures, will free researchers to pursue more fruitful investigations.

Acknowledgements

Over the years, we have had lot of fun trying to devise better ways to sort DNA molecules, and I am grateful to all the colleagues who have participated in that endeavour, especially Jean-Louis Viovy, Bob Austin and Ted Cox. Many thanks also to the organizers, fellow lecturers, and participants of the Summer School, for creating a very enjoyable environment at Les Houches. These lecture notes are partly based on a course published in *Physics of Biological Systems*, edited by H. Flyvbjerg *et al.* (1997, Springer).

References

[1] Viovy, J.L., Rev. Mod. Phys. **72** 813 (2000).

[2] Olivera, B.M., Baine, P., and Davidson, N., Biopolymers **2** 245 (1964).

[3] Southern, E.M, Anal. Biochem. **100** 319 (1979).

[4] Pluen, A., Tinland, B., Sturm, J., and Weill, G., Electrophoresis **19** 1548 (1998).

[5] Doi, M., and Edwards, S.F., The Theory of Polymer Dynamics (Oxford Univ. Press, Oxford, 1986).

[6]   de Gennes, P.G., J. Chem. Phys. **55** 572 (1971).

[7]   Lerman, L.S., and Frisch, H.L., Biopolymers **21** 995 (1982).

[8]   Lumpkin, O.J., and Zimm, B.H., Biopolymers **21** 2315 (1986).

[9]   Lumpkin, O.J, Dejardin, P., and Zimm, B.H., Biopolymers **24** 1573 (1985).

[10] Slater, G.W., and Noolandi, J., Phys. Rev. Lett. **55** 1589 (1985).

[11] Duke, T.A.J., Semenov, A.N., and Viovy, J.L., Phys. Rev. Lett. **69** 3260 (1992).

[12] Duke, T.A.J., Phys. Rev. Lett. **62** 2877 (1989).

[13] van Leeuwen, J.M.J., and Kooiman, A., Physica A **184** 79 (1992).

[14] Kolomeisky, A.B., and Widom, B., Physica A **229** 53 (1996).

[15] Barkema, G.T., Marko, J.F., and Widom, B., Phys. Rev. E **49** 5303 (1994).

[16] Duke, T.A.J., and Viovy, J.L., Phys. Rev. E **49** 2408 (1996).

[17] Salas-Solano, O., Carrilho, E., Kotler, L., Miller, A.W., Goetzinger, W., Sosic, Z., and Karger,
       B.L., Anal Chem. **70** 3996 (1998).

[18] Deutsch, J.M., Science **240** 922 (1988).

[19] Gurrieri, S., Smith, S.B., Wells. K.S., Johnson, I.D., and Bustamante, C., Nucl. Acids. Res. **24**
       4759 (1996).

[20] Schwartz, D.C., and Cantor, C.R., Cell **37** 67 (1984).

[21] Carle, G.F., Frank, M., and Olson, M.V., Science **232** 65 (1986).

[22] Schwartz, D.C., and Koval, M., Nature **388** 520 (1989).

[23] Duke, T.A.J., and Viovy, J.L., Phys. Rev. Lett. **68** 542 (1992).

[24] Birren, B.W., Lai, S.M., Clark, M., Hood, L., Simon, M.I., Nucl. Acids. Res. **16** 7563 (1988).

[25] Viovy, J.L., Miomandre, F., Miquel, M.C., Caron, F., and Sor, F., Electrophoresis **13** 1 (1992).

[26] Gurrieri, S., Smith, S.B., and Bustamante, C., Proc. Natl. Acad. Sci. USA **96** 453 (1999).

[27] Volkmuth, W.D., and Austin, R.H., Nature **358** 600 (1992).

[28] Austin, R.H., Tegenfeldt, J.O., Cao, H., Chou, S.Y., and Cox, E.C., IEEE Trans Nanotech. **1** 12
       (2002).

[29] Bakajin, O.B., Duke, T.A.J., Chou, C.F., Chan, S.S., Austin, R.H., and Cox, E.C. Phys. Rev.
       Lett. **80** 2737 (1998).

[30] Long, D., Viovy, J.L., and Ajdari, A., Phys. Rev. Lett. **76** 3858 (1996).

[31] Duke, T., Austin, R.H., Cox, E.C., and Chan, S.S., Electrophoresis **17** 1075 (1996).

[32] Southern, E.M., Anand, R., Brown, W.R.A., and Fletcher, D.S., Nucleic Acids Res. **15** 5925
       (1987).

[33] Bakajin, O., Duke, T.A.J., Tegenfeld, J.O., Chou, C.F., Chan, S.S., Austin. R.H., and Cox, E.C.,
       Anal. Chem. **73** 6053 (2001).

[34] Huang, L.R., Tegenfeld, J.O., Kraeft, J.J., Sturm, J.C., Austin, R.H., and Cox, E.C., Nature
       Biotech. **20** 1048 (2002).

[35] Duke T.A.J., and Austin, R.H., Phys. Rev. Lett. **80** 1552 (1998).

[36] Ertaz, D., Phys. Rev. Lett. **80** 1548 (1998).

[37] Chou, C.F., Bakajin, O.B., Turner, S.W., Duke, T.A.J., Chan, S.S., Cox, E.C., Craighead, H.G.,
       and Austin, R.H., Proc. Natl. Acad. Sci. USA **96** 13762 (1999).

[38] Huang, L.R., Silberzan,P., Tegenfeldt,J.O., Cox, E.C., Sturm, J.C., Austin, R.H., and Craighead,
       H., Phys. Rev. Lett. **89** 178301 (2002).

[39] Huang, L.R., Cox, E.C., Austin, R.H., and Sturm, J.C., Science **304** 987 (2004).

Course 6

SINGLE-MOLECULE STUDIES OF DNA MECHANICS AND DNA/PROTEIN INTERACTIONS

T.R. Strick

Cold Spring Harbor Laboratory, 1 Bungtown Road, Cold Spring Harbor NY 11724 USA
Present address: CNRS, Institut Jacques Monod and Universites de Paris VI et Paris VII,
2 Place Jussieu 75251 Paris Cedex 05 France

D. Chatenay, S. Cocco, R. Monasson, D. Thieffry and J. Dalibard, eds.
Les Houches, Session LXXXII, 2004
Multiple aspects of DNA and RNA: from Biophysics to Bioinformatics
© *2005 Elsevier B.V. All rights reserved*

Contents

1. Introduction

In the past ten years, physicists and biologists have been able to isolate and study individual biomolecules such as molecular motors (myosin [1], kinesin [2], polymerases [3,4] or ATPases [5]), structural proteins (titin [6–8], actin [9, 10], microtubules [11] or nucleosomes [12]) and nucleic acids (RNA [13] and DNA [14–16]). These studies, made possible by advances in the techniques of visualization (fluorophore excitation by evanescent waves [9, 17] and fluorescence resonnant energy transfer [13, 18–20]) and micromanipulation (magnetic tweezers [15, 21, 22] and optical tweezers [3]), seek to characterize the mechanical response of these systems as well as their function.

Thus in 1992 researchers first measured the response of a single DNA molecule to a stretching force [14]. These experiments not only made it possible to measure the stiffness of DNA (and along the way to verify theoretical models of polymer elasticity [23]) but also uncovered a structural transition in DNA provoked by a stretching force of about 70 picoNewtons (1 pN = 10^{-12} N) [24, 25]. This type of experiment, in which one may measure the spring constant of DNA and observe the controlled formation of mechanically stabilized structures, cannot be performed on molecules freely diffusing in solution. Therefore in the first part of this chapter we will describe how micromechanical experiments allow us to better understand the mechanical and structural properties of nucleic acids.

Another important aspect of single-molecule DNA manipulation experiments is the study of enzymatic activity. Here again, the observation and manipulation of individual molecules makes it possible to observe events which are difficult (if not impossible) to detect in "bulk" experiments. As an example, consider the experiments of Wang *et al.* [3], who anchored an RNA polymerase to a glass surface before pulling on the extremity of the three-micron long DNA molecule serving as the enzyme's substrate. These experiments showed that the enzyme is capable of working against forces of up to about 50 pN with a broad distribution of velocities. If "bulk" molecular biology allows one to know that the enzyme catalyzes RNA formation with a certain average rate, it does not allow for the measurement of the forces generated by the process, nor does it allow for a simple observation of the distribution of velocities. Thus in the second part of this chapter we will illustrate, through the examples of topoisomerases and polymerases, how such experiments are performed and interpreted to extract new information.

Before however, it is useful to briefly compare the single-molecule and traditional "bulk" methodologies and paradigms. First of all, in a single-molecule experiment, the molecule of interest is fixed at a defined point of space; in a bulk experiment it is freely-diffusing. Note that by "fixed" we mean that the molecule may be physically tethered (as in the case of a manipulation experiment [3]), or that it may be freely-diffusing but repeatedly sampled on a timescale much shorter than that required for diffusion (as in the case of fluctuation correlation spectroscopy, or FCS, experiments [26]). As a result, in the single-molecule assay one can know the position of the target molecule at any moment in time, which is impossible in the bulk system. This makes it easy to observe through time the target molecule and reactions which take place on it, by monitoring a single point in the reaction cell. From multiple single-molecule trajectories, or time-traces, one can reconstitute not just the average reaction but also the fluctuations about the average. The latter is particularly rich in information on the underlying statistical features of the reaction [27]. In the bulk assay one can measure only the average outcome of the reaction, and fluctuations about the average are lost.

In addition, the readout of the single-molecule experiment can be done as the reaction proceeds, whereas in a bulk assay one performs the readout step much after the reaction step, again leading to a loss of information and the ability to finely "tune" the system as the reaction occurs. Another problem that the single-molecule paradigm also solves is that of "synchronization" 'in a bulk assay; since one can observe the beginning and the end of a reaction at the single-molecule level, it becomes possible to align reactions according to their temporal unfolding and, through averaging, reconstitute and detect transient events which are lost in the bulk assay.

Of course, the greatest advantage of the single-molecule methodology is that it allows the researcher to study the mechanical properties of nucleic acids and the proteins that interact with them. The effect of mechanical constraints (stretching and twisting) 'on biopolymer structure illuminates their building rules. The effect of mechanical constraints on protein-DNA interactions illuminates the role of conformational changes (i.e. physical movement of protein domains) in the work of proteins.

2. The interest of physicists for DNA

2.1. Ease of handling

Physicists are interested in DNA for several reasons. First of all, it is a unique polymer characterized by incredible lengths (up to several centimeters per molecule in the case of human chromosomes) and a very high degree of monodispersity (the DNA extracted from bacteriophage-λ, for instance, always consists of the exact same sequence of 48502 base pairs). If certain artificial polymers such as polystyrene or polyethylene glycol are capable of achieving high degrees of polymerization, it is still very difficult to impose a specific length for the whole sample. Moreover, these artificial polymers are not very rigid at the scale of the monomer; we will see that DNA, locally more rigid, is (paradoxically) much easier to stretch out.

Another important point is that DNA is, for now, the only polymer which may be easily modified and observed by scientists. An ever-growing number of tools in molecular biology – restriction enzymes, ligases, PCR, electrophoresis gels – make it possible to cut, reglue, modify and purify DNA fragments in a manner which is both simple and precise. One can readily obtain large numbers of DNA molecules, each one of them bearing the exact same modifications at the exact same position. Moreover, the use of intercalating dyes such as ethidium bromide and YOYO1 make it easy to observe single stained DNA molecules in solution using standard fluorescence microscopy.

2.2. DNA as a model polymer

A polymer is characterized by two lengths: its crystallographic length l_0 and its persistence length $\xi = \frac{A}{k_B T}$. Here A represents the flexural rigidity of the material, k_B the Boltzmann constant ($k_B = 1.38 \times 10^{-23} J/K$) and T the temperature in Kelvins. THe persistence length represents the distance over which the polymer remains oriented despite thermal agitation. As the temperature increases, thermal agitation tends to decorrelate the orientation of successive segments and the persistence length of the DNA decreases. At zero temperature, the polymer's path would be a straight line. At room temperature on the other hand, the polymer essentially follows a random walk, taking on the form of a self-avoiding random coil. This fluctuating coil is capable of exploring the greatest possible number of distinct conformations, maximizing the system's entropy. On the other hand, a polymer which is completely stretched out between two points has a unique conformation – the straight line – and its entropy is zero. If one assumes that a polymer of length l_0 performs a random walk consisting of $N = l_0/b$ independent steps ($b = 2\xi$), one finds that a force $F = \frac{3}{2} \frac{k_B T}{\xi} \frac{l}{l_0}$ [28] is necessary to separate the ends of the polymer by a distance $l \ll l_0$. As one stretches the polymer, the stretching force performs work which reduces the polymer's conformational entropy; one therefore speaks of the polymer's entropic elasticity.

This helps to understand why it is easier to stretch a polymer with a long persistence length. Consider a chain of a give crystallographic length. If the material's persistence length increases, the effective number of segments N in the chain decreases. As a result, the configurational entropy of the system decreases and less work (and thus smaller forces) will be required to stretch the polymer. Thus, when a polymer's *local* rigidity increases, its entropic rigidity decreases. With a persistence length $\xi \sim 50$ nm, a DNA random coil is easily unraveled and stretched by a force of order $F \sim k_B T/\xi \sim 0.1$ pN. In comparison, a chain of polystyrene, characterized by a persistence length on the order of a few angstroms, will require stretching forces roughly one hundred times greater to obtain the same relative extension.

Thus researchers have succeeded in observing using fluorescence microscopy single DNA molecules labeled by fluorescent markers and stretched by the velocity gradient of a hydrodynamic flow. When the flow is stopped, it is possible to measure the relaxation time of the stretched polymer and compare this value to theoretical models [29,30]. DNA has also been used to verify models of polymer reptation [31], according to which a polymer snakes between physical obstacles in its environment by advancing or retreating, but without performing any lateral displacements. These experiments indicate that DNA has been widely adopted as a model system for the behavior of polymers. This is underscored by the fact that the relevance of the Worm-Like Chain model for describing biopolymer elasticity – not just of DNA but unfolded proteins [6–8] – was first shown in the context of DNA [23].

Finally, it is important to remember that DNA is one of the rare polymers which can be supercoiled, opening many new avenues of research into the statistical beavhior of twisted elastica. This remarkable property, as we will see, is the direct consequence of the double-helical nature of this polymer.

2.3. Introduction to single-molecule DNA manipulation techniques

2.3.1. Strategies and forces involved

Techniques for micromanipulating individual biomolecules have known great progress in the last decade. The problem consists of two separate issues: one must first bind the macromolecule to macroscopic "handles" on which it will afterwards be possible to pull or apply a torque. Typical "handles" include glass [32] or ferrite [14] microspheres, thin glass fibers [25, 33] or AFM cantilevers [6, 34]. It is also common to anchor one extremity of the biomolecule to a glass surface; this is employed in the experiments we describe in this these (see Figure 1). Binding of the macromolecule to these physical supports usually involves labeling the support and the biomolecule with pairs of molecules of the antibody/antigen type. This is used to guarantee the specificity of the binding. Mechanical constraints can then be applied to the handle and measured using an ever-increasing number of instruments: optical [24, 27] or magnetic tweez-

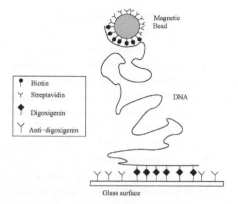

Fig. 1. Sketch of the anchoring strategy used in single-molecule DNA experiments. One extremity of the DNA is labeled with biotin, and the other is labeled with digoxigenin. The interaction between biotin and streptavidin guarantees that the biotin-labeled DNA will bind specifically to a small magnetic bead coated with streptavidin. When this bead-DNA construct is incubated on an anti-digoxigenin coated glass surface, the digoxigenin-labeled end of the DNA will bind to the surface. The presence of multiple binding points between the bead and its physical supports makes it possible to control the twisting of the double helix. The magnetic bead is manipulated using a magnetic field, allowing for both stretching and twisting of the DNA.

ers [22], AFM cantilevers [6, 34] or glass microneedles [25, 33, 35], or also by using hydrodynamic flows [14].

Let us first give an idea of the scale of forces involved. The strongest bonds in a macromolecule are covalent. They correspond to energies on the order of the eV ($1.6 \cdot 10^{-19}$ J, equivalent to $40 \, k_B T$ at room temperature or 24 kCal/mol), and act over distances on the order of an angstrom. The force necessary to break a covalent bond is therefore on the order of eV/Å $\sim$ 1 nanoNewton (10^{-9} N). To give an example, sugar chains such as dextran break at forces of 2 nN, and DNA can remain intact under a 1 nN stretching force [36]. At the other extreme, the smallest forces which can be measured are limited by the thermal agitation of the instrument used to measure the force (bead, glass fiber, AFM cantilever). This random agitation gives rise to the Langevin force, whose amplitude depends on the environment's viscosity and the instrument's friction, but also on the time-scale over which the fluctuations are averaged out. For a spherical object,

$$F_{Langevin} = \sqrt{4k_B T 6\pi \eta R \Delta f} \qquad (2.1)$$

where η represents the viscosity, R the object's radius and Δf the bandwidth. For a sphere of radius $R = 1.5 \mu$m in water ($\eta = 10^{-3} N \cdot s \cdot m^{-2}$, or 1 centipoise) at room temperature, the root-mean-square force is of order $20 f N/\sqrt{Hz}$, or 20 fN on the time-scale of one second (the mean of the Langevin force is zero). Any force measurement done on this time-scale will be limited by a 20 fN resolution, and faster measurements will be hindered by even higher thermal noise.

The typical forces involved in the interaction of biological molecules lie between the extremes of thermal forces and the forces related to covalent bonds. They are brought about by hydrogen or ionic bonds, as well as the van der Waals interactions which lend structure to nucleic acids and proteins. Unzipping DNA (as if it were a zipper) by pulling on the two strands at one extremity of the molecule, requires forces in the range of 10 to 15 pN [33, 37]. The force exerted by a molecular motor such as myosin as it tracks along an actin fiber is of the order of 3 pN [1, 9]. *E. coli* RNA polymerase translocating on its DNA substrate exerts forces reaching up to 50 pN [38, 39]. Finally, proteins such as titin – the folding of which depends on a large number of non-covalent bonds – unfold at forces of about 100 pN [6–8].

It is also important to have an idea of the forces which can be withstood by the molecules used to bind the biological object-of-interest to its macroscopic handles. In our experiment for example, one extremity of the DNA is labeled with digoxigenin, and this extremity will specifically anchor to a glass surface coated with anti-digoxigenin (Figure 1). The forces which hold together an antibody and its antigen (for example digoxigenin and anti-digoxigenin), or pairs of molecules such as biotin and streptavidin, are on the order of hundreds of pN [34, 40]. The adsorption forces of proteins onto surfaces, also due to Van der Waals interactions, are probably of the same order of magnitude. For these reasons, non-covalent anchoring strategies do not allow unambiguous measurement of forces greater than about 100 pN.

2.3.2. Measurement techniques

Cantilevers   NanoNewton-scale forces can be generated using AFM cantilevers of glass microfibers (Fig. 2a and b). In order to measure these forces, the stiffness of cantilever or the fiber must be calibrated; this can be done using a controlled hydrodynamic flow. Then, as one stretches the molecule of interest, the flex of the detector is measured so as to deduce the applied stretching force and the molecule's end-to-end extension. If these instruments allow rapid force measurements, they are limited in their resolution of low forces (we will come back to this point further on). Of these two techniques, only the glass microfiber-based technique can be used to supercoil the DNA by rotating the pipette around its axis [41, 42] (Fig. 2b).

Optical Tweezers   Optical tweezers generate forces of up to about 100 pN [24], and are used to trap micron-sized objects whose index of refraction is higher than that of water (where these experiments usually take place) [27]. One typically

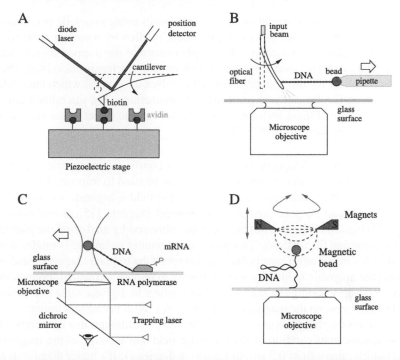

Fig. 2. A few examples of micromanipulation experiments. (A) The AFM cantilever. If the cantilever's stiffness is known, the stretching force developed by the instrument can be determined by measuring the deflection of the cantilever during the course of the stretching process. This deflection is measured by reflecting a laser beam off the cantilever's surface. (B) The same principle can be applied when one uses a thin optical fiber instead of an AFM cantilever. (C) The optical tweezer consists of a laser beam brought to a diffraction-limited focus within the sample using an oil-immersion lens. The focal point can attract and trap small objects whose index of refraction is greater than that of water. The trap stiffness must first be calibrated, which then makes it possible to measure the force by determining the position of the object with the trap. A sketch of the RNA-polymerase stretching experiment of Wang *et al.* [3] is shown. (D) Magnetic tweezers make it possible to pull on a magnetic bead, as well as cause it to rotate. The trap stiffness is calibrated by measuring the bead's Brownian fluctuations.

uses glass or latex microspheres in these experiments. The optical trap is generated by using an immersion lens to focus a laser beam at a point within the sample (Fig. 2c). Since this focal point corresponds to a maxima in the electromagnetic field's gradient, it stably attracts the bead. The stiffness of the electromagnetic trap also needs to be calibrated, either by displacing the trapped object with a controlled flow or by measuring the object's Brownian fluctuations (we will describe this technique in more detail further on). Apart from a risk of damaging photochemistry in the vicinity of the focal point, optical tweezers are non-invasive.

Their major disadvantage relative to the experiments using magnetic tweezers is that they are still complex pieces of equipment saddled by issues of short-term drift, and as such do not yet allow for *simple* control of the trapped object (both in force and in torsion) over extended periods of time (such as weeks) [43]. They also do not allow for the generation of sub-picoNewton forces, which can make them limiting for the study of protein-DNA interactions. On the other hand, a major advantage of these instruments is that they enable direct measurement of torque on supercoiled DNA [43, 44].

Magnetic tweezers Magnetic tweezers simply consist in a magnetic field gradient in one, two or three dimensions which can be used to trap small magnetic beads: the object is attracted to regions where the field is highest. The magnetic field gradient can be generated using permanent magnets [15] or electromagnets [22] (Fig. 2d). The trap stiffness can be calibrated by analyzing the trapped bead's Brownian fluctuations. This force-measurement technique is totally non-invasive. It's major advantage is that one can rotate the trapped magnetic bead by rotating the magnetic field. Moreover, it is easy to impose a constant force on the object by simply setting the magnetic field gradient to a given value (i.e. fixing the distance between the magnets and the sample). With permanent magnets one can generate stretching forces of up to about 100 pN, although this limit depends on the precise configuration of the magnetic field and the size of the magnetic bead (which ranges from 0.5 μm to 4.5 μm in diameter). It's major disadvantage is that the torque acting on the bead cannot be directly measured, but must be extracted using indirect measurements and theoretical models.

3. Force measurements

3.1. Measuring forces with Brownian motion

The force measurement technique described here is discussed in the context of the analysis of the Brownian motion of a tethered magnetic bead. The bead-DNA system subjected to a magnetic field behaves like a small gravitational pendulum: when the stretching force increases, the amplitude of the pendulum's motion decreases (see Figure 3). Consider a constant, vertical stretching force $\vec{F}$ applied to a pendulum of length l. When the pendulum's axis is displaced by a small angle θ from its vertical equilibrium position (this corresponds to a distance δx), a horizontal restoring force $F\theta$ appears which tends to return the system to its equilibrium position. In the case of the bead-DNA system, the system is displaced from its equilibrium by all the collisions with the water molecules which randomly collide with the bead. If the displacements δx are small relative to the

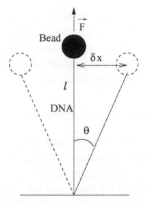

Fig. 3. The bead-DNA system behaves like a small pendulum subjected to an external tension. Stretched by a vertical force F to an extension l, the pendulum is displaced from its equilibrium position by Brownian fluctuations δx. Measuring these two lengths allows for the determination of the stretching force according to $F = \frac{k_B T l}{<\delta x^2>}$ (see text).

pendulum length, the restoring force is equal to $\frac{F}{l}\delta x$, which can also be written as $k_x \delta x$ if one uses $k_x = F/l$ as an effective spring constant. The elastic energy of this effective spring is then written as $\frac{1}{2}k_x < \delta x^2 >$, where $< \delta x^2 >$ is the mean square displacement of the pendulum's extremity. In the case of the bead-DNA system, this energy is given to the sysem by the Brownian collisions with the water molecules, and according to the equipartition theorem is worth $\frac{1}{2}k_B T$. One can rearrange these terms to obtain

$$F = \frac{k_B T l}{< \delta x^2 >}$$
(3.1)

As expected, when the stretching force increases the amplitude of the pendulum's motion decreases. Since $k_B T$ is known, this equation makes it possible to measure in a non-invasive manner the stretching force simply by measuring distances: this technique does not require any sophisticated calibrations to work. Moreover, since the amplitude of the Brownian motion increases as the force decreases, the evaluation of small forces is favored by the term $1/ > \delta x^2 >$. This force measurement technique can be used to cover a wide range of forces, ranging from a few femtoNewtons to a hundred picoNewtons (see Figure 4 for an example).

The only disadvantage of this technique is that one needs to measure the bead's motion over relatively long times so as to accurately determine the force. Indeed, the estimate of $< \delta x^2 >$ will be accurate to within 10% only if the acquisition time is on the order of one hundred times the system's characteristic time; we will return to this point further on. Finally, it is advantageous to analyze the power spectrum of the bead's Brownian fluctuations, since this enables one to remove the mechanical drift of the microscope (which essentially takes place at low frequency) and to correct for filtering effects and spectral folding due to the rate at which the bead position is sampled [45].

3.2. Advantages and disadvantages of the manipulation techniques

This force-measurement technique has several advantages. It is non-invasive, and does not require any complex calibrations. At high forces, the technique is only limited by the recording system's sampling rate and spatial resolution. At low forces, this technique is limited by the Langevin force acting on the bead. For a 3μm diameter bead, this force is of order 20 fN/$\sqrt{Hz}$. With a bandwidth of about 10^{-3} Hz (this technique has the disadvantage of being slow), this yields a noise on the order of the femtoNewton. By using small beads (0.5μm diameter for instance), one has a noise level of about 10 fN/$\sqrt{Hz}$. This noise level is excellent relative to other micromanipulation and force measurement techniques.

It is interesting to compare this technique to force measurements performed using cantilevers, glass microneedles or optical tweezers. Glass microneedles have a typical stiffness of about 10^{-5} N/m [33,41]; with a 10 nm resolution of the fiber's displacements [33] one could measure forces in the range of tens of fN. However, the large size of the glass fiber imposes a relatively large Langevin force, on the order of 0.5 pN/$\sqrt{Hz}$. Fast measurements done by sampling between 10 and 100 Hz therefore limit the force measurement to a resolution of a few pN. In the case of an AFM cantilever with a typical stiffness of 10^{-3} N/m [40], a spatial resolution at the angstrom level should allow for the measurement of forces as low as 1 pN. Once again however, the Langevin force acting on the cantilever is large, of the order of 0.1 pN/$\sqrt{Hz}$ for high-quality cantilevers, and rapid measurements done with these instruments are typically limited to a resolution of $\sim$15 pN.

In the case of optical tweezers, one generates a trap with a minimum stiffness of about 10^{-5} N/m, but the trap's dimensions are small, on the order of 0.5 μm. This corresponds to forces on the order of the pN. To generate higher forces, one needs to increase the laser power. If the power is too low, the trapping potential is too weak and does not function effectively. The spatial resolution of the bead's position within the trap can be on the order of a few angstroms, implying that one should be able to measure forces as smal as a few fN. Once again though, the thermal forces acting on the bead hinder the measurement of such low forces, and for a bead with a micron-scale diameter the limiting thermal resolution is about 10 fN/$\sqrt{Hz}$. Fast experiments done with a bandwidth of a few hundred Hz thus cannot measure forces much lower that 0.1 pN. Unfortunately, if one attempts to reduce the noise by reducing the bead size, one obtains a weaker trapping potential.

Constant position vs. Constant force apparati   An important distinction between these different instruments is the control variable which it naturally allows the experimenter to set. In the case of cantilever-based systems, the control parameter is the position of the tip; in the case of the optical tweezer, the control parameter is the position of the bead relative to the center of the optical trap. These instruments allow one to fix/measure the position of the object (the force is extracted using knowledge of the spring constant), and they are thus constant-position instruments. A feedback loop (used to modulate the trap position or the cantilever defelction) needs to be introduced so as to use the constant-position instrument into a constant-force instrument [3].

The magnetic tweezer, on the other hand, is a constant-force instrument. Here the magnetic field gradient which generates the force on the magnetic bead is on the milimeter scale. Since the extension of the DNA molecule which tethers the magnetic bead is no more than a few microns, the bead can never move enough to experience a different force if the DNA extension is changed (for instance by the action of an enzyme). Note that a feedback loop could also be used to convert this constant-force system into a constant-extension one.

4. Mechanical properties and behavior of DNA

4.1. Tertiary structures in DNA

The primary (chemical) and secondary (double helical) structures of DNA are well-known; here we will focus on the tertiary (higher-order) structure of DNA. The tertiary structure is determined by the path of the double helice's axis. If DNA is quite rigid on a scale of 50 nm, it may nevertheless curve over distances greater than its persistence length. A long DNA molecule in solution is not a rigid rod, but rather resembles a random coil constantly deformed by thermal fluctuations. According to the theory of polymer elasticity [28], the radius of gyration of a random coild of DNA with end-to-end length l_0 would be of order $R_g \sim \sqrt{2\xi l_0}$, or on average 80 μm per human chromosome (to be compared with the typical size of the cell nucleus $\sim$10 μm).

Nevertheless, more compact and organized tertiary structures may appear under the influence of proteins or torsional constraints. Electron micrographs of supercoiled circular DNAs (known as plasmids) show that such molecules exhibit shapes reminiscent of the interwindings observed on a tangled phone cord. These interwound and looped structures (see Figs. 2C, 7 and 10), known as "plectonemes", represent one way in which the molecule may store a torsional constraint.

Another tertiary structure which may be used to store a torsional constraint is the solenoidal wrap. When DNA is packaged in the cell nucleus, it is first wrapped in a solenoidal manner around small, positively charged globular proteins. The eukaryotic version of such proteins is called a histone, whereas the prokaryotic version is named HU. This type of solenoidal wrapping, which allows for the stabilisation of about two complete turns of DNA around each histone, assists in the compactification and organization of DNA. A protein partner is likely required to stabilize such structures.

4.2. Topological formalism

Two topological components are used to describe DNA. The first, the *twist* Tw of the double helix, measures the number of times the two constituent strands of the molecule wrap about each other. This is simply a measure of the number of helical steps in the double-helix, which is a way of characterizing the system's secondary structure. For linear B-DNA in the absence of any exterior constraints, the natural twist Tw_0 is equal to the number of base pairs n divided by the helical pitch h of the structure: $Tw_0 = n/h$ ($h = 10.4$ for B-DNA).

The second topological component, the *writhe* Wr of the molecule, counts the number of times the axis of the molecule crosses itself: this is simply a measure of the tertiary structure of the molecule. To measure Wr, one needs to project the molecule's axis onto a plane before counting the number of crossovers observed in that plane. A positive or negative number may be assigned to the writhe (this sign will be the same as the sign of the molecule's supercoiling). One generally assumes that on average a linear DNA molecule, in the absence of external constraints, will have neither writhe nor macroscopic curvature and thus $Wr_0 = 0$.

The sum of these two components is known as the *linking number Lk*, and represents the sum total of crossings (strand-strand and axis-axis) in the molecule. Thus, the natural linking number Lk_0 of a linear DNA in the absence of external constraints is equal to the number of helical turns of the molecule $Lk_0 = Tw_0$.

A mathematical theorem derived by White [46] shows that for a circular DNA (where the extremities of the molecule are connected), the linking number is a topological invariant:

$$Lk = Tw + Wr = constant \tag{4.1}$$

This is formally true for all circular DNAs, where Lk may only take on integer values. Nevertheless, the relation is also valid for long linear DNA molecules whose extremities are prevented from rotating freely (in which case Lk may take on non-integer values).

A DNA molecule is supercoiled when $Lk \neq Lk_0$, that is to say when the molecule has an excess or a deficit of linking number relative to its torsionally relaxed state: $\Delta Lk = Lk - Lk_0 = Tw - Tw_0 + Wr = \Delta Tw + Wr$. One may then define the degree of supercoiling σ:

$$\sigma = \frac{\Delta Lk}{Lk_0} = \frac{\Delta Tw + Wr}{Lk_0} \tag{4.2}$$

This normalization of the linking number of a DNA makes it possible to compare molecules of different lengths. σ is positive when the DNA is *over*wound and negative when it is *under*wound. Plasmids extracted from bacteria are typically underwound, or negatively supercoiled, with $\sigma \sim -0.06$ [47]. To give an example, a 50 kbp DNA molecule (16 microns long) with this degree of negative supercoiling is underwound by roughly 300 turns.

If the total number of crossings in the system is an invariant of the system, this does not prevent the molecule from redistributing its linking number between writhe Wr and twist Tw. In a sense, this means that there is a form of "communication" between the secondary and tertiary structures of a topologically closed DNA: a change in one implies a compensatory change in the other.

4.3. DNA supercoiling in vivo

4.3.1. DNA unwinding and helix destabilisation

As pointed out above, bacterial DNA *in vivo* is negatively supercoiled ($\sigma \sim -0.06$); similar values are thought to hold for eukaryotic DNA in vivo. The study of negatively supercoiled circular DNAs (plasmids) has shown that the torsional constraint can destabilize the double helix [48–53]. This destabilization can lead to local denaturation of the plasmid in its A+T rich regions [50], and may be implicated in the initiation of transcription and replication. If the sequence within this denatured region is palindromic, local denaturation can then lead to the formation of cruciform structures [51, 54]. In a similar fashion, negative supercoiling can also stabilize the formation of regions of left-handed Z-DNA [51, 54]. The exact biological role of some of these more complex transitions is still somewhat unclear.

4.3.2. DNA topoisomerases

The regulation of DNA supercoiling is handled *in vivo* by the topoisomerases (topo) [55]. The ubiquitous nature of these enzymes is a testament to the importance of supercoiling in the living cell, and their inhibition generally poses a threat to the cell's survival. It is probable that these enzymes act during the course of processes such as DNA transcription and replication. In this regard, it is interesting to note that topoisomerases are a privileged target of antibacterial drugs and a certain number of anti-cancer drugs. Four topoisomerases have been identified in *E. coli*: topo I, gyrase, topo III and topo IV. Topo I and III are capable of relaxing DNA supercoiling via the introduction of a transient single-stranded break in the moleucle. Gyrase is responsible of generating negatively supercoiled DNA, and topo IV is ultimately responsible of the complete disentangling of replicated chromosomes. We will describe a number of these enzymes in more detail further on.

4.3.3. Supercoiling and transcription

DNA supercoiling and transcription also influence one another. During transcription, an RNA polymerase tracks along one strand of the DNA, polymerizing an RNA strand which is complementary to the selected template. Transcription initiation, where RNA polymerase first binds to the beginning of the gene and unwinds it over a distance of about 13 base-pairs, is stimulated by negative torque acting on the DNA [4].

If the topology of DNA can influence transcriptional activation, it is also interesting to note that transcription can transiently affect DNA supercoiling. This is the proposal put forward by Liu and Wang's [56] twin-domain model of transcription-induced supercoiling, and it is easily visualized with a pencil and two intertwined pieces of string. Holding the ends of the intertwined string so as to prevent any rotation, slide a pencil (representing the poymerase) between the two strings. When the pencil is displaced without rotating along the axis of the braided string, one notes that excess windings accumulate in front of the pencil while a deficit of winding appears behind the pencil. This effect has been observed *in vitro* and *in vivo* in plasmids borne by bacteria [57,58]: transient waves of positive supercoiling appear to propagate in front of the transcribing complex, and waves of negative supercoiling appear in the complexe's wake. It is important to note that these observations were carried out in special situations where the RNA polymerase was tethered to the inside of the cell membrane and thereby prevented from rotating.

4.4. DNA elasticity in the absence of torsion ($\sigma = 0$)

The magnetic trap is the most commonly used instrument to study DNA supercoiling. The magnetic trap setup makes it possible to control three mechanical parameters of DNA: the DNA's degree of supercoiling (expressed in terms of the number of turns n or by the degree of supercoiling σ), the stretching force F and the system's extension l. Experimentally, it is simpler to change the degree of supercoiling or the stretching force and to measure the DNA extension which results from these mechanical constraints. Before describing the behavior of su-

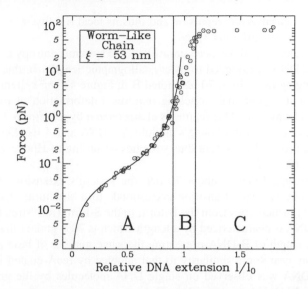

Fig. 4. Force vs. relative extension curve for torsionally relaxed DNA from 10 fN to 100 pN in a 10 mM phosphate buffer (pH 8) and at room temperature. Relative extension is the extension per base-pair (l) divided by the crystallographic length of one base-pair ($l_0 = 3.34$). Several experiments have been regrouped in one plot, which demonstrates that the force-measurement technique used can cover rougly five orders of magnitude. The different regimes of DNA elasticity are indicated by capital letters. (A) Entropic elasiticy of DNA, fit (continuous line) to the Worm-Like Chain model with a persistence length $\xi \sim 50$ nm. (B) Enthalphic regime characterized by a stiffness ~ 1000 pN. (C) B→S overstretching transition.

percoiled DNA however, we will examine the stretching of a torsionally relaxed DNA molecule.

4.4.1. Results

The force vs. extension curve $F(l)$ depicted Figure 4 shows the three regimes of DNA elasticity. At low forces (regime A), the molecule's extension increases progressively as the random DNA coil is unfurled and stretched. This regime is called the entropic regime, since stretching the molecule calls for unravelling the random coil and reducing its entropy. This regime corresponds to forces on the order of $F \sim k_B T/\xi \sim 0.1$ pN. In the first part of this regime, DNA has linear elasticity: $F = \frac{3}{2}\frac{k_B T}{\xi}\frac{l}{l_0}$, where l_0 is its crystallographic length. In ionic conditions where the concentration of monovalent cations is greater or equal to 10 mM, the persistence length of DNA is $\xi = 53$ nm ± 2 [59]. If one lowers the ionic conditions, the DNA's negative charge is not screened as well, increasing the molecule's electrostatic repulsion and thus its local rigidity: $\xi = 75nm$ in 1 mM PB (phosphate buffer pH 8) [60].

Once the random coil has been unravelled, the system's entropy is essentially zero and the DNA is extended to its crystallographic length. In this "enthalpic regime" (ranging from 10 to 70 pN, noted B in Figure 4), the system's stiffness is high since any additional stretching requires a deformation of the system's double-helical structure. This regime is characterized by a stiffness $EA \sim 1000$ pN [24, 32], where E is the Young's modulus for DNA and A its effective cross-sectional area [61]. The system thus stretches according to Hooke's law: $F = EA(\frac{l-l_0}{l_0})$, with $l > l_0$.

At a stretching force of about 70 pN, the system's extension abruptly increases (regime C). This transition corresponds to a structural change in the DNA [24, 25], which goes from the B-form to the S-form (for "stretched"). The S-form of DNA is characterized by a length which is 70% greater than the crystallographic length of B-DNA presenting the same number of base pairs. This structure may bear some similarity to that adopted by recA-coated DNA. This S-form of DNA was observed on single DNA molecules by the groups of F. Caron [25] and C. Bustamante [24]. The precise atomic structure of this form of DNA is not completely known yet. Finally, we note that DNA does not break under the action of a stretching force of one nanoNewton [36].

4.4.2. Theoretical models
The first measurements of DNA elasticity in the absence of a torsional constraint were obtained in 1992 by C. Bustamante and his co-workers [14]. Their experimental force vs. extension curve was not well described by the freely-jointed chain (or random walk) model usually considered at the time. It was not until 1994 that Marko and Siggia showed that DNA elasticity is in fact described by the worm-like chain model, which takes into account the fact that DNA is a continous, semi-flexible chain. A useful numerical approximation exists for purposes of fitting force-extension data with this model [62]. If the persistence length of the system is found to be on the order of 50 nm, it is a confirmation that only a single DNA molecule tethers the bead to the surface. If the persistence length of the system is found to be on the order of 25 nm, the bead is likely tethered to the surface by *two* DNA molecules. For now we will focus on the former, but the latter situation also can be useful in certain experiments (see the section on topoisomerase IV).

4.5. Mechanical properties of supercoiled DNA

Two types of measurements can be performed on a single, supercoiled DNA molecule [Note that, to be supercoilable, the molecule must contain no breaks along both its strands and it must also be tethered at multiple points at each end to bead and surface]:

• At a fixed force, one can measure the system's extension as a function of the number of turns $l(n)|_{F=cst}$, also written $l(n)$ or extension-supercoiling curve. These curves can be normalized by the DNA's crystallographic length l_0 to give $l_r(\sigma)$, the DNA's relative extension $l_r = l/l_0$ as a function of the normalized degree of supercoiling $\sigma = \frac{n}{Lk_0}$.

• For a given degree of supercoiling, one can measure the DNA's force-extension curve: $F(l)|_{n=cst}$, also written $F(l)$.

In what follows we will focus on the first kind of measurement, as it is the most useful in the study of how a single enzyme molecule can act on a single DNA molecule.

4.5.1. Results

The three curves shown in Figure 5A were obtained in a 10 mM phosphate buffer at room temperature and at stretching forces of 0.2, 1 and 8 pN

• For a force $F = 0.2$ pN, the molecule's extension varies rapidly with n and independently of the sign of the supercoiling. At this force, the system contracts whether one over- or under-winds the DNA, and at a rate of about 80 nm/turn. The axis of symmetry of the curve corresponds to the torsionally relaxed state of the DNA ($\Delta Lk = 0$).

• For a force $F = 1$ pN, the system's behavior depends of the sign of the supercoiling. In the range of $-n$ represented here, the system's extension does not change when the DNA is underwound. One also notes that when the system is slighly overwound, its extension also remains unchanged. It's only beyond a certain degree of overwinding that the system contracts rapidly and regularly, in a fashion analogous to that observed for $F = 0.2$ pN but with a weaker slope (~ 40 nm/turn).

• Finally, for a force $F = 8$ pN, the DNA's extension varies very litte in the range of σ discussed here.

4.5.2. Interpretation

We explain these curves by considering White's equation (Eqn. 4.1): $\Delta Lk = \Delta Tw + Wr$ (Figure 5B).

• At a low force ($F = 0.2$ pN), the rapid and regular contraction of the system is explained by the formation of plectonemes, structures which can be observed by electron microscopy on supercoiled plasmids or chromosomes [63]. These plectonemes allow the torsional constraint to be stored under the form of writhe, Wr.

• At an intermediary force ($F = 1$ pN), the system no longer contracts when it is underwound. Therefore, no plectonemic structures appear, and $Wr \sim 0$ de-

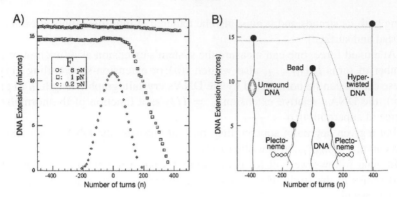

Fig. 5. (A) Extension vs. supercoiling curves $l(n)$ obtained at three different stretching forces. ($\diamond$): $F = 0.2$ pN. The system contracts whether over- or under-wound. ($\square$): $F = 1$ pN, the system's extension decreases when it is overwound, but remains essentially constant when it is underwound. ($\circ$): $F = 8$ pN, the DNA's extension does not vary noticeably in the range of supercoiling shown here. B) Interpretation of the data. At $F = 0.2$ pN, the DNA stores supercoiling under the form of plectonemic supercoils which can appear independently of the sign of supercoiling. At $F = $ 1pN, negative supercoiling generates denaturation bubbles along the DNA. At $F = 8$pN, positive supercoiling generates a new, locally hypertwisted DNA structure.

spite the fact that $\Delta Lk < 0$. By the conservation of linking number, we conclude that $\Delta Tw < 0$. In this situation localized denaturation bubbles (where $Tw \sim 0$) appear in the A+T-rich regions of the DNA [64]. On the other hand, when the DNA is overwound, one notes that the system begins to contract only after a certain number of turns have been added to the system. We interpret this by the existence of two regimes: a first regime in which the system is subjected to a pure torsional constraint $\Delta Tw > 0$, and then a second regime where the DNA contracts as plectonemes ($Wr > 0$) begin to form.

• At high forces ($F = 8$ pN), unwinding the DNA still generates denaturation bubbles. One also notices that DNA overwound at this force does not contract either, and therefore that $Wr \sim 0$. This implies that along the DNA there will appear regions where the linking number is greater that Tw_0. The existence of such a hypertwisted structure, named P-DNA, has been confirmed by the experiments of J.-F. Allemand [65].

These first experiments therefore give an idea of the two types of modifications brought about by mechanical twisting of DNA. At low forces, the supercoiling is essentially stored by the formation of interwound plectonemes. At higher forces, these tertiary structures are no longer stable, and the DNA adjusts by undergoing important changes in its secondary structure. The stretching force acts by par-

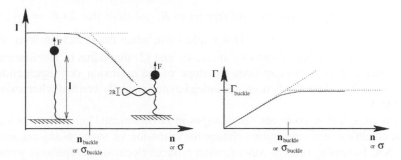

Fig. 6. (Left panel) The extension l of an elastic tube which has been stretched and overwound by n turns. Initially, the torsional constraint does not cause the system to contract. When $n = n_{buckle}$ turns have been applied to the tube, a buckling transition allows the system to relax its torsional constraint by forming a loop, causing the system to contract. As one continues to twist the tube its extension decreases linearly as the number of loops grows regularly. (Right panel) Torque Γ acting on the tube as a function of the number of turns n. As long as the system remains extended, its torque increases linearly with the number of applied turns. When the critical torque $\Gamma = \Gamma_{buckle}$ is reached ($n = n_{buckle}$), the formation of a plectoneme relaxes the torsional constraint and prevents the torque from increasing beyond Γ_b. Each additional turn added to the system further lengthens the plectonemes, preventing the torque from increasing.

titionning the supercoiling between these two types of modifications. We will now describe the appearance of plectonemic structures as a result of a buckling instability in a twisted rod.

4.5.3. The buckling instability in DNA
We begin by considering a twisted rubber tube of twist stiffness $k_B T \cdot C$, stretched by a force F to its total length l_0 (Figure 6). When one begins to twist this tube, its extension initially remains unchanged and the constraint accumulates as pure torsion. The tube's torque Γ increases linearly with the twist angle Ω, and its twist energy increases quadratically: $\Gamma = k_B T \frac{C}{l_0} \Omega$ and $E_{torsion} = \frac{1}{2} k_B T \frac{C}{l_0} \Omega^2$.

As one continues to twist the tube, one notes that after a certain number of turns n_{buckle} (written n_b and corresponding to a torque Γ_b) the system buckles and a loop of radius R is formed. The twist energy is thus transferred into bending energy, and the torque no longer increases (Figure 6B). As a result, the system's extension decreases by $2\pi R$ despite the stretching force F. Beyond this buckling instability, each turn added to the system causes the number of plectonemes to increase linearly. One writes

$$2\pi \Gamma_f = 2\pi R F + 2\pi R \frac{1}{2} \frac{B}{R^2} \tag{4.3}$$

By minimizing this energy with respect to R, one finds that $2\pi F = \frac{\pi B}{R^2}$, or $R = \sqrt{\frac{B}{2F}}$ and $\Gamma_f = \sqrt{2BF}$. This implies that when the stretching force increases (1) the number of turns n_f increases and (2) the radius of plectonemes decreases, and as a result so does the slope of the extension vs. supercoiling curves. Both of these predictions are indeed experimentally verified when twisting DNA.

It is important to point out that changes in environmental conditions such as ionic conditions and temperature change the extension vs. supercoiling behavior of DNA. Increasing the salt concentration reduces electrostatic repulsion within plectonemic structures, allowing for a smaller plectonemic radius R and leading to smaller rates of contraction per added supercoil. Increasing temperature causes the DNA pitch to increase, thereby unwinding the molecule at a rate of about $0.01°/°C/bp$. This causes a leftward shift (towards lower Lk) of extension vs. supercoiling curves. Moreover, increasing the salt concentration can also lead the DNA pitch to decrease.

Behavior of plasmids in solution   In a tethered-DNA manipulation experiment, the stretching force and degree of supercoiling can be varied independently; this is not true of circular, supercoiled plasmid DNAs in solution. However, the extension vs. supercoiling data provide a direct connection to the behavior of such DNAs in solution. The circular nature of plasmids means that their end-to-end extension is always zero. However, the fact that they are supercoiled means that they experience internal tension (imagine cutting a twisted rubber band; the two ends would rapidly separate upon cleaving). This can be intuitively understood by following the extension vs. supercoiling curve obtained for a given stretching force F until the extension of the DNA is equal to zero (this occurs at a degree of supercoiling which we note $\sigma_{plasmid}$). Thus, a plasmid DNA with a degree of supercoiling $\sigma_{plasmid}$ is subjected to an internal tension equal to F; we therefore estimate that *in vivo* the internal tension experienced by supercoiled DNA is small, on the order of $F = 0.3pN$.

4.6. Stretching single-strand DNA

The mechanical properties of single-strand DNA (ssDNA) are very different from those of double-strand DNA (dsDNA). In particular, the absence of a global helical structure means that ssDNA is much more locally flexible than ds DNA. Its persistence length is thus on the order of a few base-pairs, or about a nanometer. Because of the low aspect ratio this implies (the width of a monomer beings similar to the persistence length of the polymer, unlike in dsDNA), the statistical-mechanical properties of an ssDNA chain are best described by a worm-like chain model which incorporates self-avoidance even for a short polymer [16].

Another complication to consider is that a single strand of DNA is structurally inhomogeneous. Regions of the polymer which contain self-complementary base-pair sequences (such as palindromes) will be capable of folding into localized double-helical structures known as hairpins.

The force-extension data obtained using single-molecule techniques on single-strand DNA display these various features [16]. On one hand, the small persistence length of ssDNA means that larger forces must be generated to obtain the same relative extension as with dsDNA. Self-avoidance in ssDNA is exhibited in the low-force regime of the data as an exponential dependence of the stretching force on the relative extension. The formation of self-folded hairpins also contributes to shortening of ssDNA; these structures are resistant to forces on the order of 5–10 picoNewtons. Therefore, at forces lower than these, ssDNA is shorter than dsDNA containing the same number of bases. At force of about 5–10 pN, ssDNA and dsDNA display the same extension per base. At higher forces, the hairpins are pulled apart and the ssDNA becomes extended. Because of the absence of a helical structure in this case, the maximal length per base-pair of ssDNA is on the order of 7 Å, as compared to the 3.34 Å for dsDNA. At high forces therefore, the end-to-end extension of ssDNA is greter than that of dsDNA with the same number of bases.

The difference in mechanical properties of ssDNA and dsDNA means that polymerization of ssDNA into dsDNA (by a DNA polymerase [66, 67] or digestion of dsDNA into ssDNA (by a DNA exonuclease [68] or a DNA polymerase undergoing error-correction [67]) can be monitored by measuring changes in the extension of the system. If ssDNA is stretched by a force lower than 5–10 pN, its conversion by DNA polymerase into dsDNA will cause it to extend. If ssDNA is stretched by a 5–10 pN force, its conversion by DNA polymerase into dsDNA will cause no change in extension. Finally, if ssDNA is stretched by a force greater than those listed above, its conversion by DNA polymerase into dsDNA will cause the system to contract. By calibrating the difference in extension-per-base-pair between ssDNA and dsDNA at the force for which the experiment is done, it is possible to monitor the rate of replication by DNA polymerase by monitoring the system's overall.

4.7. Conclusions on the mechanical properties of nucleic acids

Understanding the mechanical properties of nucleic acids does more than just provide us with a better theoretical understanding of such systems; it also gives researchers a way to calibrate the DNA molecule under study. Such a calibration makes it possible to use the determination of DNA extension as a real-time transducer of enzyme activity. In what follows, we will discuss how such calibrations are used to measure gene transcription by RNA polymerase or DNA unknotting by topoisomerases.

5. RNA polymerases

5.1. An introduction to transcription

The cell must accomplish two fundamental tasks – transcription and translation – in order to produce a protein. In the process of transcription, a protein termed RNA polymerase reads the gene (DNA) encoding the protein and polymerizes a messenger RNA (mRNA) molecule whose sequence is complementary to the original gene. In the process of translation, the mRNA then serves as a blueprint for the ribosome to assemble the appropriate polypeptide (i.e. protein) chain. Thus during the course of this process genetic information flows from the gene (double-stranded DNA) to the message (single-stranded RNA) and finally to the protein (a structured polypeptide chain). Transcribing DNA into mRNA is the central function of a ubiquitous and highly-conserved class of proteins known as RNA polymerases.

Transcription initiation   Transcription by RNA polymerase is a multi-step process involving transcription initiation, transcription elongation and transcription termination. During transcription initiation, the RNA polymerase must first locate the beginning of the gene and bind to it. This is accomplished using specialized sequences known as promoters, which are located just ∼10 bp upstream of the gene and act as a binding site for the RNA polymerase. Once bound to the promoter site, the RNA polymerase mechanically unwinds about one turn of the double helix, generating a ∼13 bp "transcription bubble" between the promoter -10 element and extending slightly beyond the transcription start site. This makes the base pairs of the template strand available to the RNA polymerase for high-accuracy reading of the genetic information stored there. The transcription bubble is thought to be stabilized by interactions between the RNA polymerase σ subunit and the non-template strand of the bubble. Up to this point, the reaction is fully reversible as no source of external energy has been used to perform these tasks.

Promoter escape and elongation   Subsequent steps of transcription are less reversible. Indeed, upon addition of the full set of ribonucleotide triphophates (ATP, UTP, GTP and CTP) the RNA polymerase begins to polymerize an RNA strand complementary to the sequence read on the template DNA strand. For reasons which are not yet completely understood, it appears that the initial transcription of the messenger RNA is fraught with difficulty. This is evidenced by the repetitive synthesis of RNA fragments containing only nine to ten bases. This process of "abortive initiation" is thought to involve the initial polymerization of a short RNA fragment which is then released by the RNA polymerase as it simultaneously "backslides" on its DNA template and finds itself back at its initial position at the +1 transcription start site. Typically, tens to hundreds of these short abortive transcripts are genereated by the RNA polymerase before it is able to fully break free from the attractive interactions which tether it to the promoter site. Presumably in a stochastic manner, an RNA polymerase will eventually polymerize a messenger RNA a few bases longer than the threshold 9-10 base pairs seen in abortive initiation. This leads to a sufficient weakening of the polymerase's interactions with the promoter that it becomes committed to transcription elongation, finally performing what is known as "promoter escape". This involves full dissociation from the promoter and the engagement of the RNA polymerase to processively transcribing the gene into messenger RNA.

This phase of processive transcription, known as transcription elongation, constitutes the bulk of transcription. During elongation, RNA polymerase moves processively along the DNA at a rate on the order of twenty base-pairs per second, polymerizing RNA. Although the rate of transcription is not constant along the template, this process is characterized by the fact that a single RNA polymerase will be able to scan/read the entire gene and transcribe it into messenger RNA.

Transcription termination   Finally as the RNA polymerase reaches the end of the gene, it typically encounters a termination sequence which serves to dissociate the RNA product from the RNA polymerase, and presumably as well the RNA polymerase from the DNA. These terminators, at least in bacteria, are typically characterized by their ability to form secondary structures in the DNA such as cruciforms.

RNA polymerases are remarkably conserved across all living species. Beyond the strong similarity in the primary sequence of genes encoding for RNA polymerases, there is a strong similarity in the actual structures of RNA polymerases. For simplicity sake, we will concentrate on the RNA polymerase from *E. coli* which has been extensively studied using single-molecule techniques.

RNA polymerase: a canonical molecular motor   From this introduction, it should be apparent that RNA polymerase is a remarkable molecular motor capable of a range of large-scale mechanical interactions with its DNA substrate. During transcription initiation the RNA polymerase physically unwinds the DNA, perhaps by the application of mechanical torque to the DNA. During transcription elongation the RNA polymerase must physically translocate along the entire length of the DNA, an activity typical of molecular motors. Thus, the RNA polymerase is capable of first rotating (or torquing) the promoter DNA at transcription initiation and then moving (translating) along the DNA during elongation. These properties make RNA polymerase an excellent candidate for single-molecule experiments adressing questions such as the nature of the enzymatic steps affected by force or torque. In our discussion or RNA polymerase, we will proceed as per the historical development of the field, covering first transcription elongation before returning to studies of transcription elongation.

5.2. Historical overview: transcription elongation, or RNA polymerase as a linear motor

The first single-molecule studies of transcription elongation by RNA polymerase were reported in 1991 by Dorothy Schafer and collaborators [69]. In these experiments, a DNA molecule was prepared with a promoter site at one end and a biotin label at the other end (similar to the configuration depicted in Fig. 2C). The DNA was attached at one end to a micron-size polystyrene bead coated with streptavidin, and the promoter site (located at the other end) was oriented so as to direct transcription towards the bead. RNA polymerase was loaded onto the promoter and fixed there. The bead-DNA-polymrease complex was then attached to the glass surface. The micron-sized bead's Brownian motion could be observed under the light micrsocope by digital intereference contrast (DIC). Upon transcription, the RNA polymerase was seen to progressively reel-in the bead towards the site of attachment of the RNA polymerase on the surface. Progressive shortening of the bead's DNA tether led to an observable and quantifiable decrease in the amplitude of the beads' lateral fluctuations. These and subsequent experiments [70, 71], based on the "tethered-particle motion" method and performed under conditions of zero external load, allow researchers to detect and measure in real-time the rate of transcription by RNA polymerase. Transcription rates on the order of ten to twenty nucleotides per second are typically observed, in good agreement with results obtained by bulk biochemistry.

This work was soon followed by experiments of a similar conception, where an external force was applied on the bead using optical tweezers [3]. Indeed, the strength of the TPM technique is also its weakness: although one can measure RNA polymerase's velocity under zero external load, the weak entropic stiffness of DNA leads to significant thermal noise in the system which makes rapid and accurate determination of the bead position difficult [70]. By imposing a pico-Newton scale stretching force on the RNAP-DNA-bead assembly one extends the DNA molecule and greatly reduces the system's compliance. As the system becomes stiffer, accurate determination of the bead position becomes faster.

Transcription by the surface-bound RNA polymerase again causes the DNA tether to progressively shorten, pulling the bead away from the center of the optical trap and thereby subjecting it to higher trapping forces. Beyond a force on the order of 30 pN, RNA polymerase was no longer able to transcribe. This determined the second extremum of the RNAP's motor-like characteristics, the stalling force for which the enzyme's velocity goes to zero. Interestingly, the enzyme's rate of transcription, on the order of 20 base-pairs per second, is essentially independent of the applied load until it is but a few pN away from the stall force. A Boltzmann model for the force inhibiting a mechanically-coupled step involving displacement of RNAP against the applied force was described by Wang *et al.* [3]. This model implies that at low loads the rate-limiting steps of transcription elongation are biochemical in nature, and not mechanical (excluding the translocation step for instance). Only at high loads does the enzymatic cycle become affected by force, presumably by the effect of an external opposing force on the process of translocation.

Another interesting observation from single-molecule experiments is that RNA polymerase frequently pauses on the DNA during productive elongation. Short-lived pauses, on the order of a few seconds, occur at random along the DNA and presumably involve stuctural rearrangement of the protein rather than movement relative to the template [72]. On the other hand, long-lived pauses (ranging from tens of seconds to tens of minutes) appear to be correlated with mechanical movement, or "backtracking" of the enzyme in the direction contrary to transcription elongation [73]. The sequence-dependence of transcription elongation and pausing is now coming under scrutiny [74].

5.3. RNA polymerase as a torquing device: the case of transcription initiation

A number of other single-molecule experiments have investigated the role played in transcription by the helical structure of DNA. Indeed, it has been postulated that as RNA polymerase slides along the DNA double helix in its search for a promoter site, the enzyme tracks along the helical pitch of DNA. Similarly, it was expected that as RNA polymerase transcribed DNA into RNA it would also track along the helical path of DNA, leading either to rotation of the RNA polymerase/RNA complex around the DNA or rotation of the DNA as it is threaded through the RNA polymerase. The latter point was demonstrated in experiments by Harada *et al.* [75].

Thus historically, transcription elongation was the first to be studied using single-molecule techniques, as it beautifully demonstrated the motor-like nature of translocating RNA polymerase. Recently, single-molecule experiments using the magnetic trap have made it possible to study transcription initiation, an important early stage of transcription. As pointed out earlier, this involves local

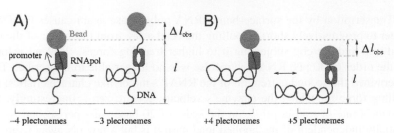

Fig. 7. Principle of detection of promoter unwinding by RNA polymerase. A) Negatively supercoiled
DNA B) Positively supercoiled DNA. See text for details.

unwinding of promoter DNA and the subsequent formation of a "transcription
bubble", a region of unpaired bases where the sequence of the template strand
can be read with high accuracy. This process, which precedes RNA polymerization and does not require an external energy source (the $\alpha - \beta$ phosphate bond
of nucleotides), is fully reversible. Transcription elongation thereafter involves
displacement of the RNA polymerase along the DNA, all the while dragging the
transcription bubble along with it.

Calibration and detection of transcription initiation   Transcription initiation is
itself a multi-step process, involving (i) search for promoter DNA, (ii) binding of
RNA polymerase to promoter DNA to form the RNAP-promoter closed complex
(or RP_c) (iii) unwinding of about one turn of promoter DNA to form the RNAP-
promoter open complex (iv).

Figure 7 shows the principle behind detection of promoter unwinding by RNA
polymerase [4]. In a topologically closed system such as this, any change in local DNA winding (ΔTw) must be compensated by an equal and opposite change
in global DNA writhing (Wr). Since RNA polymerase-induced unwinding of a
promoter site leads to a decrease in DNA twist $\Delta Tw \sim -1$, it must therefore
be accompanied by a unit increase in the DNA's writhing number $\Delta Wr = +1$.
This change in DNA writhing number can be detected in real-time by monitoring the end-to-end extension of a mechanically stretched and supercoiled DNA
molecule. Indeed as discussed earlier, the end-to-end extension of a mechanically stretched and supercoiled DNA undergoes a large change (in the conditions
used for these experiments, on the order of $\sim$56 nm) when its writhing number
changes by a unit amount (that is, when one supercoil is added to the system).
This change in DNA extension is calibrated by measuring the extension vs. supercoiling curve (Fig. 5) of the DNA prior to experiments.

This detection scheme therefore couples the local changes in DNA twist which
occur at the promoter site into a large-scale change in DNA end-to-end exten-

sion, yielding a large amplification of signal. DNA unwinding at the promoter site covers about 10 base-pairs, equivalent to about 3 nm of DNA. The actual distances over which the two DNA strands at the promoter sites move is likely much smaller. Since the topologically-coupled change in DNA writhing number leads to a change in end-to-end DNA extension on the order of 50 nm, an amplification factor of about 20 is obtained in this detection scheme.

As shown in Figure 7, the change in DNA extension observed upon unwinding of the promoter site by RNA polymerase is expected to be different for positively or negatively supercoiled DNA. In both cases, unwinding of the promoter site will lead to a unit increase in the DNA writhing number as per White's theorem. For negatively supercoiled DNA, increasing the DNA writhing number by one is equivalent to removing a negative supercoil, allowing the DNA end-to-end extension to increase. For positively supercoiled DNA, increasing the DNA writhing number by one leads to the addition of a positive supercoil in the DNA, causing the DNA end-to-end extension to decrease. This detection principle should also be applicable to other protein-DNA interactions which lead to torsional deformation of the DNA.

Results Unwinding of the negatively supercoiled lac promoter is irreversible due to the strength of the protein-DNA interaction and the stabilization of unwinding by negative torque acting on the DNA (Fig. 8A). Reversible unwinding of the positively supercoiled lac promoter by RNA polymerase gives rise to a classical "telegraphic signal" which shows in real-time the promoter site switching between a "closed" and an "unwound" state (Fig. 8B). Unwinding of the positively supercoiled lac promoter is reversible due to destabilization of unwinding by positive torque acting on the DNA. The amplitudes of the change in DNA extension ($\Delta l_{obs,-}$ and $\Delta l_{obs,+}$ for, respectively negatively and positively supercoiled DNA) can be related to the amplitude of promoter unwinding (about 13 bp are unwound) and promoter bending (a $\sim 90°$ bend in the DNA is generated upon promoter unwinding) by RNA polymerase [4].

In the case of reversible unwinding, the time interval between two successive unwinding events T_{wait} is the inverse of the rate of formation of the unwound promoter/RNAP complex. It is influenced by a wide range of basic experimental parameters, including temperature, ionicity and buffer formulation. More complex activation processes involving protein cofactors such as CAP/cAMP, fundamental to the regulation of gene expression *in vivo*, are known to enhance this rate.

The concentration dependence of the rate of formation of the unwound promoter complex essentially displays Michelis-Menten kinetics; this is expected for a bimolecular reaction where initial binding of enzyme to substrate is reversible. This makes it possible to extract the binding affinity 'of the enzyme for

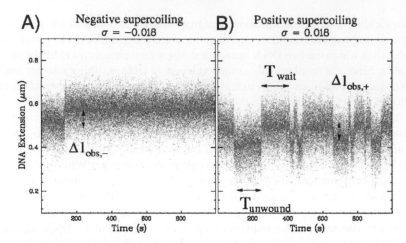

Fig. 8. Detection of lac promoter unwinding by RNA polymerase. A) Negatively supercoiled DNA B) Positively supercoiled DNA. See text for details.

the promoter site, as well as the time required for an enzyme already bound to the promoter site to unwind it.

The lifetime of an unwinding event $T_{unwound}$ is the inverse of the dissociation rate and a direct measure of the stability of the unwound promoter/RNAP complex. As with T_{wait}, it is sensitive to environmental conditions such as temperature and buffer composition; unlike T_{wait} it is independent of protein concentration. It also responds to a range of small molecule compounds believed to interact with RNA polymerase, including ppGpp, nucleotides and the antibiotic rifampicin (used in the treatment of tuberculosis). The rate of dissociation of the binary RNAP/unwound promoter complex is independent of protein concentration. Thus the lifetime of the unwound promoter/RNAP complex is a robust measurement of the stability of the unwound promoter. It can be a useful parameter for careful, quantitative measurements of the action of inhibitors of bacterial transcription such as certain antibiotics.

Action of torque on polymerase-promoter interactions   The response of the system to supercoiling was also studied using the magnetic trap. By varying the degree of positive or negative supercoiling, it was confirmed that positive supercoiling reduced the rate of formation and increased the rate of dissociation of the unwound promoter complex. Conversely, negative supercoiling increased the rate of formation and decreased the rate of dissociation of the unwound promoter complex. These kinetic characteristics of the unwound promoter complex did not vary linearly with supercoiling however. Instead, a biphasic response was observed, with an initial regime ($|\sigma| < 0.013$) where rates changed with supercoiling and a second regime ($|\sigma| > 0.013$) where rates were insensitive to changes in supercoiling. This biphasic response essentially recapitulates the biphasic response of DNA torque to supercoiling (where torque initially increases linearly with increased winding and then saturates as plectoneme supercoils are formed, see Fig. 6 and the accompanying discussion), leading us to propose that transcription initiation is under direct mechanical control. Thus in contrast to transcription elongation, transcription initiation is truly under mechanical control as mechanical unwinding of the promoter site is, even at low torque, a rate-limiting step of the process. In this picture, RNA polymerase could be viewed as a torque wrench.

Conclusions on RNA polymerase   Single-molecule experiments have provided important new insights into the mechanochemistry of transcription, both at the level of transcription initiation and transcription elongation. It is important to note that, since the cell actively regulates DNA supercoiling (and thus torque) *in vivo*, it is possible that it exploits the sensitivity of transcription initiation to supercoiling as one of several means of regulating gene expression.

6. DNA topoisomerases

Topoisomerases, discovered nearly thirty years ago, regulate the supercoiling of DNA *in vivo*. These enzymes modify the linking number Lk of the double helix by generating transient breaks in the molecule [55]. Identified in an ever-growing number of organisms (viruses [76], mesophilic [77] and archaeal [78] bacteria, eukaryotes [79]), it is likely that these enzymes are present in all forms of life. Together, they maintain the topological homeostasis of DNA during the course of processes as varied as DNA transcription, replication or recombination. They are potent targets of antibacterial and anticancer agents, since their inhibition is generally lethal for the cell [80].

6.1. Type I and Type II topoisomerases

Classified as type I or type II, topoisomerases act by generating reversible single- or double-strand breaks (respectively) on a DNA molecule. The biochemical mechanism which catalyzes these breaks is remarkably conserved across the two classes, as well as in enzymes involved in site-specific recombination [81, 82]. It consists in the nucleophilic attack by a tyrosine group on the phophodiester backbone of one of the DNA's two constituent strands. A covalent bond between

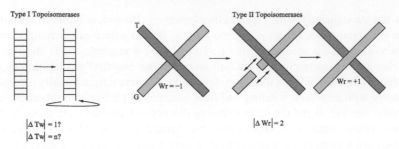

Fig. 9. Rough sketch of the action of topoisomerases. Type I topoisomerases generate single-stranded breaks in the DNA, allowing the torsional constraint ΔTw to be relaxed either by rotation around the intact strand or by its translocation through the single-stranded break. Type II topoisomerases generate a double-stranded break in the DNA, through which another DNA segment may be passed. Each enzymatic cycle causes $|\Delta Wr| = 2$. The cleaved segment is denoted G for "gate", and the transported segment is noted T.

the enzyme and one of the ends of the cleaved DNA strand is formed as a result of this attack. Before the end of the enzymatic cycle, this transesterification is reversed, sealing the bread and releasing the enzyme from the DNA. These cleavage and religation steps can therefore take place in the absence of an energetic cofactor. Type I topoisomerases are generally monomeric enzymes with only one active tyrosine, whereas type II topoisomerases are usually multimeric and have two active tyrosines.

With one active tyrosine available, type I topoisomerases are only capable of breaking one of the two strands of the double helix. The introduction of a swivel point in the DNA thereafter makes it possible to relax the molecule's degree of supercoiling (see Figure 9). It is proposed that each enzymatic cycle changes the system's twist ΔTw by increments of one in the case of type IA enzymes [83] ("constrained rotation") and by $|\Delta Tw| = n >> 1$ for type IB enzymes ("free rotation"). These enzymes do not usually require ATP, although the reverse gyrase identified in thermophilic bacteria is an exception. In the case of type II topoisomerases, two active tyrosines (one on each of the protein's functional subunits) cut the two strands of the double-helix quasi-simultaneously. This generates a temporary double-strand break (Fig. 9) in one DNA segment (named the G, or gate, segment). The transport of a second DNA segment (named the T, or transport, segment) through the break makes it possible to resolve knots or other crossovers (the junctions between plectonemes or catenanes between two molecules) between two double-stranded DNAs. The type II enzymes all hydrolyze ATP to function normally.

For the purposes of this chapter, we will focus here on the type II topoisomerases from prokaryotes (*E. coli* topo IV) and eukaryotes (*D. melanogaster* topo II). Note, however, that recent work on prokaryotic and eukaryotic type I topoisomerases has brought to light many interesting new features of these systems [84].

6.2. Eukaryotic topoisomerase II

Eukaryotic topoisomerase II is a homodimer A_2. In the presence of ATP and magnesium, eukaryotic topo II transports a segment of DNA (named T segment, for Transport) through a transient double-strand break generated in a second DNA segment (named G segment, for Gate). This enables the enzyme to relax positive or negative supercoils, and to unknot and untangle DNA [3]. The enzyme typically acts in a processive manner [85], performing many catalytic turnovers before dissociating from DNA. As with other type II topoisomerases, each enzymatic cycle reduces the DNA linking number by increments of 2, releasing two supercoils. Classical bulk experiments have determined the turnover rate (measured in cycles per second) to be on the order of $k_{cat} \sim 1s^{-1}$ [85]. This enzyme is absolutely essential to the successful completion of DNA replication [86], where the last remaining topological links between replicated chromosomes must be removed. Eukaryotic topo II (as well as prokaryotic topo IV) uses the energy stored in ATP to reduce the number of catenanes or supercoils in DNA to levels *below* those observed in thermal equilibrium [87]. The mechanism by which this happens is still unknown.

6.2.1. Enzymatic cycle

The enzymatic cycle of eukaryotic topo II is partly known, thanks to numerous studies based on the use of non-hydrolyzable ATP analogues (such as AMPPNP) or the decatenation of radiolabeled DNA circles [88–90]. The first step of the cycle corresponds to the binding of the enzyme to a DNA segment (G). Next, a second DNA segment (T) must be bound by the enzyme. When the enzyme then binds ATP, it cleaves the G segment and pushes the T segment through the transient break. The break is then resealed and the T segment released from the enzyme. So as to recover its initial conformation, the enzyme must hydrolyze at least one ATP. Topologically, the net result of this cycle is to invert the sign of a crossover between two DNA segments, leading to a change in the DNA's linking number by an increment of $\Delta Lk = -2$ [83] (see Figure 9).

A remarkable property of this type of topoisomerase II is that it uses the energy of hydrolysis of ATP to reduce the number of knots or catenanes in a DNA to levels *below* those observed at thermal equilibrium [87]. This activity is coherent with respect to the enzyme's main role *in vivo*, that is to say the complete removal of any topological links between replicated chromosomes. The mechanism by which an enzyme acting locally can measure the global topological state

of the DNA remains a mystery however. It has been recently suggested [91] that
this property could be explained by a kinetic proofreading mechanism, whereby
the enzyme would sample at *two* successive steps the equilibrium distribution of
topoisomers. If at each step the probability P of entangling the DNA is small:
$P \sim \exp(-\Delta G) \ll 1$, the probability of obtaining a DNA more strongly entangled than at the beginning goes as P^2.

6.3. Prokaryotic topoisomerase IV

Topo IV is an $(AB)(AB)$ heterotetramer which belongs to the type II family of
topoisomerases. It is therefore capable of decatenating topologically intercon-
nected DNA molecules (such as Olympic circle plasmids or replicated chromosomes). It is also highly homologous to gyrase [92]. In the presence of ATP and
$MgCl_2$, it is capable *in vitro* of removing positive or negative supercoils as well
as catenating/decatenating and knotting/unknotting DNA. Its rate of decatenation
has been measured at about 5 events/minute, whereas it would relax positive supercoils at a rate of 0.15 events/minute [93]. Topo IV is essential to the end of
DNA replication because it removes the last remaining topological links between
replicated DNA molecules [94]. Its absence at this stage of the cellular cycle is
generally lethal to the cell.

6.4. Experimental results: D. melanogaster topoisomerase II

6.4.1. Calibrating the experiment
These experiments, performed using the magnetic trap apparatus, consist in measuring in real time and at constant force the extension of a supercoiled DNA
molecule in the presence of topoisomerase (in Figure 10 we depict the catalytic

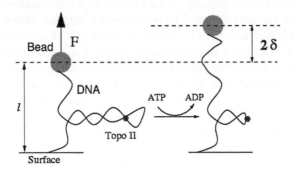

Fig. 10. Principle of detection of supercoil relaxation by topoisomerase II. By removing supercoils,
the enzyme causes the DNA's extension to increase. Monitoring the system's extension during the
course of the reaction makes real-time measurement of enzyme activity possible.

action of a topo II on a supercoiled DNA). If this supercoiling is relaxed by a topoisomerase, the molecule's extension will grow as supercoils are removed. Here the plectonemes serve as a signal amplifier: typically, a change in the linking number $\Delta Lk = \pm 2$ will lead to a detectable change in the DNA's extension: $\Delta l \sim \pm 90$ nm.

As with experiments measuring promoter unwinding by RNA polymerase, this prediction is obtained by calibrating the extension vs. supercoiling curve $l(n)$ at a given force and in the enzyme's activity buffer. Two parameters are extracted from this curve: the number of turns n_{buckle} beyond which the molecule forms plectonemes, and the change $\pm \delta$ in the system's extension which results from a change in its linking number $\Delta Lk = \pm 1$.

Thus in these experiments, for a stretching force $F = 0.7$ pN we determine that the system must be twisted by at least ~ 15 turns before plectonemes form, and that its extension changes by about 45 nm for each turn added or removed ($\Delta Lk = \pm 1$). During the course of the experiment we then measure the DNA's extension as a function of time, $l(t)$. Using the calibration described above, we can then determine the DNA's degree of supercoiling as a function of time $n(t)$.

6.4.2. Crossover clamping in the absence of ATP

Witholding ATP makes it possible to explore the initial steps of the reaction cycle, where topoisomerase II is expected to bind first to one, and then a second, DNA segment.

In these experiments, a single DNA molecule was supercoiled to the threshold of the buckling transition, where individual plectonemic loops are expected to rapidly appear and disappear as a result of thermal agitation. In the absence of ATP, topoisomerase II was seen to sufficiently stabilize these loops such that they may be detected, see Fig. 11. With additional witholding of Mg^{2+} the lifetime of the clamped state was short (~ 20 s). Adding back Mg^{2+} induced the appearance of a second, long-lived state (~ 260 s) which represented about 30% of all events.

It is important to note that, if bulk experiments may perhaps be able to detect the long-lived state, the short-lived state would be very difficult to assess by standard methods. Indeed, this had led to the erroneous conclusion that the enzyme did not significantly bind DNA crossovers in the absence of magnesium [95]. Single-molecule results, uniquely able to detect transient events, showed that in fact the enzyme stabilizes loops even in the absence of magnesium. Additional experiments confirmed that the enzyme was stabilizing the loops by binding to the two DNA segments which form a crossover at the base of the loop [96].

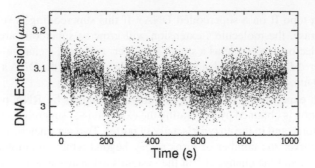

Fig. 11. In the absence of ATP and magnesium, topoisomerase II stabilizes transient loops formed at the onset of the DNA buckling transition. Reversible, $\sim$45 nm changes in DNA extension are observed, corresponding to the stabilization of one plectonemic supercoil by topo II.

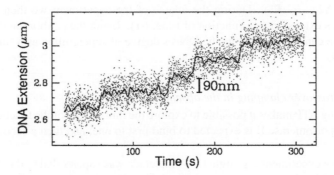

Fig. 12. Relaxation of supercoils by topo II in the presence of 10μM ATP. Measuring the extension of the system as a function of time $l(t)$ makes it possible to observe discrete, 90 nm steps in the system's extension. These steps represent single enzymatic cycles which result in $\Delta Lk = -2$. The mean time between two cycles is about 20 s. Points correspond to raw data obtained at 12.5 Hz, and the full line is a one-second average of the raw data. Over this timescale, the error on the system's extension due to the bead's Brownian fluctuations is on the order of 10 nm.

6.4.3. Low ATP concentrations: detecting a single enzymatic cycle

At a low enough ATP concentration, the topo II reaction is sufficiently slowed down so that individual catalytic cycles may be resolved by averaging the system's extension over a $\sim$ 1s timescale. Figure 12 shows the extension of a positively supercoiled DNA molecule stretched by a $F = 0.7$ pN force in the presence of topo II and ATP (10μM). The system's extension increases by discrete steps spaced by 90 nm. According to the mechanical calibration of the DNA in these experimental conditions, $\delta = 45$ nm/turn, indicating that this change in extension corresponds to a change in the system's linking number $\Delta Lk = -2$. We deduce

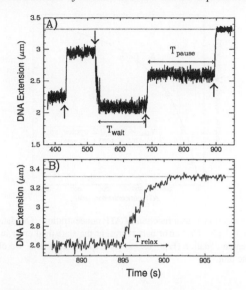

Fig. 13. (A) Monitoring the DNA's extension $l(t)$ in the presence of topo II and 300 μM ATP shows the enzymatic relaxation of supercoils ($\uparrow$), followed by the mechanical regeneration of super-coils ($\downarrow$). The dotted line represents the system's maximal extension l_{max} in the absence of supercoils ($Wr = 0$). Long waiting times $T_{wait} \sim 150s$ separate relaxation events. One also notes the exis-tence of long pauses $T_{pause} \sim 200s$ in enzyme activity. (B) Enlargement of the third relaxation from curve (A). The relaxation of supercoils is too quick ($T_{relax} \sim 5s$) to clearly resolve individual enzy-matic cycles. The fact that $T_{relax} \ll T_{wait}$ indicates that a single enzyme is responsible for supercoil removal.

that these discrete steps correspond to individual enzymatic cycles of topo II, and that each enzyme cycle indeed removes two supercoils from the system. By de-termining the time between steps, we directly measure the reaction rate in these experimental conditions: $V = 0.2$ cycles/second.

6.4.4. High ATP concentration

By increasing the ATP concentration, the reaction rate is made to increase. The time between two successive enzymatic cycles then becomes too short for us to resolve individual cycles. The curves obtained at 300 μM which show the system's extension as a function of time (see Fig. 13(A)) no longer display the discrete steps described in the previous paragraph, but rather a continous and rapid ascent. Since the time between two relaxations $T_{wait} \sim 200s$ is very long compared to the time required to relax the DNA molecule $T_{relax} \sim 5s$, we may conclude that a single topo II molecule is responsible for supercoil removal. This also allows us to say that the enzyme behaves processively, since the enzymatic

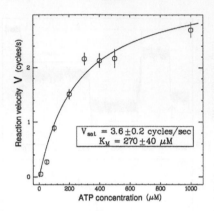

Fig. 14. Velocity of the reaction V as a function of ATP concentration. The data were obtained with a stretching force $F = 0.7$ pN. The' error bars represent the statistical error in determining V. By fitting the Michelis-Menten equation (Eqn. 6.1) to the experimental data, we obtain (solid line) the parameters shown in the plot.

cycles follow one another in rapid succession, without the enzyme detaching from its substrate. By repeating the experiment twenty times, we obtain a statistical sample of twenty curves corresponding to roughly one hundred enzymatic cycles, allowing us to determine to within about 10% the reaction velocity V.

6.4.5. Determining V_{sat}, the saturated reaction velocity

The procedure described above is then repeated for different ATP concentrations. At each ATP concentration we determine the mean reaction curve $< l(t) >$, and from this we extract the velocity of the reaction V for a given ATP concentration. It is important to note that since the experiments are performed in the constant-torque regime, the reaction at cycle $n + 1$ is the same as at cycle n. We then plot (see Figure 14) the reaction velocity as a function of ATP concentration. The Michelis-Menten equation relating reaction velocity to ATP concentration is

$$V = \frac{V_{sat}[ATP]}{k_m + [ATP]} \tag{6.1}$$

where V_{sat} represents the reaction velocity in conditions where ATP is present in saturating amounts, and k_M the ATP concentration (written [ATP]) for which $V = V_{sat}/2$. By fitting this equation to the experimental data (Figure 14), we obtain $k_M = 270 \pm 40\mu$M and $V_{sat} = 3.6 \pm 0.2$ cycles/second. These values are in good agreement with results obtained by bulk experiments on the same enzyme [85]: $k_M = 270\mu$M and $V_{sat} = 1.5$ cycles/second.

The difference in the value of V_{sat} can be explained by two factors which each demonstrates the interest of single-molecule experiments. First of all, it is impossible to determine in bulk experiments the exact number of active enzymes, and this number is typically overestimated relative to the total mass of enzyme employed. This typically leads to an under-estimate of enzyme activity. Another problem with bulk experiments is that they cannot separate out the time required for the enzyme to diffuse and find its substrate and the time it then takes to relax all the supercoils. This also leads to an under-estimate of enzyme activity. Measurements performed on single molecules are free of these particular problems: the time required for the enzyme to acquire its subtrate does not contaminate the measurment, and by diluting the enzyme sufficiently one can obtain conditions where the measurement truly represents the activity of a single, active enzyme molecule. Thus, measurements performed at the single-molecule level offer the possibility of defining without ambiguity and in a quantitative fashion the real activity of an enzyme.

6.4.6. Effect of the stretching force
The experiments described previously were performed at a force $F = 0.7$ pN on positively supercoiled DNA. We will now describe experiments performed in saturating ATP conditions (1 mM) but at different forces.

So as to determine the effect of the stretching force on the reaction velocity, we performed measurements on positively supercoiled DNA in the presence of 1 mM ATP. Using positively supercoiled DNA makes it possible to work at forces as high as $F = 5$ pN without inducing structural transitions in the DNA (remember that this would not be possible with negatively supercoiled DNA). The results of these supercoil removal experiments are shown Figure 15, and they show that V decreases as the stretching force increases.

This result is rather surprising, as on would imagine that increasing the force would favor supercoil relaxation. Indeed, increasing the force should decrease the energetic barrier to opening of the G-segment. Also, since the torque acting on the DNA increases with the force, one would have thought that the increase in torque would favor the transport of the T-segment through the G-segment. What we observe is the exact opposite: since the reaction velocity decreases as the force increases, this suggests that the rate-limiting step in the reaction corresponds to work performed *against* the stretching force. Rather than lower the activation energy of the rate-limiting step, the stretching force raises it. This suggests that the rate-limiting step of the reaction could correspond to rejoining the ends of the cleaved G-segment, as this step clearly performs work against the stretching force.

By writing that the energetic barrier to this rate-limiting step is increased by $F \cdot \Delta$, where Δ represents the distance over which the enzyme performs

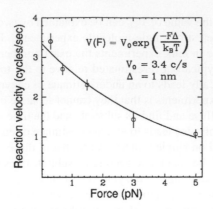

Fig. 15. Velocity of the reaction V as a function of the stretching force applied to the DNA. The experiments were done in the presence of 1 mM ATP. The error bars represent the statistical error in the determination of V. The continuous line is a fit of an exponential decay $V(F) = V_0 \exp(-F\Delta/k_B T)$ to the experimental data (see text).

work against the force, we obtain an Arhenius equation for the reaction rate: $V(F) = V_0 \exp(-F\Delta/k_B T)$. By fitting this equation to the experimental data, we obtain a good description of the experiments with $\Delta = 1$nm. This distance is reasonnable considering the hypothesis made above. Indeed, if the cleaved gate segment must allow the transported DNA segment to pass through it, it must open by at least ~ 1 nm.

6.4.7. Relaxation of negatively supercoiled DNA

So as to study the activity of topo II on a negatively supercoiled DNA, we stretched the DNA with a force sufficiently low ($F = 0.3$ pN) such that unwinding does not cause local denaturation. We then measured the maximal velocity of the reaction V in the presence of 1 mM ATP. In the experimental conditions tested, we did not detect a great difference between the rate of relaxation of positive or negative supercoils: $V^{\sigma < 0} = 2.6 \pm 0.2$ cycles/second and $V^{\sigma > 0} = 3.4 \pm 0.2$ cycles/second. As we will see, a far different situation holds for prokaryotic topo II.

6.4.8. Topo II only removes crossovers in DNA

By measuring the residual amount of supercoiling which remained after a topoisomerase II enzyme molecule had acted on the DNA, we found that the enzyme was only able to remove supercoiling which was stored as plectonemes. Indeed, the excess linking number of DNA molecules acted upon by topo II never dropped below $n_{buckle} \sim 15$, the number of supercoils observed at the buckling

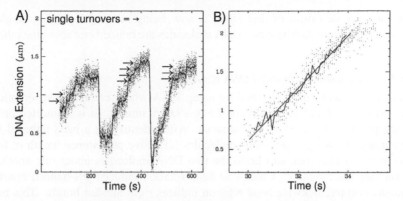

Fig. 16. Relaxation of DNA supercoils by topoisomerase IV. A) Relaxation of negative supercoils occurs in a stepwise, non-processive fashion even at high enzyme concentrations. Points correspond to the raw experimental data, and the full line to an averaging over a few seconds. The relaxation is much slower than if the DNA were positively supercoiled. Arrows indicate single cycles which were resolved. B) Relaxation of positive supercoils occurs in a highly processive fashion (as with topo II), even in single-molecule conditions. We superpose ~ 10 curves (shown as points), and the average is shown as a continous line; from this we obtain a velocity $V \sim 3$ cycles/second.

transition where plectonemic supercoils begin to appear. This implies that topo II is capable of acting on supercoiled DNA only when there are DNA crossovers present.

6.5. A comparison with E. coli topo IV

Although highly similar to eukaryotic topo II, prokaryotic topo IV display some remarkable differences which help illuminate its function and mechanism of action.

Relaxation of positive supercoils   For these experiments, we measured the maximal rate of relaxation of positive supercoils V by the method described previously. The results are presented Figure 16B. We obtain $V \sim 3$ cycles/second, very close to the rate observed on *D. melanogaster* topo II.

Relaxation of negative supercoils   The action of topo IV changes radically when its target DNA is negatively supercoiled, see Figure 16A. Indeed, the relaxation of negative supercoils only begins at enzyme concentrations of about 50 ng/ml, and is not at all processive. At this concentration, is it quite likely that multiple enzymes are in interaction with the DNA at any given moment.

   These experiments suggest that, unlike eukaryotic topo II, *E. coli* topo IV is capable of distinguishing between a positively and a negatively supercoiled DNA.

The microscopic origin of this effect is now being investigated using single-molecule experiments where two DNA molecules are braided one about the other.

6.5.1. Experiments on braided DNA molecules

These experiments are simplified by using DNA which cannot be supercoiled (i.e. contains a break along one of the two DNA strands, or is connected only at one point to the bead or the surface). After identifying a bead tethered to the surface by two parallel DNA molecules (effective persistence length of the system $\sim$ 25 nm), one can braid the two DNA molecules about one another by rotating the magnets. Clockwise bead rotation induces left-handed braids, whereas counterclockwise bead rotation induces right-handed braids. This has the advantage of allowing the researcher to independently control the topological sign and crossing angle in the braided region by controlling, respectively, the direction of braiding and the applied stretching force.

Recent experiments [97] using such a setup to investigate topoisomerase IV activty have shown that in fact the enzyme acts most efficiently when uncatenating left-handed brais. It is important to note that the topology of DNA crossovers in a left-handed braid is the same as that in positivly supercoiled DNA, whereas the topology of DNA crossovers in a right-handed braid is identical to that found in negatively supercoiled DNA. Thus, the results on both supercoiled and braided DNA are consistent with one another. In addition, by varying the force, the authors found that the enzyme worked best on positively signed crossovers presenting an acute angle. These preferences suggest a selection mechanism whereby topoisomerase IV determines the global, large-scale topology of its preferred (and biologically most important) substrate by detecting local features of this topology!

6.6. Conclusions on type II topoisomerases

Single-molecule DNA micromanipulation has made it possible to directly confirm that type II topoisomerases remove two supercoils per cycle, and also to perform quantitative studies on the nature of enzyme binding to DNA, the kinetics of enzyme binding to DNA, the effect of DNA topology (both braided and supercoiled) and the effect of force on the rate-limiting step of the reaction. Almost all of these measurements provide new, unambiguous and previously inaccessible information, thereby highlighting the usefulness of single-molecule approaches in the detailed understanding of protein-DNA interactions. The comparison between prokaryotic and eukaryotic enzymes has been particularly illuminating in this respect.

7. Conclusions and future prospects

Although single-molecule DNA manipulation experiments have provided for exciting new insights into nucleic acid structure and its interactions with proteins, new technical developments are poised to provide yet more information on these systems. In particular, the coupling of single-molecule fluorescence-based detection techniques (such as FRET and evanescent-field excitation techniques) will allow researchers to further study the temporal correlation between fast biochemical events (such as binding of molecules) and slow mechanical events (such as force generation in molecular motors) [9,98]. These advances will certainly provide for unsurpassed spatial and temporal analysis of dynamic molecular systems.

References

[1] J.T. Finer, R.M. Simmons, and J. A. Spudich. Single myosin molecule mechanics: piconewton forces and nanometre steps. *Nature*, 368:113–119, 1994.

[2] J. Gelles, B.J. Schnapp, and M. Sheetz. Tracking kinesin-driven movements with nanometre-scale precision. *Nature*, 331:450–453, 1988.

[3] M.D. Wang, M.J. Schnitzer, H. Yin, R. Landick, J. Gelles, and S. Block. Force and velocity measured for single molecules of RNA polymerase. *Science*, 282:902–907, 1998.

[4] A. Revyakin, R.H. Ebright, and T.R. Strick. Promoter unwinding and promoter clearance: detection by single-molecule DNA nanomanipulation. *Proc. Natl. Acad. Sci. (USA)*, 101:4776–4780, 2004.

[5] H. Noji, R. Yasuda, M. Yoshida, and K. Kinosita. Direct observation of the rotation of F_1-ATPase. *Nature*, 386:299–302, 1997.

[6] M. Rief, M. Gautel, F. Oesterhelt, J.M. Fernandez, and H.E. Gaub. Reversible unfolding of individual titin immunoglobulin domains by AFM. *Science*, 276:1109–1112, 1997.

[7] L. Tskhovrebova, J. Trinic, J.A. Sleep, and R.M. Simmons. Elasticity and unfolding of single molecules of the giant musle protein titin. *Nature*, 387:308–312, 97.

[8] M.S.Z. Kellermayer, S.B. Smith, H.L. Granzier, and C. Bustamante. Folding-unfolding transition in single titin molecules characterized with laser tweezers. *Science*, 276:1112–1116, 1997.

[9] A. Ishijima, H. Kojima, T. Funatsu, M. Tokunaga, H. Higuchi, H. Tanaka, and T. Yanagida. Simultaneous observation of individual ATPase and mechanical events by a single myosin molecule during interaction with actin. *Cell*, 92:161–171, 1998.

[10] Y. Arai, R. Yasuda, K. Akashi, Y. Harada, H. Miyata, K. Kinosita Jr., and H. Itoh. Tying a molecular knot with optical tweezers. *Nature*, 399:446–448, 1999.

[11] D.K. Fygenson, M. Elbaum, B. Shraiman, and A. Libchaber. Microtubules and vesicles under controlled tension. *Phys. Rev. E*, 55:850–859, 1997.

[12] B.D. Brower-Toland, C.L. Smith, R.C. Yeh, J.T. Lis, C.L. Peterson, and M.D. Wang. Mechanical disruption of individual nucleosomes reveals a reversible multistage release of DNA. *Proc. Natl. Acad. Sci. (USA)*, 99:1960–1965, 2002.

[13] T. Ha, X. Zhuang, H.D. Kim, J.W. Orr, J.R. Williamson, and S.Chu. Ligand-induced conformational changes observed in single rna molecules. *Proc. Natl. Acad. Sci. USA*, 96:9077–9082, 1999.

[14] S.B. Smith, L. Finzi, and C. Bustamante. Direct mechanical measurements of the elasticity of single DNA molecules by using magnetic beads. *Science*, 258:1122–1126, 1992.

[15] T.R. Strick, J.F. Allemand, D. Bensimon, A. Bensimon, and V. Croquette. The elasticity of a single supercoiled DNA molecule. *Science*, 271:1835–1837, 1996.

[16] M.N. Dessinges, B. Maier, Y. Zhang, M. Peliti, D. Bensimon, and V. Croquette. Stretching single stranded DNA, a model polyelectrolyte. *Phys. Rev. Lett.*, 89:248102, 2002.

[17] M. Tokunaga, K. Kitamura, K. Saito, A.H. Iwane, and T. Yanagida. Single molecular imaging of fluorophores and enzymatic reactions achieved by objective-type total internal reflection fluorescence microscopy. *Biochem. and Biophys. Res. Comm.*, 235:47–53, 1997.

[18] L. Stryer. Fluorescence spectroscopy of proteins. *Science*, 162:526–533, 1968.

[19] J.-L. Mergny and M. Djavaheri-Mergny. Le transfert d'énergie d'excitation en biologie. *Regard sur la biochimie*, 4:17–27, 1994.

[20] S. Weiss. Fluorescence spectroscopy of single biomolecules. *Science*, 283:1676–1683, 1999.

[21] F. Amblard, B. Yurke, A. Pargellis, and S. Leibler. A magnetic manipulator for studying local rheology and micromechanical properties of biological system. *Rev. Sci. Instrum.*, 67(2):1–10, 1996.

[22] C. Gosse. *Conception d'un réseau bidimensionnel d'acides nucléiques et micromanipulation et mesure de force par pinces magnétiques.* PhD thesis, Université Pierre et Marie Curie Paris VI, 1999.

[23] C. Bustamante, J.F. Marko, E.D. Siggia, and S. Smith. Entropic elasticity of λ-phage DNA. *Science*, 265:1599–1600, 1994.

[24] S.B. Smith, Y. Cui, and C. Bustamante. Overstretching B-DNA: the elastic response of individual double-stranded and single-stranded DNA molecules. *Science*, 271:795–799, 1996.

[25] P. Cluzel, A. Lebrun, C. Heller, R. Lavery, J.-L. Viovy, D. Chatenay, and F. Caron. DNA: an extensible molecule. *Science*, 271:792–794, 1996.

[26] G. Altan-Bonnet, A. Libchaber, and O. Krichevsky. Bubble dynamics in double-stranded DNA. *Phys. Rev. Lett.*, 90:138101, 2003.

[27] K. Svoboda, P.P. Mitra, and S.M. Block. Flucutuation analysis of motor protein movement and single enzyme kinetics. *Proc. Natl. Acad. Sci. (USA)*, 91:11782–11786, 1994.

[28] P.-G. de Gennes. *Scaling concepts in polymer physics.* Cornell University Press, 1979.

[29] T.T. Perkins, S.R. Quake, D.E. Smith, and S. Chu. Relaxation of a single DNA molecule observed by optical microscopy. *Science*, 264:822–826, 1994.

[30] S.R. Quake, H. Babcock, and S. Chu. The dynamics of partially extended single molecules of DNA. *Nature*, 388:151–154, 1997.

[31] T.T. Perkins, D.E. Smith, and S. Chu. Direct observation of tube-like motion of a single polymer chain. *Science*, 264:819–822, 1994.

[32] M.D. Wang, H. Yin, R. Landick, J. Gelles, and S. Block. Stretching DNA with optical tweezers. *Biophys. J.*, 72:1335–1346, 1997.

[33] B. Essevaz-Roulet, U. Bockelmann, and F. Heslot. Mechanical separation of the complementary strands of DNA. *Proc. Natl. Acad. Sci. (USA)*, 94:11935–11940, 1997.

[34] E.L. Florin, V.T. Moy, and H.E. Gaub. Adhesion force between individual ligand-receptor pairs. *Science*, 264:415–417, 1994.

[35] M.G. Poirier and J.F. Marko. Mitotic chromosomes are chromatin networks without a mechanically contiguous protein scaffold. *Proc. Natl. Acad. Sci. (USA)*, 99:15393–15397, 2002.

[36] M. Rief, H. Clausen-Schaumann, and H.E. Gaub. Sequence-dependent mechanics of single DNA molecules. *Nature Struct. Bio.*, 6:346–349, 1999.

[37] U. Bockelmann, B. Essevaz-Roulet, and F. Heslot. Molecular stick-slip revealed by opening DNA with piconewton force. *Phys. Rev. Lett.*, 79(22):4489–4492, 1997.

[38] H. Yin, M.D. Wang, K. Svoboda, R. Landick, S. Block, and J. Gelles. Transcription against a applied force. *Science*, 270:1653–1657, 1995.

[39] M.D. Wang, M.J. Schnitzer, H. Yin, R. Landick, J. Gelles, and S.M. Block. Force and velocity measured for single molecules of RNA polymerase. *Science*, 282:902–907, 1998.

[40] U. Dammer, M. Hegner, D. Anselmetti, P. Wagner, M. Dreier, W. Huber, and H.-J. Guntherodt. Specific antigen/antibody interactions measured by force microscopy. *Biophysical Journal*, 70:2437–2441, 1996.

[41] J.-F. Leger. *L'ADN: une flexibilité structurale adaptée aux interactions avec les autres macromolécules de son environnement.* PhD thesis, Université Louis Pasteur Strasbourg I, 1999.

[42] J.F. Léger, G.Romano, A. Sarkar, J. Robert, L. Bourdieu, D. Chatenay, and J.F. Marko. Structural transitions of a twisted and stretched dna molecule. *Phys. Rev. Lett.*, 83:1066–1069, 1999.

[43] Z. Bryant, M.D. Stone, J. Gore, S.B. Smith, N.R. Cozzarelli, and C. Bustamante. Structural transitions and elasticity from torque measurements on DNA. *Nature*, 424:338–341, 2003.

[44] A. La Porta and M.D. Wang. Optical torque wrench: angular trapping, rotation, and torque detection of quartz microparticles. *Phys. Rev. Lett.*, 92:190801, 2004.

[45] J.-F. Allemand. *Micromanipulation d'une molécule individuelle d'ADN.* PhD thesis, Université Pierre et Marie Curie Paris VI, 1997.

[46] J.H. White. Self linking and the gauss integral in higher dimensions. *Am. J. Math.*, 91:693–728, 1969.

[47] B.J. Peter, J. Arsuaga, A.M. Breier, A.B. Khodursky, P.O. Brown, and N.R. Cozzarelli. Genomic transcriptional response to loss of chromosomal supercoiling in escherichia coli. *Genome Biol.*, 5:R87, 2004.

[48] J. Vinograd, J. Lebowitz, and R. Watson. Early and late helix-coil transitions in closed circular DNA. The number of superhelical turns in polyoma DNA. *J. Mol. Biol.)*, 33:173–197, 1968.

[49] A.V. Vologodskii, A.V. Lukashin, V.V. Anshelevich, and M.D. Frank-Kamenetskii. Fluctuations in superhelical DNA. *Nucleic Acids Research*, 6:967–982, 1979.

[50] D. Kowalski, D.A. Natale, and M.J. Eddy. Stable DNA unwinding, not breathing, accounts for the nuclease hypersensitivity of A+T rich regions. *Proc. Natl. Acad. Sci. (USA)*, 85:9464–9468, 1988.

[51] E. Palecek. Local supercoil-stabilized structures. *Crit. Rev. Biochem. Mol. Biol.*, 26:151–226, 1991.

[52] C.J. Benham. Sites of predicted stress-induced DNA duplex destabilization occur preferentially at regulatory loci. *Proc. Natl. Acad. Sci. (USA)*, 90:2999–3003, 1993.

[53] K.L. Beattie, R.C. Wiegand, and C.M. Radding. Uptake of homologous single-stranded fragments by superhelical DNA. *J. Mol. Biol.*, 116:783–839, 1977.

[54] A. Murchie, R. Bowater, F. Aboul-ela, and D. Lilley. Helix opening transitions in supercoiled DNA. *Biochimica and Biophysica Acta*, 1131:1–15, 92.

[55] J.C. Wang. DNA topoisomerases. *Annu. Rev. Biochem.*, 65:635–692, 1996.

[56] L.F. Liu and J.C. Wang. Supercoiling of the DNA template during transcription. *Proc. Natl. Acad. Sci. USA*, 84:7024–7027, 1987.

[57] P. Dröge. Transcription-driven site-specific DNA recombination *in vitro. Proc. Natl. Acad. Sci. (USA)*, 90:2759–2763, 1993.

[58] H.-Y. Wu, S. Shyy, J.C. Wang, and L.F. Liu. Transcription generates positively and negatively supercoiled domains in the template. *Cell*, 53:433–440, 1988.

[59] C. Bouchiat, M.D. Wang, S. M. Block, J.-F. Allemand, T.R. Strick, and V.Croquette. Estimating the persitence length of a worm-like chain molecule from force-extension measurements. *Biophys. J.*, 76:409–413, 1999.

[60] T.R. Strick, J.-F. Allemand, D. Bensimon, and V. Croquette. The behavior of supercoiled DNA. *Biophys. J.*, 74:2016–2028, 1998.

[61] M.E. Hogan and R.H. Austin. Importance of DNA stiffness in protein-DNA binding specificity. *Nature*, 329:263–266, 1987.

[62] C. Bouchiat and M. Mézard. Elasticity theory of a supercoiled DNA molecules. *Phys. Rev. Lett.*, 80:1556–1559, 1998.

[63] L. Postow, C.D. Hardy, J. Arsuaga, and N.R. Cozzarelli. Topological domain structure of the escherichia coli chromosome. *Genes Dev.*, 18:1766–1779, 2004.

[64] T.R. Strick, V. Croquette, and D. Bensimon. Homologous pairing in streched supercoiled DNA. *Proc. Nat. Acad. Sci. (USA)*, 95:10579–10583, 1998.

[65] J.-F. Allemand, D. Bensimon, R. Lavery, and V. Croquette. Stretched and overwound DNA form a Pauling-like structure with exposed bases. *Proc. Natl. Acad. Sci. USA*, 95:14152–14157, 1998.

[66] B. Maier, D. Bensimon, and V. Croquette. Replication by a single DNA polymerase of a stretched single-stranded DNA. *Proc. Natl. Acad. Sci. (USA)*, 97:12002–12007, 2000.

[67] G.J.L. Wuite, S.B. Smith, M. Young, D. Keller, and C. Bustamante. Single-molecule studies of the effect of template tension on T7 DNA polymerase activity. *Nature*, 404:103–106, 2000.

[68] A.M. van Oijen, P.C. Blainey, D.J. Crampton, C.C. Richardson, T. Ellenberger, and X.S. Xie. Single-molecule kinetics of lambda exonuclease reveal base dependence and dynamic disorder. *Science*, 301:1235–1238, 2003.

[69] D.A. Schafer, J. Gelles, M.P. Sheetz, and R. Landick. Transcription by single molecules of RNA polymerase observed by light microscopy. *Nature*, 352:444–448, 1991.

[70] H. Yin, R. Landick, and J. Gelles. Tethered particle motion method for studying transcript elongation by a single RNA polymerase molecule. *Biophys. J.*, 67:2468–2478, 1994.

[71] S.F. Tolic-Norrelykke, A.M. Engh, R. Landick, and J. Gelles. Diversity in the rates of transcript elongation by single rna polymerase molecules. *J. Biol. Chem.*, 279:3292–3299, 2004.

[72] K.C. Neuman, E.A. Abbondanzieri, R. Landick, J. Gelles, and S.M. Block. Ubiquitous transcriptional pausing is independent of RNA polymerase backtracking. *Cell*, 115:437–447, 2003.

[73] J.W. Shaevitz, E.A. Abbondanzieri, R. Landick, and S.M. Block. Backtracking by single RNA polymerase molecules observed at near-base-pair resolution. *Nature*, 426:684–687, 2003.

[74] A. Shundrovsky, T.J. Santangelo, J.W. Roberts, and M.D. Wang. A single-molecule techniques to study sequence-dependent transcription pausing. *Biophys J.*, 87:3945–3953, 2004.

[75] Y. Harada, O. Ohara, A. Takatsuki, H. Itoh, N. Shimamoto, and Jr. K. Kinosita. Direct observation of DNA rotation during transcription by escherichia coli RNA polymerase. *Nature*, 409:113–115, 2001.

[76] L.F. Liu, C.C. Liu, and B.M. Alberts. T4 DNA topoisomerase: a new ATP-dependent enzyme essential for initiation of T4 bacteriophage DNA replication. *Nature*, 281:456–461, 1979.

[77] J.C. Wang. Interaction between DNA and an *escherichia coli* protein ω. *J. Mol. Biol.*, 55:523–533, 1971.

[78] O. Guipaud, E. Marguet, K.M. Noll, C. Bouthier de la Tour, and P. Forterre. Both DNA gyrase and reverse gyrase are present in the hyperthermophilic bacterium *thermotoga maritima*. *Proc. Natl. Acad. Sci. (USA)*, 94:10606–10611, 1997.

[79] E.R. Shelton, N. Osheroff, and D.L. Brutlag. DNA topoisomerase II from *drosophila melanogaster*: purification and physical characterization. *J. Biol. Chem.*, 258:9530–9535, 1983.

[80] S.J. Froelich-Ammon and N. Osheroff. Topoisomerase poisons: harnessing the dark side of enzyme mechanism. *J. Biol. Chem.*, 270:21429–21432, 1995.

[81] F. Guo, D.N. Gopaul, and G.D. Van Duyne. Structure of Cre-recombinase complexed with DNA in a site-specific recombination synapse. *Nature*, 389:40–46, 1997.

[82] C. Cheng, P. Kussie, N. Pavletich, and S. Shuman. Conservation of structure and mechanism between eukaryotic topoisomerase I and site-specific recombinases. *Cell*, 92:841–850, 1998.

[83] P.O. Brown and N.R. Cozzarelli. Catenation and knotting of duplex DNA by type i topoisomerases: a mechanistic parallel with type 2 topoisomerases. *Proc. Natl. Acad. Sci. (USA)*, 78:843–847, 1981.

[84] N.H. Dekker, V.V. Rybenkov, M. Duguet, N.J. Crisona, N.R. Cozzarelli, D. Bensimon, and V. Croquette. The mechanism of type IA topoisomerases. *Proc. Natl. Acad. Sci. (USA)*, 99:12126–12131, 2002.

[85] N. Osheroff, E.R. Shelton, and D.L. Brutlag. DNA topoisomerase II from *drosophila melanogaster*: relaxation of supercoiled DNA. *J. Biol. Chem.*, 258:9536–9543, 1983.

[86] T. Uemura and M. Yanagida. Mitotic spindle pulls but fails to separate chromosomes in type II DNA topoisomerase mutants: uncoordinated mitosis. *EMBO Journal*, 5:1003–1010, 1986.

[87] V.V. Rybenkov, C. Ullsperger, A.V. Vologodskii, and N.R. Cozzarelli. Simplification of DNA topology below equilibrium values by type II topoisomerases. *Science*, 277:690–693, 1997.

[88] J. Roca and J.C. Wang. The capture of a DNA double helix by an ATP-dependent protein clamp: a key step in DNA transport by type II DNA topoisomerase. *Cell*, 71:833–840, 1992.

[89] J. Roca and J.C. Wang. DNA transport by a type II DNA topoisomerase: evidence in favor of a two-gate model. *Cell*, 77:609–616, 1994.

[90] J. Roca, J.M. Berger, S.C. Harrison, and J.C. Wang. DNA transport by a type II topoisomerase: Direct evidence for a two-gate mechanism. *Proc. Natl. Acad. Sci. (USA)*, 93:4057–4062, 1996.

[91] J. Yan, M.O. Magnasco, and J.F. Marko. Kinetic proofreading mechanism for disentanglement of DNA by topoisomerases. *Nature*, 401:932–935, 1999.

[92] J. Kato, Y. Nishimura, R. Imamura, H. Niki, S. Hiraga, and H. Suzuki. New topoisomerase essential for chromosome segregation in *e. coli*. *Cell*, 63:393–404, 1990.

[93] H. Hiasa and K.J. Marians. Two distinct modes of strand unlinking during θ-type DNA replication. *J. Biol. Chem.*, 271:21529–21535, 1996.

[94] H. Hiasa, R.J. DiGate, and K.J. Marians. to find. *J. Biol. Chem.*, 269:2093–2099, 1994.

[95] J. Roca, J.M. Berger, and J.C. Wang. On the simultaneous binding of eukaryotic DNA topoisomerase II to a pair of double-stranded DNA helices. *J. Biol. Chem.*, 268:14250–14255, 1993.

[96] T.R. Strick, V. Croquette, and D. Bensimon. Single-molecule analysis of DNA uncoiling by a type II topoisomerase. *Nature*, 404:901–904, 2000.

[97] G. Charvin, D. Bensimon, and V. Croquette. Single-molecule study of DNA unlinking by eukaryotic and prokaryotic type-II topoisomerases. *Proc. Natl. Acad. Sci. (USA)*, 100:9820–9825, 2003.

[98] M.J. Lang, P.M. Fordyce, and S.M. Block. Combined optical trapping and single-molecule fluorescence. *J. Biol.*, 2:1–4, 2003.

Course 7

INTRODUCTION TO SINGLE-DNA MICROMECHANICS

John F. Marko

Department of Physics, University of Illinois at Chicago,
Chicago, IL 60607-7059, USA

D. Chatenay, S. Cocco, R. Monasson, D. Thieffry and J. Dalibard, eds.
Les Houches, Session LXXXII, 2004
Multiple aspects of DNA and RNA: from Biophysics to Bioinformatics
© *2005 Elsevier B.V. All rights reserved*

211

Contents

1. Introduction

Over the past ten years new 'single-molecule' techniques to study individual bio-molecules have been developed. Many of the new approaches being used are based on micromanipulation of single DNAs, allowing direct study of DNA, and enzymes which interact with it. These lectures focus on mechanical properties of DNA, crucial to the design and interpretation of single-DNA experiments, and to the understanding of how DNA is processed and therefore functions, inside the cell.

A seminal example of a single-DNA experiment was the measurement of the force exerted by RNA polymerase [1], done by Jeff Gelles, Steve Block and co-workers. Gene sequences in DNA are 'read' by RNApol, which synthesizes an RNA copy of a DNA sequence. The experiment (Fig. 1) revealed that as a RNApol moves along a DNA, it is able to pull with up to 30×10^{-12} Newtons, or 30 piconewtons (pN) of force. In the single-molecule world, this is a hefty force: the motor proteins which generate your muscle contractions, called *myosin* generate only about 5 pN.

RNApol is an example of a *processive enzyme* which works rather like a macroscopic engine, using stored chemical energy to catalyze not only the synthesis of RNA, but also converting some of that energy to mechanical work. This mechanical work is absolutely necessary for RNApol's function: it must move 'processively' along the DNA double helix in order to make a faithful copy of DNA. Another important DNA-processing enzyme is *DNA polymerase* which is able to synthesize a copy of a DNA strand; this is important in cell division, since in order to make a copy of itself, a cell must faithfully copy its chromosomal DNAs. Proper understanding of this kind of DNA-processing enzyme machinery requires us to first understand the mechanical properties of DNA itself.

DNA has extremely interesting and unique polymer properties. In double helix form it is a water-soluble, semiflexible polymer which can be obtained in gigantic lengths. We often measure DNA length in 'bases' or 'base pairs'; each DNA base of nm dimensions encodes one of four 'letters' (A, T, G or C) in a genetic sequence. A human genome contains 3×10^9 bases divided into 23 chromosomes. Each chromosome therefore contains a DNA roughly 10^8 bases long; chromosomal DNAs are the longest linear polymers known. Furthermore, the

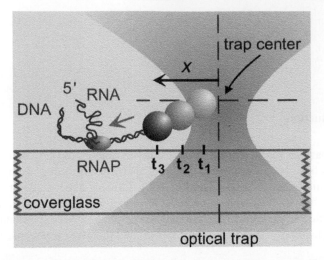

Fig. 1. Sketch of single-DNA experiment of Wang et al. to measure force generated by RNA polymerase (reproduced from Ref. [1]). The polymerase is attached to the glass, and the DNA is pulled through it. A bead at the end of the DNA is held in a laser trap; deflection of the bead in the trap indicates the applied force.

base-paired complementary-strand structure of the double helix offers up new types of polymer physics problems, which we will explore in these lectures.

I note some physical scales relevant to these lectures. The fundamental length scale of molecular biology is the nanometer (nm); this is a distance several atoms long, the size of a single nucleic acid (DNA or RNA) base (the basic unit of information in molecular biology), or a single amino acid (the elementary unit of proteins). Cells must maintain their nm-scale organizational structure at room temperature: this requires that components be acted on by forces of roughly

$$1 \, k_B T / \text{nm} = 4 \times 10^{-21} \, \text{J} / 10^{-9} \, \text{m} = 4 \times 10^{-12} \, \text{N} = 4 \, \text{pN}.$$

We can expect the forces generated by single mechanoenzymes to be on the pN scale. If RNApol generated smaller forces than this, it would get pushed around by thermal forces, and would be unable to read DNA sequence in a processive manner.

Problem 1: Consider a molecule localized by a harmonic force $f = -kx$. What force constant is necessary to have $\langle x^2 \rangle = 1 \, \text{nm}^2$? What is the typical (root-mean-squared) force applied to the molecule in this case? Repeat this calculation if the localization is done to 1 Å (atomic) accuracy.

Problem 2: Consider a nanowire made of some elastic material, with circular cross-section of diameter d. In any cross-section of the wire, what will be the typical elongational stress (force per area) due to thermal fluctuations? What does this suggest about the Young modulus of the material that you might try to use to make a nanowire?

Problem 3: Consider a random sequence of DNA bases 48502 bases long. How many times do you expect to find the sequences AATT, ACTAGT and GGC-CGGCC?

2. The double helix is a semiflexible polymer

The double helix (sometimes called the 'B-form') is taken by DNA most of the time in the cell. This form of DNA has a regular helical structure with remarkably uniform mechanical properties. This section will focus on the bending flexibility of the double helix, which gives rise to polymer elasticity effects which are of biological importance, and accessible in biophysical experiments.

2.1. Structure

The double helix is made of two DNA polymer molecules. Each DNA polymer is a string of four interchangable types of 'monomers', which can be strung together in any sequence. The monomers each carry a *sugar-phosphate backbone* element: these are covalently bound together in the polymer. However, each monomer also carries, attached to the sugar (which is deoxyribose), one of four possible 'bases': either adenine (A), thymine (T), guanine (G) or cytosine (C).

The length of each backbone unit is about 0.7 nm when extended. The bases are each about 1 nm wide, and 0.3 nm thick.

The structure of each polymer gives it a definite 'polarity'. It is conventional to report DNA sequence along each strand in the direction read by RNA polymerase, from 5′ to 3′ (the number refer to carbon atoms in the deoxyriboses). Often people just omit the leading 5′: in this case it is almost always in 5′ to 3′ order.

The bases have shapes and hydrogen-bonding sites which make A-T and G-C bonds favorable, under the condition that the two strands are anti-aligned (see sketch). Such *complementary strands* will bind together, making inter-strand hydrogen bonds, and intra-strand *stacking interactions*. The stacking of the bases drives the two strands to twist around one another to form a helix, since each base is only about 0.3 nm thick while the backbones are roughly 0.7 nm long per base.

We can roughly estimate the helix parameters of the double helix, assuming that the backbones end up tracing out a helical path on the surface of a cylinder of radius 1 nm (the bases are 1 nm across). Since each base is 0.34 nm thick, and traces a helix contour length of 0.7 nm, the circumference occupied by each base is $\sqrt{0.7^2 - 0.34^2}$ nm = 0.61 nm. Dividing the total circumference (6.3 nm) by

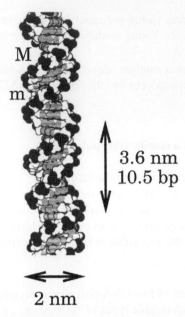

M

m

3.6 nm
10.5 bp

2 nm

Fig. 2. DNA double helix structure. The two complementary-sequence strands noncovalently bind together, and coil around one another to form a regular helix. The two strands can be seen to have directed chemical structures, and are oppositely directed. Note the different sizes of the major (M) and minor (m) grooves. The helix repeat is 3.6 nm, and the DNA cross-sectional diameter is 2 nm. DNA image reproduced from Ref. [2].

this indicates that the double helix contains about 10.3 base pairs (bp) per helical turn. This is very close to the number usually quoted of 10.5 bp/turn; the double helix therefore makes one turn for every $10.5 \times 0.34 = 3.6$ nm.

The B-form double helix is *right-handed*, with the two backbones oriented in opposite directions. This means that there are two types of 'grooves' between the backbones: these are in fact rather different in size in B-DNA, and are called the 'major groove' and the 'minor groove'.

We should remember the following conversion factor for the double helix, length: 1 bp = 0.34 nm, thus each micron (1000 nm) worth of DNA contains about 3000 bp = 3 kilobp (kb); one whole human genome is thus close to 10^9 nm = 1 m in length.

Problem 4: Consider a hypothetical form of double helix formed of two *parallel-orientation* strands. Describe the grooves between the backbones.

Problem 5: A student proposes that for two complementary-sequence biological DNA strands, there must be an equivalent form of double helix, of free energy equal to B-DNA, which is instead left-handed. Explain under what circumstances of symmetry of the monomers this conjecture can be expected to be true. Based on textbook pictures of the base and backbone chemical structures, what is your conclusion?

Problem 6: Do you expect the average helix repeat (base-pairs per turn) of the double helix to increase or decrease with increased temperature?

Problem 7: Estimate the 'Young modulus' of the double helix, using the assumption that the single-base helix parameters described above apply to room-temperature structure of DNA to roughly 1 Å precision.

2.2. DNA bending

Although the structure of DNA is often presented in books as if it is static, at room temperature and in solution the double helix undergoes continual thermally excited changes in shape. Per base pair, the fluctuations are usually small displacements (a few degrees of bend, 0.03 nm average separations of the bases) but over long stretches of double helix, they build up to significant, thermally excited random bends. Note that rarely, more profound thermally-excited disturbances of double helix structure (e.g., transient unbinding of base pairs) can be expected to occur.

2.2.1. Discrete-segment model of a semiflexible polymer

We can make a simple one-dimensional lattice model of thermally excited bending fluctuations. If we describe our DNA with a series of tangent vectors $\hat{\mathbf{t}}_j$ that indicate the orientation of the *center axis* of the molecule, then the bending energy associated with two adjacent tangents is $E/(k_B T) = -a\hat{\mathbf{t}}_j \cdot \hat{\mathbf{t}}_{j+1}$.

The dimensionless constant a describes the molecule's bending rigidity: $a >> 1$ means very rigid (adjacent tangent vectors point in nearly the same direction); $a < 1$ means very floppy. We'll talk more about a below, but just to give a rough idea of the stiffness of the DNA double helix, if we consider adjacent base pairs to be described by successive tangents, the value of a to use is about 150.

Problem 8: Estimate the bend between two adjacent tangent vectors excited thermally in the limit $a >> 1$; your result should be of the form

$$\langle |\hat{\mathbf{t}}_j - \hat{\mathbf{t}}_{j+1}|^2 \rangle \propto a^p$$

where p is a power. Hint: $\frac{1}{2}|\hat{\mathbf{t}}_j - \hat{\mathbf{t}}_{j+1}|^2 = 1 - \hat{\mathbf{t}}_j \cdot \hat{\mathbf{t}}_{j+1}$.

What is the typical single-base bending angle (in degrees) if we take $a = 150$?

We write the (unnormalized) probability distribution of a given conformation of an $N + 1$-tangent-vector-long chunk of molecule using the Boltzmann distribution:

$$P(\hat{t}_0, \cdots, \hat{t}_N) = \prod_{j=0}^{N-1} e^{a\hat{t}_j \cdot \hat{t}_{j+1}} \tag{2.1}$$

Now we compute the thermal correlation of the ends of this segment of polymer:

$$\langle \hat{t}_0 \cdot \hat{t}_N \rangle = \frac{\int d^2 t_0 \cdots d^2 t_N \, \hat{t}_0 \cdot \hat{t}_N \, P(\hat{t}_0, \cdots, \hat{t}_N)}{\int d^2 t_0 \cdots d^2 t_N \, P(\hat{t}_0, \cdots, \hat{t}_N)} \tag{2.2}$$

This calculation is not too hard to do using the formula (recall decomposition of plane waves into spherical waves):

$$e^{a\hat{t}\cdot\hat{t}'} = \sum_{l=0}^{\infty} 4\pi i^l j_l(ia) \sum_{m=-l}^{l} Y_{lm}(\hat{t}) Y_{lm}^*(\hat{t}') \tag{2.3}$$

and if you write the dot product $\hat{t}_0 \cdot \hat{t}_N$ as an $l = 1$ spherical harmonic, and place $\hat{t}_N$ along the $\hat{z}$ axis.

The orthogonality of the spherical harmonics leads to a 'collapse' of the many sums over l's and m's into one sum. In the numerator only the $l = 1$ term (from the dot product) survives; in the denominator only the $l = 0$ term contributes. The result is:

$$\langle \hat{t}_0 \cdot \hat{t}_N \rangle = \left(\frac{i j_1(ia)}{j_0(ia)} \right)^N = e^{N \ln[\coth(a) - 1/a]} \tag{2.4}$$

The function $\coth(a) - 1/a$ is less than 1 for positive a. Therefore the correlation of direction falls off simply exponentially with contour distance N along our polymer. Small local fluctuations of bending of adjacent tangents build up to big bends over the 'correlation length' of $-1/\ln[\coth(a) - 1/a]$ segments.

Problem 9: For the $a \gg 1$ limit, how many segments long is the tangent-vector correlation length?

Problem 10: Explain the relation between the discrete-tangent model discussed above and the one-dimensional Heisenberg (continuous-spin) model of classical statistical mechanics. Suppose a magnetic field is added: what would that correspond to in the polymer interpretation?

2.2.2. *Bending elasticity and the persistence length*

We can connect this discrete model to the continuous model for bending of a thin rod, from the theory of elasticity. We note that the bending energy of two adjacent tangents was, in $k_B T$ units, $-a\hat{\mathbf{t}}_j \cdot \hat{\mathbf{t}}_{j+1}$, which up to a constant is $\frac{a}{2}|\hat{\mathbf{t}} - \hat{\mathbf{t}}'|^2$.

The bending of a thin rod can be described in terms of tangent vectors $\hat{\mathbf{t}}$ distributed continuously along the rod contour. A bent rod has energy which is locally proportional to the square of its bending curvature $d\hat{\mathbf{t}}/ds$ (s is contour length):

$$E = \frac{B}{2} \int_0^L ds \left| \frac{d\hat{\mathbf{t}}}{ds} \right|^2 \tag{2.5}$$

where B is the rod bending modulus. For a rod of circular cross section of radius r made of an isotropic elastic material, $B = \frac{\pi}{4} Y r^4$ where Y is the Young modulus [4].

Problem 11: By considering a simple circular arc, find the contour length along a thin rod for which a one-radian bend has energy cost $k_B T$.

Problem 12: Pretend that dsDNA is made of a plastic material of Young modulus 3×10^8 Pa. Predict the bending constant B.

We can now connect our discrete and continuous models of bending, if we introduce the length b of the segments in our discrete model:

$$\begin{aligned} E &= -k_B T a \sum_{j=1}^N \hat{\mathbf{t}}_j \cdot \hat{\mathbf{t}}_{j+1} = \frac{k_B T a b}{2} \sum_{j=1}^N b \left| \frac{\hat{\mathbf{t}}_j - \hat{\mathbf{t}}_{j+1}}{b} \right|^2 \\ &\rightarrow \frac{B}{2} \int_0^L ds \left| \frac{d\hat{\mathbf{t}}}{ds} \right|^2 \end{aligned} \tag{2.6}$$

where a constant energy shift has been dropped. The final term represents the limit where we make b small, while making a big, keeping the product ab constant. This continuum limit turns the finite difference into a derivative, and the sum into an integral.

The bending elastic constants a and B are related by $k_B T a b = B$, and the rod length corresponds to the number of tangents through $Nb = L$. So, for a rod with bending modulus B, if we wish to use a discrete tangent vector model with segment length b, we need to choose $a = B/(k_B T b)$.

If we now go back to the correlation function (2.4), we can write it in the continuum limit where a becomes large, replacing $\ln[\coth a - 1/a] \rightarrow -1/a$

and obtaining

$$\langle \hat{\mathbf{t}}(s) \cdot \hat{\mathbf{t}}(s') \rangle = e^{-k_B T |s-s'|/B} = e^{-|s-s'|/A} \qquad (2.7)$$

The final term introduces the continuum version of the correlation length of (2.4) $A = B/(k_B T)$, called the *persistence length*. For the double helix, a variety of experiments show that $A = 50$ nm (150 bp) in physiological aqueous solution [5] (this term 'physiological solution' usually means water containing between 0.01 and 1 M univalent salt, and with pH between 7 and 8, at temperature between 15 and 30 C).

Problem 13: Starting with the persistence length $A = 50$ nm, estimate the bending modulus B, and the 'effective Young modulus' Y of the DNA double helix.

Both a and B represent effective elastic constants, and the bending energies being discussed here are really free energies (as in the theory of elasticity, we consider deformations at fixed temperature [4]). The 'real', microscopic internal energy must include thermal energy and chemical binding energies of the atoms, but as in many other areas of condensed matter physics we'll choose to ignore atomic details and use coarse-grained models, since I will focus on phenomena at length scales of nm and larger (double helix deformations, DNA-protein interactions). This is not to say that atomic detail is not important: many important questions about the stability of the double helix and DNA-protein interactions require information at atomic scales, and can be theoretically approached only via numerical simulation of all the atoms involved [6] (Fig. 3).

2.2.3. End-to-end distance

The tangent vector $\hat{\mathbf{t}}(s)$ can be used to compute the distance between two points on our polymer, using the relation $\mathbf{r}(L) - \mathbf{r}(0) = \int_0^L ds \hat{\mathbf{t}}(s)$. This relation can be used to compute the mean-square distance between contour points a distance L apart:

$$\langle |\mathbf{r}(L) - \mathbf{r}(0)|^2 \rangle = 2AL + 2A^2 \left(e^{-L/A} - 1 \right) \qquad (2.8)$$

In the limit where we look at points closer together than a persistence length, $L/A << 1$, we have a mean-square distance $= L^2 + \mathcal{O}(L/A)$; in this limit, the polymer doesn't bend very much, so its average end-to-end distance is just L.

In the opposite limit of a polymer many persistence lengths long, $L/A >> 1$, we have a mean-square-distance of $2AL$, just the size expected for a random-walk of $L/(2A)$ steps each of length $2A$. We sometimes talk about the *statistical segment length* or *Kuhn segment length* in polymer physics: for the semiflexible

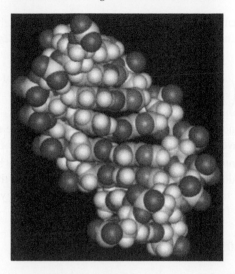

Fig. 3. Molecular-dynamics snapshot of typical DNA conformation for a short 10 bp molecule in solution at room temperature. Reproduced from Ref. [3].

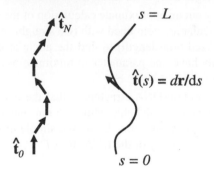

Fig. 4. Discrete-tangent and continuous-tangent models for DNA bending (see text).

polymer this segment length is $2A$. For the double helix, $2A$ is about 100 nm or 300 bp [5].

2.2.4. *DNA loop bending energies*
We'll hear in Section 5 about proteins which stabilize formation of DNA loops. Often, looping of DNA occurs so that sequences roughly 10 to 1000 bp away from the start of a gene can regulate (repress or enhance) that gene's transcription [7]. Formation of such a loop requires DNA bending, and now we can estimate the associated free energy.

Suppose we form a loop of length L. The simplest model is a circle of circumference L, with radius $L/(2\pi)$ and bending curvature $2\pi/L$, and bending energy

$$\frac{E_{\text{circle}}}{k_B T} = \frac{A}{2}L\left(\frac{2\pi}{L}\right)^2 = 2\pi^2\frac{A}{L} \tag{2.9}$$

For $L = 300$ bp and $A = 150$ bp, this is a big energy - close to $10\,k_B T$. A 100 bp circle would have a bending free energy three times larger than this!

You might be interested in the *lowest* energy necessary to bring two points a contour length L along a rod together. The optimal shape of the rod is of course not circular, but is instead tear-drop-shaped. The exact energy can be computed in terms of elliptic functions to be [8]

$$\frac{E_{\text{teardrop}}}{k_B T} = 14.055\frac{A}{L} \tag{2.10}$$

about 71% of the energy of the circle. In either teardrop or circle case, the energy of making a loop diverges as $1/L$ for small L.

Problem 14: Carry out an approximate calculation of the tear-drop shape and energy, by using a circular arc combined with two straight segments. Use energy minimization (with fixed total length) to find the angle at the base of the tear-drop (you should only have one parameter to minimize over) and the tear-drop configuration energy.

Problem 15: In a protein-DNA structure called the nucleosome, 146 bp of DNA make 1.75 helical turns with helix radius of 5 nm, and helical pitch (spacing of turns along the helix axis) of 3 nm. Using the simple models of this section, estimate the bending free energy of the DNA in $k_B T$.

2.2.5. Site-juxtaposition probabilities

These bending energies are not by themselves enough to accurately predict the probability that a DNA segment of length L forms a loop; we must also sum over bending fluctuations, thermally excited changes in shape. For the simple bending model described above, sophisticated calculations have been done for the probability of forming a loop.

Calculations of Stockmayer, Shimada and Yamakawa [8,9] tell us the probability density for finding the two ends of a semiflexible polymer brought smoothly together (with the same orientation):

$$J_{\text{circle}} = \frac{\pi^2}{(2A)^3}\left(\frac{2A}{L}\right)^6 e^{-E_{\text{circle}}/k_B T + 0.257L/A} \tag{2.11}$$

If the condition that the ends come together smoothly is relaxed, the same authors found

$$J_{\text{teardrop}} = \frac{28.01}{(2A)^3} \left(\frac{2A}{L} \right)^5 e^{-E_{\text{teardrop}}/k_B T + 0.246 L/A} \tag{2.12}$$

The units of these expressions are density (inverse volume), i.e., concentration of one end, at the position of the other.

For DNA, since the double helix can bend only over roughly 100 nm, the natural scale for J is very roughly $J \approx (100 \text{ nm})^{-3} = 10^{-6} \text{ nm}^{-3} \approx 10^{-6}$ Mol/litre (1 Mol/litre, or M, is 0.6 nm^{-3}).

The empirical results provide an accurate interpolation between the two limits where bending energy ($L/A < 1$), and entropy ($L/A > 1$) dominate, including the experimentally- and numerically-established result that the *peak* probability of juxtaposition occurs for molecules about $L = 170$ nm (500 bp) long [5, 11].

For $L \gg A$ we reach the long-distance limit, where we may estimate the probability of finding the two ends of a long DNA close together, using the average end-to-end distance (2.8), which is $\approx \sqrt{2AL}$ in this limit. For $L \gg A$, the two ends are somewhere in a volume $\approx (AL)^{3/2}$. Therefore, the probability of finding the two ends together for $L/A \gg 1$ decays as $J \approx 1/(AL)^{3/2}$.

This formula does not account for self-avoidance, but because the double helix has a segment length $2A$ so much longer than its diameter (only 3 nm even when electrostatic repulsion in physiological solution is taken into account) self-avoidance effects can be neglected for molecules as large as 10^4 bp in length.

2.2.6. *Permanent sequence-driven bends*

We've focused on thermally excited bends, using a model which has as its 'ground state' a perfectly straight conformation. The average shape of any DNA molecule depends on its sequence: different sequences have slightly different average distortions. A remarkable discovery is that it is possible, by 'phasing' sequences that generate kinks, one can obtain DNAs with strong permanent bends along them [14]. Some of these strong permanent bends are implicated in biological processes, for example facilitation of the binding of proteins that bend or wrap DNA.

2.3. *Stretching out the double helix*

One type of single-molecule experiment which has become widely studied is the stretching of DNAs using precisely calibrated forces. Early experiments showed that the double helix displayed polymer stretching elasticity of exactly what was

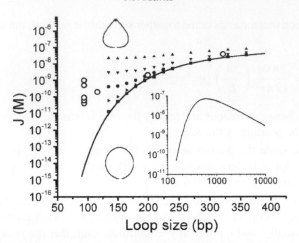

Fig. 5. Juxtaposition probability (J_{circle} in Mol/litre) for double helix. Solid lines show theoretical result for simple semiflexible polymer model of DNA double helix. Inset shows theoretical J_{circle} for large distances, showing peak near 500 bp and $L^{-3/2}$ decay. Main figure focuses on energetically-dominated small-L behavior, showing strong suppression of probability in the simple semiflexible polymer model (solid line). Open circles show recent experimental data of Cloutier and Widom for short DNAs [12]; there is an anomalously large probability of juxtaposition for 94 bp. Filled symbols correspond to a theory of DNA site juxtaposition including the effect of thermally excited 'hinges' [13]: we'll hear more about this in Sec. 3.1.3.

expected from the semiflexible polymer model introduced above. This has been important to the design of many experiments focusing on the effects of proteins or other molecules binding to, or moving along DNA. This subsection reviews the basic polymer stretching elasticity of a long ($L \gg A$) double helix DNA.

A force f applied to a single DNA molecule of length L appears in the Boltzmann factor coupled to the end-to-end vector along the force direction (which we take to be z). Our energy becomes:

$$E = \frac{k_B T A}{2} \int_0^L ds \left| \frac{d\hat{\mathbf{t}}}{ds} \right|^2 - f\hat{\mathbf{z}} \cdot [\mathbf{r}(L) - \mathbf{r}(0)] \qquad (2.13)$$

We can turn the end-to-end vector into an integral over $\hat{t}$ as before, giving

$$\beta E = \int_0^L ds \left[\frac{A}{2} \left| \frac{d\hat{\mathbf{t}}}{ds} \right|^2 - \beta f \hat{\mathbf{z}} \cdot \hat{\mathbf{t}} \right] \qquad (2.14)$$

A single parameter $\beta A f$ controls this energy (to see this, write Eq. 2.14 using contour length in units of A). We therefore have two regimes to worry about: forces below, and above the characteristic force $k_B T / A$.

For the double helix, $A = 50$ nm, so $k_B T/A = 0.02\, k_B T/$nm $= 0.08$ pN. This is a low force due to the long persistence length of the double helix.

Ideally, we want to calculate the partition function

$$Z(\beta A f) = \int \mathcal{D}\hat{\mathbf{t}} e^{-\beta E} \tag{2.15}$$

and then calculate the end-to-end extension, using

$$\langle \hat{\mathbf{z}} \cdot [\mathbf{r}(L) - \mathbf{r}(0)] \rangle = \frac{\partial \ln Z}{\partial \beta f} \tag{2.16}$$

This can be done in general numerically, but we can find the low- and high-force limits analytically.

2.3.1. Small forces ($< k_B T/A = 0.08$ pN)

For small forces, we can calculate the end-to-end extension using linear response, since we know the zero-force fluctuation of the mean-square end-to-end distance: recall that this was $2AL$. This counted three components; by symmetry we have

$$\left\langle \left(\hat{\mathbf{z}} \cdot [\mathbf{r}(L) - \mathbf{r}(0)] \right)^2 \right\rangle = \frac{2AL}{3} \tag{2.17}$$

The linear force constant will be $k_B T$ divided by this fluctuation, giving a small-extension force law:

$$f = \frac{3k_B T}{2AL} z + \cdots \tag{2.18}$$

where we use the shorthand $z = \langle \hat{\mathbf{z}} \cdot [\mathbf{r}(L) - \mathbf{r}(0)] \rangle$ to indicate the average end-to-end extension in the force direction.

This is just the usual ideal (Gaussian) low-extension force law familiar from polymer physics. The spring constant of the polymer is inversely proportional to the persistence length, and to the total chain length.

2.3.2. Larger forces ($> k_B T/A = 0.08$ pN)

The linear force law shows that our ideal DNA will start to stretch out when forces of $\approx k_B T/A$ are applied to it. We can also calculate the very nonlinear elasticity associated with the nearly fully stretched polymer, using an expansion in $1/\sqrt{f}$.

Suppose that the polymer is quite stretched out, so that $\hat{\mathbf{t}}(s) = \hat{\mathbf{z}} t_\parallel + \mathbf{u}$, where $\mathbf{u}$ is in the xy plane, and has magnitude $<< 1$. Since $\hat{\mathbf{t}}^2 = 1$, $t_\parallel = \sqrt{1 - |\mathbf{u}|^2} =$

$1-\frac{1}{2}|\mathbf{u}|^2+\cdots$ Plugging this into the Hamiltonian (2.14) and expanding to leading order in $|\mathbf{u}|^2$ gives:

$$\beta E = -\beta f L + \frac{1}{2}\int_0^L ds \left[A\left|\frac{d\mathbf{u}}{ds}\right|^2 + \beta f |\mathbf{u}|^2 \right] \tag{2.19}$$

In this limit, the fluctuations can be seen to slightly reduce the length, generating the final energy cost term.

Introducing Fourier modes $\mathbf{u}_q = \int_0^L ds e^{iqs}\mathbf{u}(s)$ diagonalizes the Hamiltonian:

$$\beta E = -\beta f L + \frac{1}{2L}\sum_q (Aq^2 + \beta f)|\mathbf{u}_q|^2 \tag{2.20}$$

where $q = \pm 2\pi n/L$ for $n = 0, \pm 1, \pm 2, \cdots$. The fluctuation amplitude of each mode is therefore

$$\langle|\mathbf{u}_q|^2\rangle = \frac{2L}{Aq^2 + \beta f} \tag{2.21}$$

where the leading 2 comes from the two components (x and y) of $\mathbf{u}$. Now we can compute the real-space amplitude:

$$\langle|\mathbf{u}(s)|^2\rangle = 2\int_{-\infty}^{\infty}\frac{dq}{2\pi}\frac{1}{(Aq^2 + \beta f)} = \frac{1}{\sqrt{\beta A f}} \tag{2.22}$$

and finally the extension in the force direction

$$z = L\langle t_\|\rangle = L\left(1 - \frac{1}{2}\langle|\mathbf{u}|^2\rangle + \cdots\right) = L\left(1 - \frac{1}{\sqrt{4\beta A f}} + \cdots\right) \tag{2.23}$$

The semiflexible polymer shows a distinct $1/\sqrt{f}$ behavior as it is stretched out. Also note that the energy expressed in wavenumbers shows that there is a force-dependent correlation length for the bending fluctuations, given by $\xi = \sqrt{k_B T A/f}$. Experiments on double-helix DNAs show this relation [15–20].

The asymptotic linear relation between z and $1/\sqrt{f}$ is quite useful. It turns out this holds well theoretically for the exact solution of the semiflexible polymer model under tension, for $z/L > 0.5$. If you have experimental data for stretching a semiflexible polymer, you can plot z versus $1/\sqrt{f}$ and fit a line to the $z/L > 0.5$, the z-intercept of the linear fit estimates the molecular length L, and the $1/\sqrt{f}$ intercept gives an estimate of $\sqrt{4\beta A}$, i.e., a measurement of persistence length. The agreement between different kinds of single-DNA experiments gives strong evidence that for long molecules, most of the elastic response comes from thermal bending fluctuations.

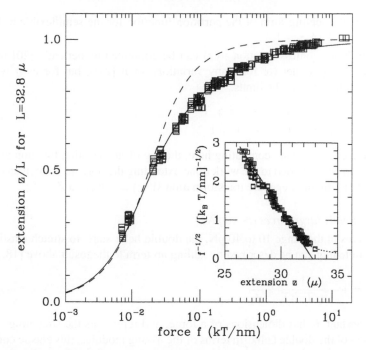

Fig. 6. Experimental data and models for stretching of the double helix, from Ref. [20]. Main figure shows experimental data (squares) of Smith et al. [15] and a fit to the semiflexible-polymer model (solid line), for a persistence length $A = 53$ nm. The units of force are $k_B T$/nm; recall $1k_B T$/nm$= 4.1$ pN, for $T = 300$ K. Inset shows a plot of extension versus inverse square root of force, showing the linear relation between these two quantities. Dashed line shows result for freely-jointed polymer model with segment length 100 nm; this model describes the low-force polymer elasticity, but fails to describe the high-force regime of the experiment.

2.3.3. Free energy of the semiflexible polymer

It is useful to compute the free energy difference between unstretched and stretched polymer from the extension in the force direction, by integrating (2.16):

$$\ln Z(f) = \beta \int_0^f df' z(f') + \ln Z(0) \tag{2.24}$$

We'll drop the constant $\ln Z(0)$, which amounts to taking the relaxed random coil as a 'reference state' with free energy defined to be zero. For the semiflexible polymer the free energy takes the form

$$\ln Z(f) = \frac{L}{A} \gamma(\beta A f) \tag{2.25}$$

which is the scaling form of the partition function for the semiflexible polymer in the limit $L/A >> 1$.

The dimensionless function $\gamma(x)$ can be computed numerically [20] for the semiflexible polymer (or for many variations of it [21]), but for us it will be sufficient to consider the limits:

$$\gamma(x) = \begin{cases} x - \sqrt{x} + \cdots & x > 1 \\ 3x^2/4 + \cdots & x < 1 \end{cases} \tag{2.26}$$

The free energy we are computing here, the log of the partition function at fixed force, can be converted to the work done extending the polymer to a given extension, $W(z)$, by the Legendre transformation $W(z) = -k_B T \ln Z + fz$.

2.3.4. Really large forces (> 10 pN)

For forces in the range 10 to 40 pN, the double helix starts to stretch elastically. This stretching can be described by adding an term to the result above [18, 20]:

$$\frac{z}{L} = 1 - \frac{1}{\sqrt{4\beta A f}} + \frac{f}{f_0} \tag{2.27}$$

The constant f_0 has dimensions of a force, and represents the stretching elastic constant of the double helix. In terms of the Young modulus, this elastic constant for a rod of circular cross-section of radius r is $f_0 = \pi r^2 Y$. Experimental data indicate $f_0 \approx 1000$ pN [22, 23].

Finally, at about 60 to 65 pN, depending a bit on salt concentration, there is an abrupt transition to a new double helix state about 1.7 times longer than B-form. This is sometimes called the S-form of DNA; there is at present some controversy over whether this form is base-paired or not [24].

Fig. 7 shows some experimental data for the high-force response of dsDNA (squares and diamonds) from two groups. Note the elongation of the double helix above the fully extended double helix value of 0.34 nm/bp, and the sharp 'overstretching' transition force 'plateau' near 63 pN.

Problem 16: Above we saw that $B = (\pi/4)Yr^4$ where r is the cross-sectional radius of an elastic rod. Compare the Y values inferred from B and f_0. Are they consistent?

Problem 17: Consider longitudinal stretching fluctuations of adjacent base pairs. Compute the energy of a fluctuation of amplitude (length) δ: what is the root-mean-square value of the single-base-pair longitudinal fluctuation $\sqrt{\langle \delta^2 \rangle}$?

Problem 18: Under some conditions, a *single strand* of DNA will behave like a flexible polymer of persistence length $A_{ss} \approx 1$ nm. Find the characteristic force at which you might expect a single-stranded DNA to become 50% extended.

Problem 19: Consider the Hamiltonian (2.14) generalized so that it contains a *vector* force **f** coupled via dot product to end-to-end extension.

Show $\partial_{\beta f_i} \ln Z = \langle x_i(L) - x_i(0) \rangle$ and $\partial^2_{\beta f_i} \ln Z = \langle [x_i(L) - x_i(0)]^2 \rangle$ where the indices i label the three spatial coordinates.

Now, assuming the force to be in the z direction, verify the following formula relating the average extension and the 'transverse' end-to-end vector fluctuations:

$$\frac{\langle z(L) - z(0) \rangle}{\langle [x(L) - x(0)]^2 \rangle} = \frac{f}{k_B T} \tag{2.28}$$

Hint: use the fact that the partition function is a function of only the *magnitude* of **f**.

This exact, nonperturbative relation is used in magnetic tweezer experiments to infer forces applied to single DNA molecules [25]. This does not depend on the details of the polymer part of the Hamiltonian - even if it contains long-ranged interactions - as long as it is invariant under space rotation.

Problem 20: For the semiflexible polymer, consider the approximate force-extension relation $\beta A f = z/L + 1/[4(1 - z/L)^2] - 1/4$. Show that this function reproduces the high- and low-force limiting behaviors derived above (it is not a terribly accurate representation for the exact behavior of $f(z)$). Compute the free energy $W(z)$ using this relation. Hint: integrate (2.16).

Problem 21: Consider the 'freely jointed chain' obtained by setting $a = 0$ in the segment model. Calculate the extension, and free energy ($\ln Z$) as a function of force. Also calculate the transverse mean-squared fluctuations as a function of force, and verify Eq. 2.28 for this model.

3. Strand separation

In the previous section we didn't say much about a feature of the double helix of paramount biological and biophysical importance: it consists of two covalently bonded *single-stranded* DNAs (ssDNAs) which are relatively weakly stuck to one another. The weakness of the binding of the two strands makes it possible for the two strands of a double helix to be separated from one another, either permanently as occurs in vivo during DNA replication, or transiently as occurs during DNA transcription (reading of DNA by RNApol) and DNA repair.

Conversion of dsDNA to ssDNA can be accomplished in a few ways:
Elevated temperature: The double helix is stable in 'physiological' buffer (pH near 7, univalent salt in the 10 mM to 1 M range) for temperatures below about 50 C. Over the range 50 to 80 C, the double helix 'melts', with AT-rich sequences

falling apart at the low end of this range, and highly GC-rich sequences holding together until the high end of this temperature range.

Denaturing solution conditions: Too little salt ($<$ 1 mM NaCl), which increases electrostatic repulsion of the negatively charged strands, or pH too far from 7, destabilizes the double helix, lowering its melting temperature.

Sufficient 'unzipping force' applied to the two strands: If you pull the two strands apart, they will separate at forces in the 10 to 20 pN range, with force variations reflecting the sequence composition.

Below I will discuss the last of these three modes of strand separation, unzipping by force. A process similar to idealized 'forced-unzipping' is carried out in the cell to generate single-stranded DNA for DNA repair and replication; specialized motor enzymes called *DNA helicases* track along the double helix, pushing the two strands apart. The function of helicases can be precisely studied using single-DNA methods [26–28], and models for their activity require us to understand unzipping by force. To do that, we'll need to learn about the strength of the base-pairing interactions, and the polymer elasticity of ssDNA.

3.1. Free-energy models of strand separation

In the simplest picture of DNA melting, we ignore base sequence entirely, and consider simply the average free energy difference g per base pair between isolated, relaxed ssDNAs and dsDNA, at room temperature and in physiological solution conditions. Then, for an N-base-pair-long molecule, the free energy difference between ssDNAs and dsDNAs would be just $G_{\text{ssDNAs}} - G_{\text{dsDNA}} = Ng$. For random DNA sequences, this $g \approx 2.5 k_B T$; its positive value reflects the fact that the double helix is more stable than isolated single strands: very roughly, the probability of observing melted single strands is $e^{-\beta Ng}$.

Thermal melting can be most simply thought about by considering the temperature dependence of the base-pairing free energy, breaking it into 'enthalpy' h and 'entropy' s per base pair, i.e., $g = h - sT$. At the melting temperature $T_m = h/s$, the free energy of isolated ssDNAs is equal to free energy of double helix, making these two states equally probable.

Problem 22: Random-sequence DNA has $g \approx 2.5 k_B T$ at 25 C, and melts near 70 C ($T = 343$ K). Estimate h and sT at 25 C ($T = 298$ K).

Problem 23: For the simple model where the strand separation free energy per base is a constant $g = h - sT$, calculate the probability of finding separated single strands as a function of temperature (you may consider this to be a two-state system). How does the *width* of the melting transition as a function of temperature scale with N? You may want to plug in some numbers from the previous problem.

3.1.1. Sequence-dependent models

A number of groups are working on accurate algorithms to predict the melting temperatures of dsDNAs as a function of sequence. One of these classes of models assign a contribution to base-pairing free energy for each *pair* of bases, the idea being that stacking interactions of adjacent base pairs play an important role in determining the stability of the double helix. The raw data behind such models are melting temperature data for a set of different-sequence, short (10 to 20 bp) dsDNAs.

Table 1 lists a set of free energies due to Santalucia [10] for the ten different oriented pairs of bases that occur along a DNA strand. All remaining pairs of bases can be obtained from considering the complementary sequence on the adjacent strand, e.g., the contribution of $5'$-GA is the same as that of $5'$-TC found in the table. The free energy of strand separation for a long N−bp molecule is obtained by adding the $N − 1$ adjacent-base contributions together. In addition, there are contributions for the ends which we won't discuss – all though they are significant when considering melting of short molecules.

The key point of Table 1 is that AT-rich sequences are lower in strand-separation free energy (the values for AT, AA and TA are all less than $1.7k_B T$), while GC-rich sequences are higher (GG, GC and CG are $3k_B T$ or more). Models of this type are not infallible – in reality, double-helix structure and energy depends on longer than nearest-neighbor sequence correlations – but they do give some idea of sequence dependence of base-pairing free energy.

The data of Table 1 are for the physiological ionic strength of 150 mM NaCl; lower ionic strengths reduce the base-pairing free energy. An ionic-strength correction for the base-pairing free energy has been given by Ref. [10]): $\Delta g_i = 0.2 \ln(M/0.150)$ where M is the molarity of NaCl.

Table 1

Base-pairing-stacking free energies of Santalucia [10]. Free energies are in $k_B T$ units, and are for 25 C, 150 mM NaCl, pH 7.5. For other temperatures and salt concentrations the values must be corrected (see text).

Base i and $i + 1$ ($5' \rightarrow 3'$)	Free energy g_i (150 mM NaCl, pH 7.5, 25 C)
AA	1.68
AT	1.42
AG	2.19
AC	2.42
TA	0.97
TG	2.42
TC	2.12
GG	3.00
GC	3.75
CG	3.68

Problem 24: Calculate the free energy differences between separated ssDNAs and double helicies, for the following sequences: 5'-AATTAATTAATT, 5'-GCGCGCGGCCGG, 5'-AGCTCCAAGGCT. You may want to consult reference [10] to include the end effects.

Problem 25: In Table 1 you can see that AT-rich sequences have roughly $2k_B T$ less free energy holding them together than do GC-rich sequences. For a random N-base sequence, there will therefore be a mean free energy of strand separation, and fluctuations of that free energy. Calculate the mean free energy per base pair, and estimate the fluctuations.

3.1.2. Free energy of internal 'bubbles'

The above discussion suggests that thermal melting might be described by a one-dimensional Ising model with sequence-dependent interactions, i.e., with some quenched 'randomness'. However, this would ignore an important physical effect that acts to suppress opening of bubbles in the interior of a long double helix. This effect is the *entropic cost* of forcing an internal 'bubble' to close [29]. This cost is not included in the strand separation free energy models described above which are fit to data obtained from melting of short double helicies.

This loop free energy is easy to roughly understand – we have already discussed it above indirectly in our discussion of juxtaposition of DNA sequences. We mentioned that the long-molecule limit for DNA juxtaposition probability should be $J \approx N^{-3/2}$ simply from considering the fact that the two molecule ends should be found in a volume of radius $R \approx N^{1/2}$. If we think about this probability in terms of a free energy cost of constraining the ends to be near one another, we obtain the loop free energy cost

$$\Delta G_{\text{loop}} = \frac{3}{2} k_B T \ln N \qquad (3.1)$$

Since ssDNA has a persistence length of roughly one base (0.7 nm), the N relevant here is simply the number of bases in the loop. For an internal ss-DNA bubble formed by opening N base pairs, we should use $2N$ as the loop length.

This additional free energy discourages opening of internal bubbles, eliminating the use of the simple Ising model with short-ranged interactions to describe DNA melting. In fact, the logarithmic interaction of (3.1) is sufficiently long-ranged to kill the usual argument against a phase transition in a 1d system. A real phase transition occurs in the 'pure' DNA melting model including the logarithmic loop effect; however, variations in local melting temperatures due to sequence variations along long real DNAs wash out a sharp phase transition [29].

We can estimate the total free energy cost of an N-base-pair internal bubble, adding the base-pairing/stacking free energy to the loop free energy cost:

$$\sum_{i=1}^{N} g_i + \frac{3}{2} k_B T \ln(2N) \tag{3.2}$$

The sequence-dependent term ranges from about $N k_B T$ to $4 N k_B T$, making the price of a large, 10 bp bubble from roughly 20 to 45 $k_B T$: i.e., very rare except for the most AT-rich sequences. Larger bubbles are even more costly, making them exceedingly rare excitations.

3.1.3. Small internal bubbles may facilitate sharp bending

Small internal bubbles are not impossibly costly excitations: a 3 bp bubble costs 8 to 15 $k_B T$. Short AT-rich 3 bp sequences are by this reckoning, open roughly 0.1% of the time, and can be expected every few hundred base pairs (e.g., the particularly weak sequence TATA appears once every 256 base pairs in random-sequence DNA).

These small, thermally excited bubbles suggest an explanation for the recent results of Cloutier and Widom [12] (see Fig. 5) showing that the cyclization (loop-formation) probabilities of dsDNAs less than 300 bp long are far larger than we would expect from the simple elastic bending model 2.5. The experimental data indicate that tight bends of the double helix can occur via an alternative, lower-free-energy mechanism. One possibility is that via separation of a few base pairs, a 'flexible joint' might appear that could reduce the bending energy of formation of a loop. Although the free energy cost of generating a few-base-pair 'joint' is roughly $10 k_B T$, for short DNAs this becomes similar to the bending free energy saved by concentrating much of the bending into a localized, highly distorted defect in the double helix.

Problem 26: Consider Fig. 5, which shows experimental data indicating that circular closure of 94 bp DNAs occurs with probability far above the expected value J_{circle}. Suppose that for some free energy ϵ we can form a small bubble, and *kink* the DNA so that it can still close smoothly, but now if one bubble is excited, with the tear-drop shape which minimizes the bending energy. If *two* bubbles are excited, no bending is required. Estimate the probabilities of the zero-bubble, one-bubble and two-bubble closure states (use the Yamakawa-Stockmayer-Shimada loop formation probabilities 2.11 and 2.12; don't forget that the kink can appear at any base pair position along the molecule). Estimate what ϵ should be to explain the 94 bp data.

Fig. 5 shows how the juxtaposition probability is affected by the inclusion of flexible joints with energy cost 9, 10, 11 and 12 $k_B T$, via a detailed calculation [13]. The experimental data are described well by joints which cost 11 $k_B T$, close to the value expected for localized strand separation of a few base pairs.

3.2. Stretching single-stranded nucleic acids

Single-stranded DNA has also been studied in single-molecule stretching experiments, and shows polymer elasticity distinct from that of double-stranded DNA:
ssDNA has twice the contour length per base of the double helix since the helical backbones of the double helix contain about 0.7 nm per base, about half of the double helix contour length of 0.34 nm per base pair,
ssDNA has a persistence length of roughly a nanometer since the stiffness of the double helix is generated by the base pairing and stacking; once isolated, the ssDNA backbone is very flexible,
ssDNA can stick to itself by base-pairing and stacking interactions between bases along the same molecule.
These features are illustrated in Fig. 7 which plots experimental data for double helix and ssDNA side by side. The double helix, with a persistence length of 50 nm, is extended to its full contour length of about 0.34 nm/bp by forces of a few pN, and then shows a stiff force response, and finally the $\approx$ 60 pN force plateau. By comparison, ssDNA (open circles) only gradually stretches out, showing no stiff response near 0.34 nm/bp, and no force plateau. The force required to half-extend ssDNA is more than 3 pN; this reflects its short persistence length $\approx$ 1 nm (recall that the force needed to stretch out a polymer is roughly $k_B T / A$).

Fig. 7 also shows the strong dependence of ssDNA on salt concentration (open circles, left and right branches). At 150 mM NaCl ('physiological' salt, left set of data), ssDNA sticks to itself at low extensions, leading to an $\approx$ 1 pN force threshold to start opening the molecule. At low salt concentration (10 mM NaCl, right set of data) electrostatic self-repulsion eliminates this sticking effect, and the force threshold for initial extension.

For low salt concentration, the extension is well described by a logarithmic dependence on force, $\ln f/f_0$. This behavior can be understood in terms of a scale-dependent persistence length resulting from electrostatic effects [20,34,36]. At low forces, electrostatic self-repulsion effectively stiffens the polymer, helping to stretch it out; at higher forces, this effect is less pronounced (the monomers are farther away from one another) and the chain becomes harder to stretch. This effect is much more pronounced for ssDNA than for dsDNA since the backbone persistence length $\approx$ 1 nm is comparable to, or even less than, the screening length for electrostatic interactions (recall the Debye screening length is $\lambda_D = 0.3$ nm $/\sqrt{M}$ for NaCl at M Mol/litre).

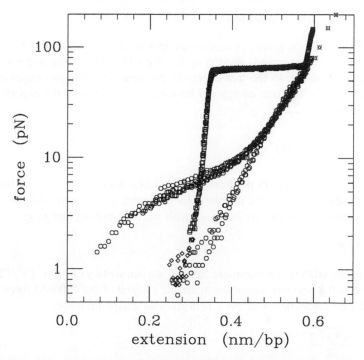

Fig. 7. Force versus extension of double helix and ssDNA. Squares show experimental dsDNA data of Léger et al. [30,31] for 500 mM NaCl buffer, diamonds show experimental dsDNA data of Smith et al. [23] for 1 M NaCl buffer. Data for physiological salinity (150 mM NaCl) are similar, but have a plateau shifted a few pN below the 500 mM result, see Refs. [24,31]. Circles show experimental data of Bustamante et al. [32] for ssDNA; stars show high-force ssDNA data of Rief et al. [33]. The left, lower-extension curve is for 150 mM NaCl, while the right, higher-extension curve is for 2.5 mM NaCl. The two ssDNA datasets converge at high force, to the behavior $x \approx \ln f$.

Problem 27: Force-extension data of Fig. 7 at low ionic strength are described by $x(f) \approx x_0 \ln(f/f_0)$ where x_0 and f_0 are constants. Compute the force-extension response in the high-force limit using Eq. 2.22, given the scale-dependent persistence length

$$A(q) = \begin{cases} A_0 q_0/q & q < q_0 \\ A_0 & q > q_0 \end{cases} \qquad (3.3)$$

A more realistic model of scale-dependence of persistence length, based on Coulomb self-interactions, gives rise to similar behavior; see Refs. [20,34–36]. A recent experiment by Visscher et al. [37] on a poly-U RNA, eliminating base-pairing, shows scale-dependent persistence length behavior rather clearly.

3.3. *Unzipping the double helix*

We now have all the pieces to analyze unzipping of the double helix by a force which pulls the two strands apart (Fig. 8). We will compare the free energy of two paired bases, g, to the free energy at constant force for two unpaired and extended bases. The free energy per base can be found from the experimental elasticity data via 2.24:

$$\gamma(f) = \int_0^f df' x(f') \tag{3.4}$$

where $x(f')$ is the length per base of the ssDNA data of Fig. 7. The function $\gamma(f)$ increases with f. The threshold for unzipping occurs when this two times this free energy – for the two bases – equals the base-pairing energy g:

$$2\gamma = g \tag{3.5}$$

Treating the ssDNA as a harmonic 'spring' we can write $\gamma(f) \approx (\ell f)^2/(2k_B T)$ where $\ell \approx 0.4$ nm (this roughly matches the integral of the 150 mM force curve of Fig. 7 for forces below 20 pN). This gives an unzipping force:

$$f = \frac{\sqrt{k_B T g}}{\ell} \tag{3.6}$$

Plugging in g from 1 to 4 $k_B T$, we see that the unzipping force varies from 10 to 20 pN, depending on sequence. Experiments of Bockelmann and Heslot on genomic molecules find fluctuations around 15 pN, the average of this range

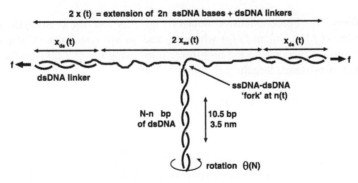

Fig. 8. Unzipping of DNA by force. Note that a torque can be applied to the end of the dsDNA region, coupled to the rotational angle θ.

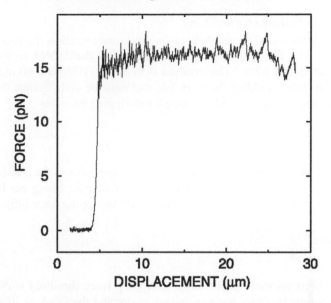

Fig. 9. Experimental data of Bockelmann et al. [38] for unzipping of DNA at 0.02 μm/sec. Sequence-dependent variations in force occur, around an average force of about 15 pN.

(see Fig. 9) [38]. The full range of unzipping forces from 10 to 20 pN has been observed by Rief et al. [33, 39] in experiments on pure AT and GC DNAs.

Problem 28: For the harmonic model of ssDNA extensibility, calculate the force-extension relation. Compare the results for forces between 1 and 20 pN with the 150 mM NaCl ssDNA data in Fig. 7.

Problem 29: We can alternately describe unzipping using *extension* as a control parameter. Suppose one has a partially unzipped dsDNA, where n base pairs have been separated. The free energy is made up of elastic stretching energy, and base pairing energy:

$$F = \frac{k_B T (2x)^2}{2(2n)b^2} + ng \tag{3.7}$$

Note that opening n base pairs results in a $2n$-base-long ssDNA (see Fig. 8). Note also that $n \geq 0$. Find the equilibrium number of base pairs unzipped, as a function of extension. For a partially unzipped molecule, also calculate the *fluctuation* in the number of bases that are unzipped. What are the corresponding *extension* fluctuations?

3.3.1. Effect of torque on dsDNA end

As unzipping proceeds, the dsDNA region must rotate to allow the two ssDNAs to be pulled out. If a torque is applied at the end of the dsDNA region, it can affect the unzipping force. This rotation is $\theta_0 = 2\pi/10.5 = 0.60$ radians per base pair unzipped. Adding the work $\tau\theta_0$ that must be done against the torque for each unzipped base pair, the equation for unzipping becomes

$$2\gamma = g - \tau\theta_0 \tag{3.8}$$

For the sign convention of Fig. 8, right-handed torque reduces the stability of the double helix, while left-handed torque acts to stabilize it. Using our harmonic approximation, we can obtain a torque-dependent unzipping force [40]:

$$f = \frac{\sqrt{k_B T(g - \tau\theta_0)}}{\ell} \tag{3.9}$$

As torque becomes more positive, the unzipping force threshold is decreased. When the torque becomes positive enough to unwind the DNA on its own, the unzipping force threshold becomes zero: this point is given by $\tau = g/\theta_0$, which ranges from $1.6k_B T$ for weakly bound (AT-rich) sequences, to $7k_B T$ for the most strongly bound (GC-rich) sequences.

If unzipping is done rapidly, the rotation of the dsDNA will generate a drag torque. In the simplest model for this where the DNA is supposed to spin around its axis, the drag torque is roughly

$$\tau = -4\pi\eta r^2 L_{\mathrm{ds}} \frac{d\theta}{dt} \tag{3.10}$$

where L_{ds} is the length of the dsDNA region, $r \approx 1$ nm is the dsDNA cross-section hydrodynamic radius, and viscosity $\eta = 10^{-3}$ Pa·sec for water and most buffers. Effects of the drag associated with dsDNA rotation have been observed in experiments of Bockelmann and Heslot (see Fig. 10) [43]; the above model is in fair agreement with the experiment [41]. Note that P. Nelson has argued that there is an additional and large contribution to the rotational drag by permanent bends along the DNA contour [42]. The shape of the DNA gives rise to an effective increase in its cross-section radius r and thus the rotational drag coefficient.

Problem 30: Estimate the number of base-pairs per second that should be unzipped in order that rotational drag can push the unzipping force up by 5 pN (assume a uniform molecule with $g = 2.5k_B T$).

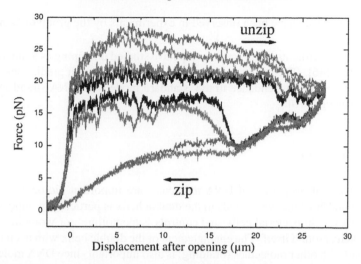

Fig. 10. Experimental results of Ref. [43] showing unzipping force rate-dependence. The two ssDNA ends are forced apart at velocities of 4, 8, 16 and 20 μm/sec. Force versus ssDNA extension (see Fig. 8) is plotted. During unzipping, higher velocities generate higher unzipping forces.

3.3.2. Fixed extension versus fixed force for unzipping

In single-molecule stretching experiments, like any experiment on a small system, choosing whether force or extension are controlled can be critical to the results. For example, laser tweezers and atomic force microscopes essentially control the position of the end of a molecule; magnetic tweezer setups by contrast provide fixed force. Unzipping of DNA provides a very good example of how these two types of experiments give different kinds of data. Fixed-extension unzipping experiments push the ssDNA-dsDNA 'fork' along, and observe jagged force 'stick-slip' events. Each stick event corresponds to the momentary stalling of the fork at a GC-rich 'barrier': the force then increases to a level where a 'slip', or barrier-crossing event occurs (see Fig. 9).

Conversely, in a fixed-force experiment, one observes the increase of extension as a function of time. For unzipping, this typically takes the form of a series of extension *plateaus*. These plateaus again correspond to the stalling of the fork at a GC-rich barrier region; however, now the force is constant, and one must wait for a thermal fluctuation for unzipping to proceed. If one is well below the maximum unzipping force for GC-rich sequences (see 3.6), the barriers can be immense: even a fraction of a $k_B T$ per base pair required to cross a long, slightly GC-rich region can give rise to an immense barrier. This effect has been theoretically emphasized by D. Lubensky and D. Nelson [44] and the constant-force extension plateaus have been observed in experiments by the group of Danilow-icz *et al.* [45].

Recent experiments by the same group have studied unzipping as a function of temperature [46]. The results are in surprisingly discord with predictions based on the temperature dependence in the 'standard models' of DNA strand separation free energy [10].

4. DNA topology

The topological properties of DNA molecules are important biologically. The linking number of the two strands in the double helix is particularly important to DNA structure in bacterial cells, and controls 'supercoiling', or wrapping of the double helix around itself. The entanglement of the double helix with itself (knotting), and with other molecules (braiding) is also important since DNA molecules (chromosomes) must be separated from one another during cell division.

4.1. DNA supercoiling

The phenomenon of supercoiling is familiar from dealing with twisted strings or wires: twist strain in a string can be relaxed by allowing the string to wrap around itself. For DNA molecules, description of this behavior requires one more ingredient, thermal fluctuation of the molecule conformation.

The physical feature of the double helix that gives rise to supercoiling is the wrapping of the two strands around one another. Neglecting bending for the moment, the relaxed double helix has one link between strands for each 10.5 bp along the molecule. This 'relaxed linking number' can be expressed as $\text{Lk}_0 = N/10.5$ bp for an N-bp double helix. The relaxed helix repeat of 10.5 bp can be expressed as a length $h = 3.6$ nm, allowing us to also write $\text{Lk}_0 = L/h$.

4.1.1. Twist rigidity of the double helix

Still avoiding bending, if we twist the double helix so that one end is rotated by an angle Θ relative to the other, the number of links between the strands will be changed by an amount $\Theta/(2\pi)$. In this case where there is no bending, the change in linking number of the double helix, ΔLk, equals the change in twist, ΔTw.

It costs some energy for this twist distortion: a simple harmonic model is

$$\frac{E}{k_B T} = \frac{C}{2L}\Theta^2 = \frac{2\pi^2 C}{L}(\Delta\text{Tw})^2 \tag{4.1}$$

This 'twist' energy is controlled by an elastic constant C with dimensions of length. This *twist persistence length* is about 100 nm for double helix DNA based on recent single-molecule experiments [47]; note that this is appreciably larger than the estimate of ≈ 75 nm that is the result of a number of solution-phase experiments. We'll see a possible explanation for this disagreement later when we discuss twist rigidity of DNA.

Problem 31: Consider the harmonic twist energy. Calculate the thermal expectation value of Θ^2: your result will depend on the molecule length L. Why is C called the twist persistence length?

Problem 32: Assuming the double helix to be composed of a uniform isotropic elastic medium, use A and C to determine the two Lamé coefficients, and equivalently the Young modulus and the Poisson ratio (you will likely want to review Landau and Lifshitz' *Theory of Elasticity* [4] unless you are really an expert in elasticity theory; also recall that we have already figured out the Young modulus from both the bending persistence length *and*, independently, from the stretching force constant).

Problem 33: What torque is necessary to twist a DNA of length L by angle Θ? For left-handed twisting, for what angle Θ will the twisting build up enough torque to start unwinding AT-rich sequences (see Sec. 3.3.1)?

4.1.2. Writhing of the double helix

When we allow bending of the double helix to occur, the linking number is no longer equal to the twisting number. However, as long as the bending radius is large compared to the radius of the double helix, there is a simple relation between twisting and bending contributions to the total linking number of the double helix:

$$\Delta Lk = \Delta Tw + Wr \tag{4.2}$$

The quantity Wr, or 'writhe', is dependent only on the bending of the double helix backbone. Very roughly, Wr measures the signed number of crossings of the molecule axis over itself, when the molecule shape is projected onto a plane.

Formally, linking number of the two strands can only be defined if the double helix is *circular*, i.e., if both of the strands are closed circles. I will be slightly loose with this, and sometimes talk about linking number of *open* molecules. If you want to make linking number of a linear molecule precise, you can just imagine extending the strand ends straight off to infinity, and closing them there. This will not lead to large corrections in the situations we will be interested in.

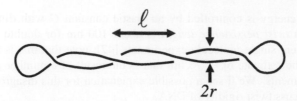

Fig. 11. Plectonemic supercoiled form of circular DNA, showing length between crossings ℓ, and cross-sectional radius r. Note that the only appreciable DNA bending occurs at the ends; also note that the line indicates double-helix DNA.

We consider the situation where as bending occurs, the linking number remains fixed. This is most relevant to *circular* double helix molecules (with no breaks or 'nicks' in their backbones), the linking numbers of which are constant. Circular DNAs are found in bacteria: both the large 4.5 Mb chromosome and small plasmids (typically 2 to 15 kb in circumference) are normally found in closed circular form.

A second situation where ΔLk can be considered constant is when one is holding onto the two ends of a DNA molecule, and forcing them to be parallel and unable to rotate. This case can be studied experimentally in single-DNA micromanipulation experiments, most notably in elegant magnetic tweezer experiments [25].

By rearranging 4.2 to ΔTw $= \Delta$Lk $-$ Wr we can see the mechanism for buckling of a twisted wire: twisting without bending will change ΔTw away from zero, costing twist energy. However, now if the wire is allowed to buckle so that it wraps around itself, the Wr from the wrapping can cancel the ΔLk, and reduce the twisting energy. By braiding the molecule with itself, the bending energy can be small as well. This self-wrapping of DNA is called *plectonemic supercoiling*.

For the plectonemic structure shown above, the magnitude of the writhe is and equal to the number of crossings: $|\text{Wr}| \approx L/(2\ell)$. The sign of the writhe for the right-handed coiling shown in Fig. 11 is negative; for a left-handed plectonemic supercoil, the writhe would be positive. For achiral conformations, Wr $= 0$.

4.1.3. Simple model of plectonemic supercoiling

We can write down a simple model for the free energy of the plectoneme:

$$\frac{F}{k_B T} = \frac{2\pi^2 C}{L}\left(\Delta\text{Lk} \pm \frac{L}{2\ell}\right)^2 + \frac{AL}{2}\left(\frac{r}{\ell^2}\right)^2 + \frac{L}{(Ar^2)^{1/3}} \tag{4.3}$$

The first term is just 4.1 with 4.2 rearranged and plugged in, using the plectonemic writhe $\mathrm{Wr} = \mp L/2\ell$; the top sign is for a right-handed plectnome, the bottom is for left-handed. The second term is the bending energy 2.5, using r/ℓ^2 as the curvature.

The third term arises from confinement of the DNA inside the 'tube' of the supercoil, of radius r. We can think about this in terms of a correlation length λ for thermally excited bending fluctuations: the smaller this wavelength, the smaller the transverse fluctuations. For bending with transverse displacement r over wavelength λ, the curvature is r/λ^2; the energy of this bend is $k_B T A r^2/\lambda^3$. Using the equipartition theorem this energy will be $k_B T$, giving us the relation $\lambda = A^{1/3} r^{2/3}$. Finally, the confinement free energy density will be $k_B T/\lambda$, giving the third term of 4.3.

The free energy model 4.3 needs to be minimized to determine the equilibrium values of r and ℓ. First, we can determine r:

$$r \approx \frac{\ell^{3/2}}{A^{1/2}} \qquad (4.4)$$

Then we can plug this result in to 4.3; simplifying some numerical factors we have

$$\frac{F}{k_B T L} = 2\pi^2 C \left(\frac{|\Delta\mathrm{Lk}|}{L} - \frac{1}{\ell} \right)^2 + \frac{1}{\ell} \qquad (4.5)$$

The sign has been chosen so that the writhe has the same sign as $\Delta\mathrm{Lk}$, which always reduces the free energy. Minimizing this with respect to $1/\ell$ gives the result:

$$\frac{1}{\ell} = \frac{|\Delta\mathrm{Lk}|}{L} - \frac{1}{4\pi^2 C} \qquad (4.6)$$

There is no solution for positive ℓ when linking number is too small: when $|\Delta\mathrm{Lk}| < L/(4\pi^2 C)$, the confinement free energy is too expensive, so the DNA does not supercoil. Then, as $|\Delta\mathrm{Lk}|$ is increased beyond this limit, $1/\ell$ becomes gradually smaller and the supercoil tightens up. This threshold indicates that until the added linking number exceeds one per twist persistence length, the DNA molecule will not supercoil.

Linking number is often expressed intensively using $\sigma \equiv |\Delta\mathrm{Lk}|/\mathrm{Lk}_0$ which just normalizes the change in linking to the relaxed linking number. In a more careful calculation where numerical factors and geometrical details are accounted for carefully, the threshold for supercoiling is at $\sigma \approx h/(2\pi C)$; plugging in

$h = 3.6$ nm and $C = 100$ nm gives a threshold σ of roughly 0.01. Another feature of the more complete theory is that the transition is 'first-order': the minimizing ℓ jumps from $\ell = \infty$ to a finite value. Electron microscopy experiments [48] indicate that plectonemic supercoiling requires about this level of σ (see Fig. 12); calculations of structural parameters of plectonemes also are in accord with the results of EM studies. In eubacteria such as *E. coli*, the chromosome and small circular 'plasmid' DNAs have nonzero ΔLk, with a $\sigma \approx -0.05$. This undertwisting is thought to play a role in gene regulation, since AT-rich promoter regions will be encouraged to open by the torsional stress associated with this amount of unlinking.

An important feature of plectonemically supercoiled DNA is its *branched* structure. Branch points can be thought of as defects in the plectonemic supercoil structure: like the ends, there is some energy cost associated with them. However, there is an entropy gain $\approx k_B \ln L/A$ of having a branch point, since it can be placed anywhere in the molecule. Balance of branch point energy and entropy determines the observed density of a Y-shaped branch point for every 2 kb along a supercoil with $\sigma = -0.05$. Branching is also very important to the internal 'sliding' of DNA sequence around in the interior of a plectonemically supercoiled DNA, is important to some enzymes which bind to two sequences simultaneously, often across a plectnomemic superhelix [49].

Problem 34: Find the dependence of r on ΔLk for the model of plectonemic supercoiling discussed above. At what value of σ does r reach $r_0 = 2$ nm, roughly the point at which the double helix will run into itself?

This effect is important as when the double helix starts to run into itself, twist compensation can no longer occur.

The following three problems will work out well best using the slightly more detailed models for the writhe and for the bending energy of the plectoneme discussed in Refs. [50, 51].

Problem 35: For the plectonemic supercoil *including* the constraint $r > r_0$, find the free energy $F/k_B T$ (note the two regimes where r is free and where r is constrained to be r_0).

Problem 36: For the plectonemic supercoil model discussed above, find the dependence of ΔTw/ΔLk on ΔLk and σ.

Problem 37: Calculate the *torque* in a DNA double helix of length L, as a function of ΔLk, for the plectnonemic supercoil described above. For $\sigma < 0$, at what value of σ does unwinding of AT-rich sequences in the double helix start to occur?

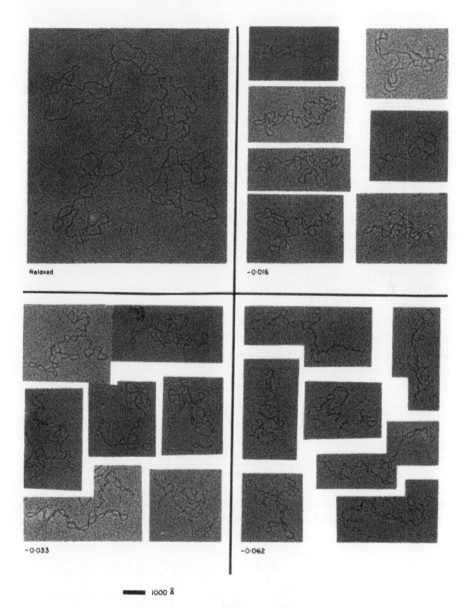

Fig. 12. Electron micrographs of supercoiled DNA at a few different σ values. Scale bar is 100 nm (300 bp); molecules are all 7 kb (2300 nm) in length. Reproduced from Ref. [48].

4.2. *Twisted DNA under tension*

It is possible to carry out single-DNA experiments as a function of force, and linking number [25]. The description of this situation along the framework of Sec. 2.3 is straightforward. At a fixed force, changing σ away from zero compacts a DNA molecule. If enough torsional stress is placed on a DNA molecule under tension, buckling will occur and plectonemic supercoils will appear along its length, leading to strong reduction in molecule extension [50,51], as has been observed experimentally [25]. However, if σ is not so large that plectonemic coils appear, a milder compaction occurs which can be treated using a small-fluctuation approach, discussed by Moroz and Nelson [47], and by Bouchiat and Mezard [52]. This region of relatively mild compaction by writhing is a good regime in which to measure the double helix torsional modulus.

To carry out the treatment of this mild compaction analytically, we need the writhe of a nearly straight DNA, in terms of tangent vector fluctuations. As long as the tangent vector stays in the hemisphere around $\hat{\mathbf{z}}$, we have:

$$\text{Wr} = \int_0^L \frac{ds}{2\pi} \frac{\hat{\mathbf{z}} \cdot \hat{\mathbf{t}} \times \partial_s \hat{\mathbf{t}}}{1 + \hat{\mathbf{z}} \cdot \hat{\mathbf{t}}} \tag{4.7}$$

Now we can write the Hamiltonian for a DNA subjected to tension, plus held at fixed linking number, by just adding the twist energy 4.1 to the stretching Hamiltonian 2.14. The White formula 4.2 allows us to express the twist in terms of linking number and writhe:

$$
\begin{aligned}
\frac{E}{k_B T} &= \int_0^L ds \left[\frac{A}{2} \left(\frac{d\hat{\mathbf{t}}}{ds} \right)^2 - \frac{f}{k_B T} \hat{\mathbf{z}} \cdot \mathbf{t} \right] \\
&+ \frac{2\pi^2 C}{L} \left(\Delta \text{Lk} - \int_0^L \frac{ds}{2\pi} \frac{\hat{\mathbf{z}} \cdot \hat{\mathbf{t}} \times \partial_s \hat{\mathbf{t}}}{1 + \hat{\mathbf{z}} \cdot \hat{\mathbf{t}}} \right)^2
\end{aligned}
\tag{4.8}
$$

We'll do a harmonic calculation, expanding 4.8 to quadratic order in $\mathbf{u}$, the transverse (xy) components of the tangent vector:

$$
\begin{aligned}
\frac{E}{k_B T} &= \frac{2\pi^2 C}{L} (\Delta \text{Lk})^2 - \frac{Lf}{k_B T} \\
&+ \int_0^L ds \left[\frac{A}{2} \left(\frac{d\mathbf{u}}{ds} \right)^2 + \frac{f}{2k_B T} \mathbf{u}^2 - \frac{2\pi C}{h} \sigma \hat{\mathbf{z}} \cdot \mathbf{u} \times \partial_s \mathbf{u} \right] + \mathcal{O}(\mathbf{u}^3)
\end{aligned}
\tag{4.9}
$$

Here the ΔLk in the cross term of the twist energy has been converted to the intensive linking number density σ. For $\sigma = 0$, we return to the high-extension limit of the stretched semiflexible polymer; for nonzero σ, the cross-product term will generate chiral fluctuations.

The fluctuation (**u**-dependent) part of the quadratic Hamiltonian 4.9 can be rewritten in terms of Fourier components of **u**:

$$\int \frac{dq}{2\pi} \left[\frac{1}{2} \left(Aq^2 + \beta f \right) |\mathbf{u}_q|^2 + \frac{2\pi C\sigma}{h} iq\hat{\mathbf{z}} \cdot \mathbf{u}_q^* \times \mathbf{u}_q \right] + \cdots \tag{4.10}$$

Problem 38: Show that the quadratic part of the twisted-stretched DNA energy 4.10 has an instability (a zero eigenvalue) at the point $(2\pi C\sigma/h)^2 = 4Af/k_B T$.

This result is just the classical buckling instability of a beam of bending rigidity B subjected to compressive force f and torque τ, which occurs when $\tau^2 = 4Bf$ [53].

Problem 39: Show that the log of the partition function $\ln Z(f, \sigma)$ for the quadratic-u fluctuations, including the non-fluctuation contributions, has the form, in an expansion in inverse powers of force:

$$\frac{\ln Z}{L} = \frac{f}{k_B T} - \frac{2\pi^2 C\sigma^2}{h^2} - \left(\frac{f}{k_B T A} \right)^{1/2}$$
$$+ \frac{1}{4} \left(\frac{2\pi C\sigma}{h} \right)^2 \left(\frac{k_B T}{4A^3 f} \right)^{1/2} + \cdots \tag{4.11}$$

You will need to find the normal modes of the fluctuations in order to compute this partition function. Note that the result can be written in a way similar to Eq. 2.26: $\frac{A}{L} \ln Z = \gamma(x, y)$ where $x = \beta Af$ and $y = 2\pi C\sigma/h$ are dimensionless variables characterizing the applied force and the linking number, respectively.

The extension as a fraction of the total molecular length follows from 4.11 and 2.16, as:

$$\langle \hat{\mathbf{z}} \cdot \hat{\mathbf{t}} \rangle = 1 - \left(\frac{k_B T}{4Af} \right)^{1/2} - \frac{1}{2} \left(\frac{2\pi C\sigma}{h} \right)^2 \left(\frac{k_B T}{4Af} \right)^{3/2} + \cdots \tag{4.12}$$

For this model, twisting the DNA ($\sigma \neq 0$) leads to a reduction in extension. This can be seen in the data of Fig. 13; however, note that the quadratic twist-dependence occurs only quite near to $\sigma = 0$, and is clearest in the 0.2 pN data of the figure.

The free energy 4.11 can be used to find the relation between the torque applied at the end of the chain, and the linking number. Since linking number is

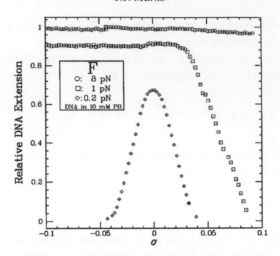

Fig. 13. Extension of DNA as a function of linking number σ, for a few fixed forces. Reproduced from Ref. [25].

controlled by rotating the end of the chain, the torque applied at the end of the chain is obtained from the linking number derivative of 4.11:

$$\frac{\tau}{k_B T} = -\frac{1}{2\pi} \frac{d}{d\Delta\text{Lk}} \ln Z = \left[1 - \frac{1}{2}\frac{C}{A}\left(\frac{k_B T}{4Af}\right)^{1/2}\right] \frac{2\pi C\sigma}{h} \qquad (4.13)$$

Moroz and Nelson have emphasized this result: the effective twist modulus of the double helix goes down as tension in the chain goes down. This effect occurs because for lower forces, more writhing can occur, allowing the twist energy and therefore the torque to be reduced.

The linearized calculation of the Wr fluctuations in Eq. 4.9 cannot account for plectoneme-type crossings. Recently, Rossetto has argued that these fluctuations modify the $\mathcal{O}(\sigma^2)$ term in Eq. 4.12 [54]. This effect may be behind discrepancies between C determinations by different theoretical approaches, thanks to either over- or under-estimation of writhe fluctuations.

4.3. Forces and torques can drive large structural reorganizations of the double helix

The previous calculations have assumed that the forces and torques were not able to cause structural phase transitions in the double helix. We've already seen that at zero force an unwinding torque of about $-2k_B T$ is sufficient to start unwinding

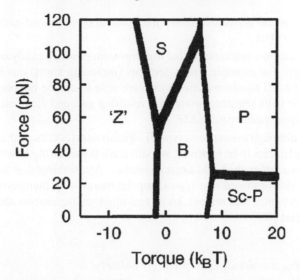

Fig. 14. 'Phase diagram' of double helix as function of external torque and force. Reproduced from Ref. [55].

AT-rich regions of the double helix. Experiments show that there may be as many as five different structural states of the double helix which can be accessed by twisting and pulling on DNA [55], and those experiments allow one to predict a 'force-torque phase diagram' (Fig. 14). For forces in the < 50 pN range the double helix is stable roughly over the torque range $-2k_BT$ to $+7k_BT$ (the B-DNA region of Fig. [55]).

4.4. DNA knotting

Above we've discussed the effect of supercoiling, which is controlled by the 'internal' linking number of the two ssDNAs inside the double helix. A separate and important property of the double helix is the 'external' entanglement state of the double helix backbone, the more usual case of topology discussed in usual polymer physics. A single circular DNA can carry a knot along its length; alternately two or more circular DNAs can be linked together.

When an initially linear DNA is closed into circular form, there is some possibility that a knot is generated. You might be wondering why all the molecules of Fig. 12 are all *unknotted*. There are two reasons for this, the first one biological and the second one biophysical.

4.4.1. Cells contain active machinery for removal of knots and other entanglements of DNA

Cells contain enzyme machines called *topoisomerases* which catalyze changes in DNA topology. For example, entanglements (including knots) can be removed or added by *type-II topoisomerases* which are able to cut the double helix, pass another double helix segment through the resulting gap, and then seal the gap up. Type-II topoisomerases require ATP.

However, although existence of type-II topoisomerases tells us that it is *possible* for entanglements to be removed, we still are left wondering how they 'know' how to remove rather than add entanglements. Astonishingly, it has been experimentally demonstrated that type-II topoisomerases by themselves have the capacity to recognize and remove knots and other entanglements along circular DNA molecules [56].

4.4.2. Knotting a molecule is surprisingly unlikely

Let's suppose we take an ensemble of linear DNA molecules of length L at low enough concentration that they do not interact with one another. Now, let's add a small quantity of an enzyme which catalyzes closure of the molecules into circles, and the reverse process (this is possible). Then we can ask what the probability $P_{unknot}(L)$ is that the molecule is unknotted, as a function of L.

We can argue that $P_{unknot} \approx \exp[-L/(N_0 A)]$, for some constant N_0. For small L, there will be a large free energy cost of closing a molecule into a circle making $P_{knot} \to 1$. However, for larger molecules, the probability of an unknotted configuration should go down. The exponential decay reflects the fact that over some length (N_0 persistence lengths) the probability of having no knot drops to $1/e$: applying this probability to each L_0 along a DNA of length L gives us $P_{unknot}(L) \approx (1/e)^{L/(N_0 A)}$. This rough argument can be made mathematically rigorous [57].

It turns out that even for an 'ideal' polymer which has no self-avoidance interactions, $N_0 \approx 600$. For a slightly self-avoiding polymer like dsDNA in physiological buffer, $N_0 \approx 800$. What this means is that to have a $1/e$ probability to find even one knot along a dsDNA, it has to be $800 \times 150 = 120,000$ bp long! (the long persistence length of DNA helps make this number impressive). Even more incredibly, for a self-avoiding polymer, $N_0 \approx 10^6$! This remarkable fact is theoretically understood only on the basis of numerical simulations: see Ref. [58].

Experiments on circular DNAs are in good quantitative agreement with statistical mechanical results for the semiflexible polymer model including DNA self-avoidance interactions. For example, it is found that the probability of finding a knot generated by thermal fluctuations for a 10 kb dsDNA is only about 0.05. [59, 60]. Topoisomerases are by themselves (using ATP) able to push this probability down, by a factor of between 10 and 100 [56].

4.4.3. Condensation-resolution mechanism for disentangling long molecules

Although topoisomerases seem to be able to help get rid of entanglements, there must be other mechanisms acting in the cell to completely eliminate them. Here I'll mention a very simple model that might give you an idea of how this machinery works.

Suppose we have a long dsDNA of length L, in the presence of some proteins which act to fold DNA up along its length. We imagine that these proteins cannot 'cross-link' DNA segments, but that they can only compact the molecule along its length. As these proteins bind, we imagine that they modify the total contour length to be $L' < L$, and the effective persistence length to be $A' > A$.

If these proteins bind slowly in the presence of type-II topoisomerases so that the knotting topology can reach close to equilibrium, then the unknotting probability will have the form:

$$P_{\text{unknot}} = \exp\left(-\frac{L'/A'}{N_0}\right) \tag{4.14}$$

As you can see, gradually compacting (decreasing L') while stiffening (increasing A') DNA can drive knotting out of it; unknotting ('entanglement resolution') will occur on progressively larger length scales as this condensation process proceeds.

Problem 40: Consider a condensation process which gradually condenses DNA a DNA of length L by folding it along its length, to make a progressively thicker fiber, of length L' and cross-section radius r'. If volume is conserved during condensation, and if the effective Young modulus of the fiber is a constant, find the unknot probability as a function of L'/L.

This simple model gives some idea of how proteins which structure DNA can play a role in controlling its entanglement at short enough scales where we can use polymer statistical mechanics. However, at the large scale of a whole chromosome cross-links do occur: they are necessary to fit the chromosome into the cell! It is also possible to envision a process where chromosome condensation by cross-linking also can drive out entanglements as long as the cross-links are transient [61].

5. DNA-protein interactions

So far we mostly talked about DNA by itself in buffer, focusing on the double helix's physical properties. In the cell, DNA is covered with proteins. You can think of DNA as a long string of thickness 2 nm, and the proteins as little particles of diameters of 1 to 10 nm plastered all along the string's length. The DNA plus all the proteins bound to it make up the biologically active chromosome.

Some proteins which bind to DNA are primarily *architectural*, folding and wrapping DNA so as to package it inside the cell. Other proteins have primarily *genetic* functions, interacting with particular sequences, usually between 4 and 40 bp in length. Of course, these two functions can be mixed: as examples proteins which tightly fold up DNA will likely repress gene expression; the expression of genes likely cannot occur without changes in DNA folding architecture.

Proteins which interact with DNA tend to be sorted into two groups:
Sequence-nonspecific proteins that stick anywhere along the double helix;
Sequence-specific proteins that bind to particular sequences very strongly, and to other sequences only relatively weakly.

Most proteins involved in chromosome architecture have a mainly nonspecific interaction with DNA. Such interactions are often electrostatic in character, and can be disrupted with high salt concentrations. Examples are the histone proteins of the nucleosome, and nonspecific DNA-bending proteins such as HU from *E. coli* or HMG proteins from eukaryote cells. Note that nonspecifically-interacting proteins usually bind better to some sequences than others, but not a whole lot better. Under physiological conditions, nonspecifically interacting proteins usually bind to DNA once their concentration is in the 10 to 1000 nM range.

Sequence-specific interactions occur via chemical interactions which depend on the structure of the bases. These are not usually electrostatic in character (most of DNA's charge is on the phosphates, which are common to all the bases) although most sequence-specific proteins also have a nonspecific interaction with DNA which is to some degree electrostatic. Examples of sequence-specific interactions include transcription factors and restriction enzymes. Sequence-specific proteins can often bind their targets at concentrations well below 1 nM. This high level of affinity is necessary: the concentration of transcription factors in *E. coli* can be as little as one per cubic micron (i.e., one per *E. coli* cell), which in molar units is $(1/6 \times 10^{23} \text{ Mol})/(10^{-15} \text{ litre}) = 1.6$ nM. In the human cell with a nucleus of volume $\approx 10^3 \ \mu m^3$, affinities in the picomolar range are needed to bind sequence-specific proteins to their targets with reasonably high probability.

5.1. How do sequence-specific DNA-binding proteins find their targets?

How long does it take for a protein to find a single specific target in a large DNA molecule? There is a long history of test-tube experiments studying this classic biophysics problem: one model system has been the *E. coli* protein lac repressor. Let's follow one protein of diameter d as it moves by diffusion to a target of size a; we'll suppose that the targets are present in solution at concentration c.

5.1.1. Three-dimensional diffusion to the target

In the absence of any nonspecific interaction effects, the protein will diffuse through space until it hits the target. To analyze how long it takes for the protein to find one target, let's divide space up into boxes of volume $V = 1/c$ each containing one target. We'll further divide each box up into 'voxels' of volume a^3; one voxel is the target. Since the protein moves by diffusion, its trajectory will be a random walk in the box. The number of steps at scale a that must be taken to move across the box (of edge $V^{1/3}$) is $V^{2/3}/a^2$; the probability of finding the target before the protein leaves the box is therefore $a/V^{1/3}$ (this result is often called 'diffusion to capture'). The time that this search occurs in is just the diffusion time $V^{2/3}/D$, where $D = k_B T/(3\pi\eta d)$ is the protein diffusion constant.

Once the protein leaves one box of volume V, the same search starts over in an adjacent box: this will occur $V^{1/3}/a$ times before a target is found. So, the total time required for our protein to find a target by simple three-dimensional diffusion is

$$\tau_{3d} = \frac{V^{1/3}}{a} \times \frac{V^{2/3}}{D} = \frac{V}{Da} \tag{5.1}$$

Note that this formula could apply in solution where there is one target per solution volume V, or to targeting in a cell compartment of volume V. In the latter case, the same volume is searched over and over until the target is found.

If we convert V to concentration c, our result is $1/\tau_{3d} = 4\pi Dac$. The factor of 4π comes from a more detailed calculation of diffusion to capture, originally due to Smolochowski [62]. This rate is proportional to concentration; biochemists usually describe the rate of this type of bimolecular reaction by normalizing the actual rate by concentration, leaving the *association rate*:

$$k_a = 4\pi Da = \frac{4\pi k_B T}{3\pi\eta} \frac{a}{d} \tag{5.2}$$

Since the target size is less than the protein size, we find that the maximum association rate is the prefactor $4\pi k_B T/(3\pi\eta) \approx 4 \times 10^{-18} \text{ m}^3/\text{s} \approx 10^8 \text{ M}^{-1}\text{s}^{-1}$. This rate is often referred to as the 'diffusion-limited reaction rate'.

5.1.2. Nonspecific interactions can accelerate targeting

Experiments in the 1970s showed that lac repressor binds its target at closer to $k_a \approx 10^{10} \text{ M}^{-1}\text{s}^{-1}$ which was initially quite a puzzle. However, Berg, Winter and von Hippel proposed and experimentally supported a solution to this paradox [63]. They realized that lac repressor also had a *nonspecific* interaction with DNA, and that this nonspecific interaction could make the target effectively larger than the protein and the binding site!

The picture is that when lac repressor first hits DNA, the nonspecific interaction allows it to 'slide' randomly back and forth along the DNA, exploring a region of length ℓ_{sl} before it dissociated. Now, the target size is increased to be ℓ_{sl}, so the time we will have to spend doing three-dimensional diffusion will be reduced to $1/(4\pi D_{\text{sl}}\ell_{\text{sl}}c)$, where D_{sl} is the diffusion constant for the 'sliding' motion.

However, it is not yet clear if this will really accelerate k_a, since each sliding event will eat up a time of roughly $\ell_{\text{sl}}^2/D_{\text{sl}}$, and to be sure to find the target, L/ℓ_{sl} sliding events have to occur, requiring a total one-dimensional diffusion time of $L\ell_{\text{sl}}/D_{\text{sl}}$.

So, the total time required to find the target by this 'facilitated diffusion' process is

$$\tau_{\text{fac}} = \frac{1}{4\pi D \ell_{\text{sl}} c} + \frac{L\ell_{\text{sl}}}{D_{\text{sl}}} \tag{5.3}$$

If we write the ratio of this and the three-dimensional result 5.2, we obtain

$$\frac{k_{a,\text{fac}}}{k_{a,\text{3d}}} = \frac{\tau_{\text{3d}}}{\tau_{\text{fac}}} = \frac{\ell_{\text{sl}}/a}{1 + 4\pi \frac{D}{D_{\text{sl}}} \ell_{\text{sl}}^2 L c} \tag{5.4}$$

As long as ℓ_{sl} is not too long, the nonspecific interaction does accelerate the reaction rate. This basic model and its experimental study were introduced by Berg, Winter and von Hippel [63].

A key feature of 5.3 is that for fixed total DNA length L and target concentration $c = 1/V$ there is an *optimal* ℓ_{sl}:

$$\ell_{\text{sl}}^* \approx \sqrt{V/L} \tag{5.5}$$

where we have dropped the 4π and the ratio of diffusion constants. For the *E. coli* cell, $V \approx 10^9 \text{ nm}^3$ and $L \approx 10^6 \text{ nm}$, indicating $\ell_{\text{sl}}^* = 30 \text{ nm} = 100 \text{ bp}$. This is the sliding length inferred for lac repressor from biochemical experiments on facilitated diffusion [63]. This suggests that ℓ_{sl} for lac repressor is optimized to facilitate its targeting *in vivo* [64].

Note that the above arguments are independent of the conformation of the large target-containing DNA [64]. At the level of scaling, the association rate constant should be the same for either a relaxed random coil, or a fully stretched molecule. This conclusion, and the dependence of k_a on concentration and total DNA length (see Eq. 5.4), could be checked using individual micromanipulated DNA molecules, given a way to detect when a protein binds its target along a long DNA. Recently Wuite and coworkers have demonstrated micromechanical detection of target recognition by the restriction enzyme EcoRV, which induces a DNA bend when it binds to its target sequence [65].

5.2. Single-molecule study of DNA-binding proteins

A number of groups are working at present on experiments looking at protein-DNA interactions using single-DNA micromanipulation, which we touched on in the last paragraph. The basic idea is to study proteins which change DNA mechanical properties, for example, by putting bends or loops into the double helix. Then, the binding of the proteins can be monitored via the force-extension response of the molecule. To give an idea of what can be obtained from this kind of study, I'll describe a few simple models.

5.2.1. DNA-looping protein: equilibrium 'length-loss' model

Consider a protein which binds to a double helix under tension f, resulting in a reduction in contour length available for extension by amount ℓ. If we suppose that the binding energy of the protein in $k_B T$ units is ϵ and its bulk concentration is c, we can write down a simple two-state model for the free energy of the protein-DNA complex as a function of the occupation of the protein [66]

$$\ln Z_n = \frac{(L - n\ell)}{A} \gamma(\beta A f) + n(\epsilon + \ln vc) \tag{5.6}$$

where $n = 1$ for protein bound, $n = 0$ for protein free in solution. The first term is the DNA stretching free energy: when the protein binds, the contour length of extensible DNA is reduced by ℓ. The last two terms give the free energy for a bound protein including the entropy cost of its removal from free solution. A factor of protein volume v is included to make the inside of the log dimensionless.

We can immediately find the probability for the protein to be bound:

$$\frac{P_{\text{on}}}{P_{\text{off}}} = \frac{c}{K_d} e^{-\frac{\ell}{A}\gamma(\beta A f)} \approx e^{-\beta \ell f + \ln(c/K_d) + \cdots} \tag{5.7}$$

where we have defined $K_d = e^{-\epsilon}/v$, the 'dissociation constant' or concentration at which the protein is half-bound for zero force; the final term gives the large-force limit (recall $\gamma(x) = x + \sqrt{x} + \cdots$). If the complex is bound at low force ($c > K_d$), it will stay bound until a force threshold is reached. Roughly, this characteristic force is $f^* \approx \ln c/K_d/\ell$; beyond this force, the DNA-protein complex will open up. In an experiment where extension versus force can be used to monitor the stability of a protein-DNA complex, one can therefore make a measurement of the zero-force K_d – if one can observe equilibrium (on-off fluctuations) at the force-induced dissociation point.

At zero force, Finzi and Gelles have observed such opening-closing fluctuations using lac repressor, which is able to bind two targets so as to form a loop [67]. Experiments of this type at nonzero force have recently been done for the gal repressor protein, which is also able to trap a DNA loop [68].

Problem 41: A single nucleosome involves 146 bp of dsDNA wrapped around a 10 nm-diameter octamer of 'histone' proteins. The total binding free energy of this structure is thought to be roughly $30\,k_B T$ (see [66] and references therein). Using the length-loss model, estimate the force at which nucleosomes ought to be released by applied force. You might like to compare your result with the 15 pN force observed in many 'nucleosome-pop-off' experiments (see, e.g., Bennink et al. [69], and Brower-Toland et al. [70]). What do you think is the origin of the discrepancy?

Problem 42: An elegant loop-formation experiment in which equilibrium fluctuations can be observed uses a single piece of RNA under force [71], which can fold into a hairpin 'helix-loop' structure, held together by base-pairing and base-stacking interactions. When under force, opening and closing kinetics can be observed on roughly 1 sec timescales. Using the single-strand unzipping results, construct a two-state model for the loop opening and closing (recall the free energy per base stabilizing the coil is is roughly $2.5\,k_B T$, while the persistence length of the single-stranded backbone is roughly $b = 1$ nm), and estimate how $P_{\text{closed}}/P_{\text{open}}$ varies with applied force.

5.2.2. *Loop formation kinetics*

The previous section supposes that one can observe on-off fluctuations in the presence of force. However, it is possible, given strong protein-DNA interactions, that spontaneous dissociation will be essentially unobservable. In this case, one will observe the on-kinetics only. In the case of a small DNA loop, the complex formation rate will involve a barrier made of two components: the free energy cost of pulling in a length ℓ of DNA as discussed above, plus the free energy cost of making the DNA loop [72] (recall the 'J-factors' of Sec. 2.2.5). Putting these together gives a simple estimate of the loop formation rate

$$k_{\text{loop}} \approx k_0 \exp\left[-\beta\ell f + \ln v J(\ell)\right] \tag{5.8}$$

where J is the juxtaposition 'J-factor' relevant to the reaction (see 2.11, 2.12), and where v is the 'reaction volume', the volume of the region of space in which the reaction can proceed.

The exponential dependence on force will effectively shut off the reaction for forces significantly larger than $f^* \approx (k_B T / \ell) \ln J$. The drastic force-quenching of irreversible loop formation by force has been verified using an essentially exact transfer-matrix calculation [73]; experimental data that quantitatively test this model have not yet been reported, although moderate forces have been observed to completely eliminate looping by the two-site-binding restriction enzyme BspMI [74]. Finally, if a series of loop binding sites are available on a long DNA, the rate (5.8) will be proportional to the velocity at which the DNA end retracts due to loop formation [72].

For large loops ($\ell > A$) and low forces ($f < k_B T / \ell < k_B T / A$) random-coil fluctuations will reduce the severity of the force-quenching effect, giving $\ln k_{\text{loop}}(f) \approx \ln k_{\text{loop}}(0) - A\ell(\beta f)^2$. This regime is one of quite low force, since $k_B T / A \approx 0.1$ pN.

5.2.3. DNA-bending proteins

In vivo, sequence-specific interactions such as the DNA-looping examples discussed in the previous section, take place on DNA which is in large part covered by other proteins. Much of this protein serves to package the DNA in the cell, and has sequence-nonspecific interaction with DNA. One of the classes of DNA-packaging proteins that has already been shown to be quite interesting to study using micromechanical approaches are *DNA-bending proteins*: examples are HU (a primary DNA-bending protein from *E. coli*) [75] and various HMG proteins (the prevalent nonspecific DNA-bending proteins in eukaryotes) [76], both of which are present at high copy number (roughly one per 200 bp of DNA) in vivo. These two proteins generate roughly 90° bends where they bind DNA.

An important chromosome-packaging role of DNA-bending proteins is well established for bacteria [77]. In eukaryotes, the compaction of chromosomal DNA occurs in part via wrapping in nucleosomes (the histones in which can be considered to bend DNA); however, many non-histone proteins including the DNA-bending HMGs act to establish 'higher-order chromatin structure', a term which is often invoked as a cover for our ignorance of the chromosome structure at supra-nucleosomal scales.

Traditional biochemical approaches to the study of DNA-bending proteins include study of changes in electrophoretic mobility resulting from binding of protein, among other bulk solution phase methods. Single-DNA experiments offer a complementary view of such proteins: in the presence of nonspecifically binding DNA-bending proteins we can expect a modification of the force-extension

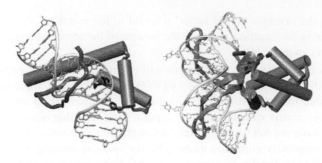

Fig. 15. Two examples of DNA-bending proteins with known structures. Left: NHP6A, an HMG protein from yeast, bound to DNA. Right: HU, a DNA-bending protein from *E. coli*, bound to DNA. Both bend DNA by roughly 90°, over roughly 10 bp. Recall the 2 nm diameter of the double helix. Image courtesy of Prof. R.C. Johnson, UCLA.

behavior of bare DNA we have discussed above. Qualitatively, we can antici-pate that a DNA will be more difficult to extend if it contains sharp bends due to bound proteins along its length; we expect a *shift* of the force-extension curve, to higher forces, or a *compaction* of DNA relative to the 'bare', protein-free case.

This force-shift effect, similar to that discussed for the two-state 'lost length' model [66] discussed in the previous section, has been observed for a few DNA-bending proteins. The first experiment of this type was carried out with the *E. coli* protein IHF [78]; the DNA-bending protein HU has been similarly stud-ied [79, 80].

Eukaryote DNA-bending proteins have also been studied and show similar effects [80]. Fig. 16 shows force-shift data for the HMG-type DNA-bending protein NHP6A from yeast. As more protein is added, the force curve shifts to larger forces. Something apparent in those NHP6A data is that at moderately high forces ≈ 10 pN, even when the DNA is covered with protein, the protein-DNA complex length can be forced to near the original bare DNA contour length (about 16 microns in this experiment on λ-DNA). In these experiments it was determined that the protein was not being dissociated by force; therefore the DNA-protein complexes are *flexible*. The *E. coli* DNA-bending proteins IHF and HU also show this effect [79, 80].

Since thermally excited DNA bends take place over molecule segments of about the persistence length $A \approx 150$ bp long, the force-extension response of a single DNA should be quite strongly perturbed when there is more than one protein bound per persistence length $A = 150$ bp. This indicates that strong force shifts such as the 33 nM data shown in Fig. 16 correspond to at least several proteins bound per persistence length, i.e., quite dense 'coverage'.

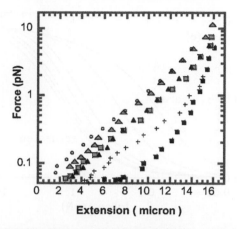

Extension (micron)

Fig. 16. Experimental force-extension data for individual λ-DNA molecules in the presence of the protein NHP6A (the HMG protein shown in Fig. 15). Black filled squares show the force-extension response of bare DNA with no NHP6A present. The other symbols show the force-extension response of single DNAs at various NHP6A concentrations: +, 3 nM; gray filled squares, 5 nM; black filled triangles, 10 nM; gray filled triangles, 33 nM; open circles, 75 nM. Reproduced from Ref. [80].

We can construct a model for this kind of experiment by supposing that where proteins bind, they generate localized (and flexible) bends. The discrete-tangent model introduced in Sec. 2 can be modified to include the effect of DNA-bending proteins [21]:

$$
\beta E = \sum_k \left[(1 - n_k)\frac{a}{2}|\hat{\mathbf{t}}_{k+1} - \hat{\mathbf{t}}_k|^2 + n_k\frac{a'}{2}(\hat{\mathbf{t}}_{k+1} \cdot \hat{\mathbf{t}}_k - \cos\psi)^2 \right.
$$
$$
\left. -\mu n_k - \beta f\hat{\mathbf{z}} \cdot \hat{\mathbf{t}}_k \right] \tag{5.9}
$$

which contains, in addition to tangent vectors $\hat{\mathbf{t}}_k$, two-state occupation degrees of freedom n_k which are either 0 or 1. The n_k are controlled by the 'chemical potential' μ; for simple, independent equilibrium binding, $\mu = $ constant $+ \ln c$ where c is the bulk protein concentration [21,66]. When $n_k = 0$, the joint between segments k and $k + 1$ has the usual semiflexible polymer bending energy. However, when $n_k = 1$, the joint has a preferred angle ψ, and a modified rigidity a'.

This type of model has as degrees of freedom, in addition to the orientation vector $\hat{\mathbf{t}}_k$, an additional 'scalar field' n_k describing protein binding. Models of this general type have been introduced to describe DNA overstretching (in this case the scalar variable describes conformational change of the double helix) [81–83] as well as protein-DNA interactions [36,84]. The idea common to these works is the use of additional variables (n_k) to account for DNA conformational

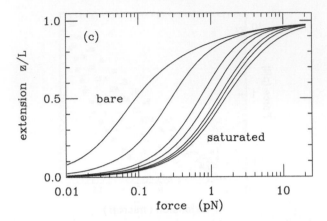

Fig. 17. Theoretical force-extension behavior for discrete-semiflexible polymer model including DNA-bending protein effects. Leftmost curve shows bare DNA (no additional bends), while the curves shifted progressively to the right show the shift to higher forces generated by a series of higher protein concentrations (chemical potential $\mu = -3.91, -2.30, -1.61, -0.69, 0$ and 3.00; corresponding low-force binding site occupation of 22%, 60%, 72%, 89%, 93% and 100%). Reproduced from Ref. [21].

change, and to examine how the coupling of those variables to the tangent vectors $\hat{\mathbf{t}}_k$ modifies force-extension behavior.

Fig. 17 shows the force-shift effect for a numerical transfer-matrix calculation of the partition function for this model, using segment length $b = 1$ nm, double helix stiffness $a = 50$ (persistence length 50 nm), a bend angle $\psi = 90°$, and a protein-bent-DNA bending constant $a' = 10$ (quite flexible). The result is, as the binding chemical potential μ is varied, a gradual shift of the force curve to larger forces qualitatively similar to that seen experimentally. The segment length b defines the maximum density of proteins that can bind, i.e., one per 3 bp for the parameters used in Fig. 17.

5.2.4. Analytical calculation for compaction by DNA-bending proteins

I now present a simple computation which gives some insight into the numerical solution of the DNA-bending-protein model (Eq. 5.9 and Ref. [21]). The aim here is to look at a 'weakly distorting' limit, where we can see how the basic semiflexible-polymer model is modified by the presence of DNA-bending proteins in the limit that they only slightly deflect $\hat{\mathbf{t}}$.

The portion of Eq. 5.9 that describes how one of the protein occupation variables n is coupled to two adjacent tangent vectors $\hat{\mathbf{t}}$ and $\hat{\mathbf{t}}'$ is

$$\frac{a}{2}|\hat{\mathbf{t}}' - \hat{\mathbf{t}}|^2 (1 - n) + \left[\frac{a'}{2} (\hat{\mathbf{t}}' \cdot \hat{\mathbf{t}} - \cos \psi)^2 - \mu \right] n \tag{5.10}$$

Using $1 - |\hat{\mathbf{t}}' - \hat{\mathbf{t}}|^2 = 2\hat{\mathbf{t}}' \cdot \hat{\mathbf{t}}$ and dropping constants and terms beyond quadratic order in $\hat{\mathbf{t}}' - \hat{\mathbf{t}}$ we have

$$\frac{a}{2}|\hat{\mathbf{t}} - \hat{\mathbf{t}}'|^2 - \frac{a + a'(1 - \cos\psi)}{2}n|\hat{\mathbf{t}}' - \hat{\mathbf{t}}|^2 - \mu n \tag{5.11}$$

where the chemical potential μ has been shifted by a constant. Keeping only the quadratic terms supposes that the DNA-bending effect of any protein present is a weak perturbation to the straightening generated by the external tension f.

We will now rewrite this model in continuum form. The discrete nature of n may be enforced by adding an n^2 term to make the amplitude of n-fluctuations well-defined. The model including force is now:

$$\int ds \left[\frac{A}{2}\left|\frac{\partial\hat{\mathbf{t}}}{\partial s}\right|^2 + \frac{1}{2}n^2 - \mu n - \beta f\hat{\mathbf{z}} \cdot \mathbf{t} - gn\left|\frac{\partial\hat{\mathbf{t}}}{\partial s}\right|^2 \right] \tag{5.12}$$

The parameter $g \propto a + [1 - \cos\psi]a'$ describes the coupling of the protein density to bending, and should be positive for DNA-bending proteins like HU, HMGB1 and NHP6A, and negative for DNA-stiffening proteins.

For fixed μ, fluctuations of n can be integrated, to obtain

$$\int ds \left[\frac{1}{2}(A - 2g\mu)\left|\frac{\partial\hat{\mathbf{t}}}{\partial s}\right|^2 - \beta f\hat{\mathbf{z}} \cdot \mathbf{t} \right] \tag{5.13}$$

A shift of chemical potential (binding free energy) causing binding of protein (positive shift of μ) causes a shift in the effective persistence length, to $A_{\text{eff}} = A - 2g\mu$. The reduction in persistence length generated by DNA-bending proteins ($g > 0$) causes a shift in the force-extension curve to larger forces, similar to that obtained in the transfer-matrix solution of the discrete model (Fig. 17).

Problem 43: Show that the expectation value of the 'bound protein density' along the DNA takes the form

$$\langle n \rangle = \mu + g\langle|\partial\hat{\mathbf{t}}/\partial s|^2\rangle$$

If you want to compute the derivative term, you must include an upper wavenumber cutoff q_{max} (equivalently, a finite lattice constant $b = 2\pi/q_{\text{max}}$). You will find that $\langle n \rangle$ diverges as you take the cutoff away (the limit $q_{\text{max}} \to \infty$).

Problem 44: The previous calculations supposed fluctuating protein occupation n. Alternately, DNA-bending proteins may be able to bind to DNA *permanently*. This may be treated in an averaged fashion by supposing n in Eq. 5.12 to take a fixed, positive value. For this case, use Eq. 5.12 to show $A_{\text{eff}} = A - 2gn$.

Problem 45: A more complete treatment of the n fluctuations would replace n^2 in Eq. 5.12 with $\frac{1}{2}\left[mn^2 + v(dn/ds)^2\right]$, where m and v are positive parameters. Find the form of A_{eff} in this case.

5.2.5. *Effects of twisting of DNA by proteins*

Proteins which bend DNA often untwist it (e.g., HU [75]), and this effect might be probed using single-DNA experiments. This has been explored for the discrete-tangent model [21], but the simple continuum calculation presented above can also be generalized to the case where DNA is twisted:

$$\int ds \left[\left(\frac{A}{2} - gn\right) \left|\frac{\partial \hat{\mathbf{t}}}{\partial s}\right|^2 - \beta f \hat{\mathbf{z}} \cdot \mathbf{t} + \frac{1}{2}n^2 - \mu n + \frac{C}{2}(\Omega - \chi n)^2 \right] \quad (5.14)$$

The last term accounts for the twist energy including a shift in its zero due to binding of protein [21]. The new field $\Omega(s)$ is just the twist density [51]; the total twist is $\int ds\,\Omega(s) = 2\pi\,\Delta\text{Tw}$. The DNA-twisting parameter χ is negative for a protein which underwinds DNA, since in this case bound protein ($n > 0$) shifts the minimum of the twisting energy to $\Omega < 0$.

Expanding the last term in (5.14) gives

$$\int ds \left[\frac{A}{2}\left|\frac{\partial \hat{\mathbf{t}}}{\partial s}\right|^2 - \beta f \hat{\mathbf{z}} \cdot \hat{\mathbf{t}} + \frac{m}{2}n^2 + \frac{C}{2}\Omega^2 \right.$$
$$\left. -\left(\mu + g\left|\frac{\partial \hat{\mathbf{t}}}{\partial s}\right|^2 + C\chi\Omega\right)n \right] \quad (5.15)$$

where $m = 1 + \chi^2 C$. Completing the square and integrating out $n = (\mu + g|\partial \hat{\mathbf{t}}/\partial s|^2 + C\chi\Omega)/m$,

$$\int ds \left[\frac{1}{2}\left(A - \frac{2g}{m}[\mu + C\chi\Omega]\right)\left|\frac{\partial \hat{\mathbf{t}}}{\partial s}\right|^2 - \beta f \hat{\mathbf{z}} \cdot \hat{\mathbf{t}} \right.$$
$$\left. +\frac{C}{2}\left(1 - \frac{C\chi^2}{m}\right)\Omega^2 - \frac{C\mu\chi}{m}\Omega \right] + \cdots \quad (5.16)$$

where constants and terms beyond quadratic order in $\hat{\mathbf{t}}$ have been dropped.

We use $\Delta\text{Tw} = \Delta\text{Lk} - \text{Wr}$ in the form $\int ds\,\Omega/2\pi = \sigma L/h - \text{Wr}$ to eliminate the twist (recall h is the DNA helix repeat). The terms of quadratic order in $\hat{\mathbf{t}}$

(recall from Eq. 4.9 that Wr is $\mathcal{O}(\hat{t}^2])$ are:

$$\int ds \left\{ \frac{1}{2} \left(A - \frac{2g}{m} \left[\mu + \frac{2\pi C}{h} \chi \sigma \right] \right) \left| \frac{\partial \hat{t}}{\partial s} \right|^2 - \beta f \hat{z} \cdot t \right\}$$

$$- \frac{2\pi C}{h} \left(\left[1 - \frac{C\chi^2}{m} \right] \sigma - \frac{h}{2\pi m} \chi \mu \right) 2\pi \text{Wr} + \cdots \tag{5.17}$$

This energy describes Gaussian fluctuations of $\hat{t}$ and has the same form as Eq. 4.10, but with shifted persistence length A and twist density σ values:

$$A \rightarrow A_{\text{eff}} = A - \frac{2g}{m} \left[\mu + \frac{2\pi C}{h} \chi \sigma \right]$$

$$\sigma \rightarrow \sigma_{\text{eff}} = \left[1 - \frac{C\chi^2}{m} \right] \sigma - \frac{h}{2\pi m} \chi \mu \tag{5.18}$$

These effective quantities can be used to write the extension in the high-force expansion form of Eq. 4.12:

$$\langle \hat{t} \cdot \hat{z} \rangle = 1 - \frac{1}{\sqrt{4\beta A_{\text{eff}} f}} - \frac{1}{2} \left(\frac{2\pi C \sigma_{\text{eff}}}{h} \right)^2 \left(\frac{k_B T}{4 A_{\text{eff}} f} \right)^{3/2} + \cdots \tag{5.19}$$

At very high force, the $1/\sqrt{f}$ term dominates, and will cause a shift of the peak of extension as a function of σ, to where $\chi\sigma < 0$. In this regime, binding of protein contracts the polymer, so the peak extension is found where the bound protein is the *least*. For a protein which untwists DNA ($\chi < 0$), the peak of extension occurs at very high force for $\sigma > 0$.

At lower, but still high force, the $1/f^{3/2}$ term can compete. This term vanishes at the zero of σ_{eff}, where the torsion in the molecule, and therefore its chiral coiling, are minimized. In the DNA-untwisting-protein case, this is for $\sigma < 0$. Therefore, for a DNA-untwisting protein ($\chi < 0$), the shift of the peak in extension as a function of force can changes from $\sigma > 0$ for large force, to $\sigma < 0$ for lower force. This gives some analytical insight into the same effect found in numerical calculations of Ref. [21].

Problem 46: Starting from Eq. 5.14 compute A_{eff} and σ_{eff} for the case that the bound protein density does not fluctuate (consider n a fixed, positive constant as in Problem 44).

5.2.6. Surprising results of experiments

Surprises have come out of the experimental studies of DNA-bending proteins. The E. coli HU protein, also known to bend DNA, has also been studied using single-DNA micromanipulation [79, 80]. While a compaction effect (shift to higher forces) has been observed for HU, at very high concentrations the force curve shifts *back* to lower forces than for bare DNA! In Ref. [79] it was suggested that at high concentration, HU is capable of essentially polymerizing along the double helix, and that HU-covered DNA is essentially straight and actually stiffer than bare DNA. This explained the shift of the force-extension curve to lower forces than for bare DNA (the result of increasing the persistence length A in Eq. 2.23). This 'bimodal' behavior of HU is a good example of a feature of a protein that was not well established using standard biochemical assays, but which was rather straightforward to infer using single-DNA methods.

A second experimental surprise associated with DNA-bending proteins is that once these proteins reach a certain high coverage on the double helix, they no longer will spontaneously dissociate, even when the protein in solution is removed (established for HU, HMGB1 and NHP6A in Ref. [80]). This high level of stability of protein densely assembled onto DNA may be a result of cooperative interactions between adjacent proteins, i.e., a term $n_k n_{k+1}$ in 5.9. While such stability due to cooperativity is expected for some proteins such as RecA, which is well known to form a polymerized 'coat' on the double helix (and which has been studied using single-DNA methods [85]), this behavior was not expected for HU, HMGB1 and NHP6A.

These results suggest that there may be appreciable restrictions on the applicability of the widely assumed 'two-state' equilibrium binding models for interaction of these non-sequence-specific proteins with DNA. Proteins like HU, HMGB1 and NHP6A are in part responsible for folding up DNA in vivo, and their ability to form highly stable complexes may be important to chromosome 'architecture'.

6. Conclusion

This course has focused on the micromechanics of individual DNA molecules studied in single-DNA micromanipulation experiments. A major focus has been on the use of statistical-mechanical models to describe how spontaneous thermal fluctuations and external applied stresses change the conformation of large DNAs.

With the double helix we have a favorable situation since the atomic and even base-pair scales (< 1 nm) are well separated from the ≈ 100 nm scale over which thermal fluctuations are able to bend the double helix. This separation of scales allows us to describe the elastic response of a dsDNA from essentially zero force, up to 10 pN or so, using only one 'effective' elastic constant, the persistence length $A = 50$ nm [20]. As force is increased beyond 0.1 pN, we have seen that a single 'correlation length' $\sqrt{k_B T A/f}$ describes the essentially Gaussian fluctuations of the tangent vector away from the force direction.

However, this course has also emphasized how quickly this pleasant long-wavelength description can fall by the wayside. Unzipping the double helix (Sec. 3) requires only about 15 pN forces, and converts the staid double helix into two single-stranded DNAs, which can by no means be described using a single elastic constant! [86] Under some circumstances ssDNA is a polymer with a strongly wavenumber- (and therefore force-) dependent persistence length [20,36]. The interplay of the backbone polymer elasticity, Coulomb interactions, and self-adhesion via attractive base-base interactions [35] makes single-stranded DNA impossible to describe using a single persistence length. Theories to describe ssDNA by necessity have a richer structure.

Something similar happens when one twists the double helix: at low forces (roughly < 0.3 pN), the response of the double helix to twisting can be captured using the twist-persistence-length model discussed in Sec. 4.2, which really just adds one more elastic constant to describe twist deformations. However, for even rather low forces, underwinding can rather easily drive strand separation.

Description of the opening of the double helix, and other abrupt force- and torque-driven structural transitions of the double helix such as 'overstretching' [22,23], need additional degrees of freedom on top of the elastic degrees of freedom used to describe relatively smooth bends and twists. For example, conversion of the double helix to ssDNA requires at least a 'helix-coil' Ising-type degree of freedom.

A few of the examples discussed in these notes have used additional degrees of freedom of this type to describe localized 'defects' along the double helix. Localized strand-separated regions excited thermally may be responsible for the anomalously large cyclization probability of short DNAs observed by Cloutier and Widom [12,13]. Similar models can be used to describe proteins which generate local bends or other distortions of the double helix [21]. Inevitably we can expect these 'local defects' to interact with one another along the double helix, and to possibly facilitate self-organization of stable protein-DNA complexes, as observed in Ref. [80]. Those experiments show that even at the level of a single species of protein interacting with a single DNA, self-assembled protein-DNA complexes can strongly resist disassembly. Fully understanding the nonequilibrium phenomena in experiments such as Ref. [80] requires at the least, theories that include kinetics.

Acknowledgments

These lectures include results of research done jointly with Eric Siggia, Didier Chatenay, Jean-Francois Légèr, Abhijit Sarkar, Simona Cocco, Sumithra Sankararaman, Dunja Skoko, Jie Yan, Michael Poirier, Steven Halford, Monte Pettitt, and Michael Feig, whose many insights I gratefully acknowledge. I am also grateful for advice and help, including experimental data, from David Bensimon, Vincent Croquette, Terence Strick, Jean-Francois Allemande, Carlos Bustamante, Steve Smith and Nick Cozzarelli. I thank NATO and the CNRS for making it possible for me to visit Les Houches to present these lectures. This work was supported in part by the US National Science Foundation, through Grants DMR-0203963 and MCB-0240998.

References

[1] M.D. Wang, M.J. Schnitzer, H. Yin, R. Landick, J. Gelles, S.M. Block, *Science* **282** 902–907 (1998).

[2] D.S. Goodsell, *The machinery of life* (Springer-Verlag, New York, 1992).

[3] M. Feig, J.F. Marko, M. Pettitt, Microscopic DNA fluctuations are in accord with macroscopic DNA stretching elasticity without strong dependence on force field choice, *NATO ASI Series: Metal Ligand Interactions*, ed. N. Russo (Kluwer Academic Press), 193–204 (2003).

[4] L.D. Landau, E.M. Lifshitz, *Theory of Elasticity* (Pergamon, NY, 1986) Ch. II.

[5] P.J. Hagerman *Ann. Rev. Biophys. Biophys. Chem.* **17**, 265 (1988).

[6] V. Makarov, B. Pettitt, M. Feig, *Acc. Chem. Res.* **35**, 376 (2002).

[7] K. Ogata, K. Sato, T.H. Tahirov, *Curr. Opin. Struct. Biol.* **13** 40 (2003).

[8] H. Yamakawa, W.H. Stockmayer, *J. Chem. Phys.* **57**, 2843 (1972).

[9] J. Shimada, H. Yamakawa, *Macromolecules* **17**, 689 (1984).

[10] J. Santalucia Jr., *Proc. Natl. Acad. Sci. USA* **95** 1460 (1998).

[11] D. Shore, J. Langowski, R.L. Baldwin, *Proc. Natl. Acad. Sci. USA* **78** 4833 (1981).

[12] T.E. Cloutier, J. Widom, *Mol. Cell* **14** 355 (2004).

[13] J. Yan, J.F. Marko, *Phys. Rev. Lett.* **93** 108108 (2004).

[14] D.M. Crothers, T.E. Haran, J.G. Nadeau, *J. Biol. Chem* **265**, 7093 (1990).

[15] S.B. Smith, L. Finzi, C. Bustamante, *Science* **258**, 1122 (1992).

[16] C. Bustamante, S. Smith, J.F. Marko, E.D. Siggia, *Science* **265**, 1599 (1994).

[17] A. Vologodskii, *Macromolecules* **27**, 5623 (1994).

[18] T. Odijk, *Macromolecules* **28**, 7016 (1995).

[19] T.T. Perkins, D.E. Smith, R.G. Larson, S. Chu, *Science* **268**, 83 (1995).

[20] J.F. Marko, E.D. Siggia, *Macromolecules* **28** 8759 (1995).

[21] J. Yan, J.F. Marko, *Phys. Rev. E* **68** 011905 (2003).

[22] P. Cluzel, A. Lebrun, C. Heller, R. Lavery, J.-L. Viovy, D. Chatenay, F. Caron, *Science* **271**, 792 (1996).

[23] S.B. Smith, Y.J. Cui, C. Bustmante, *Science* **271**, 795 (1996).

[24] J.R. Wenner, M.C. Williams, I. Rouzina, V.A. Bloomfield, *Biophys. J.* **82**, 3160 (2002).

[25] T.R. Strick, J.-F. Allemand, D. Bensimon, V. Croquette, *Biophys. J.* **74** 2016 (1998).

[26] K.M. Dohoney KM, J. Gelles, *Nature* **409** 370 (2001).

[27] P.R. Bianco, L.R. Brewer, M. Corzett, R. Balhorn, Y. Yeh, S.C. Kowalczykowski, R.J. Baskin, *Nature* **409** 374 (2001).

[28] M.N. Dessinges, T. Lionnet, X.G. Xi, D. Bensimon, V. Croquette, *Proc. Natl. Acad. Sci. USA* **101** 6439 (2004).

[29] A. Yu. Grosberg, A. R. Khokhlov, *Statistical Physics of Macromolecules*, pp. 302–312, (AIP Press, Woodbury NY, 1994).

[30] J.-F. Leger, G. Romano, A. Sarkar, J. Robert, L. Bourdieu, D. Chatenay, J.F. Marko, *Phys. Rev. Lett.* 83, 1066 (1999).

[31] J.-F. Leger, Ph.D. Thesis, l'Université Louis Pasteur Strasbourg I, Strasbourg, France (2000).

[32] C. Bustamante, D. Smith, S. Smith, *Curr. Opin. Struct. Biol.* **10**, 279 (2000).

[33] M. Rief, H. Clausen-Schaumann, H.E. Gaub, *Nat. Struct. Biol.* **6**, 346 (1999).

[34] J.-L. Barrat, J.-F. Joanny, *Europhys. Lett.* **24** 333 (1993).

[35] Y. Zhang, H.J. Zhou, Z.C. Ou-Yang, *Biophys. J.* **81**, 1133 (2001).

[36] J. Yan, A. Sarkar, S. Cocco, R. Monasson, J.F. Marko, *Eur. Phys. J. E* **10**, 249–263 (2003).

[37] Y. Seol, G.M. Skinner, K. Visscher, *Phys. Rev. Lett.* **93**, 118102 (2004).

[38] U. Bockelmann, B. Essevaz-Roulet, F. Heslot, *Biophys. J.* **82**, 1537 (2002).

[39] H. Clausen-Schaumann, M. Rief, C. Tolksdorf, H.E. Gaub, *Biophys. J.* **78**, 1997 (2001).

[40] S. Cocco, J.F. Marko, R. Monasson, *Proc. Nat. Acad. Sci. USA* **98**, 8608 (2001).

[41] S. Cocco, R. Monasson, J.F. Marko, *Phys. Rev. E* **66**, 051914 (2002).

[42] P. Nelson, *Phys. Rev. Lett.* **80**, 5810 (1998).

[43] P. Thomen, U. Bockelmann, F. Heslot, *Phys. Rev. Lett.* **88**, 248102 (2002).

[44] D. Lubensky, D.R. Nelson, *Phys. Rev. Lett.* **85**, 1572 (2000); D. Lubensky, D.R. Nelson, *Phys. Rev. E* **65**, 031917 (2002).

[45] C. Danilowicz, V.W. Coljee, C. Bouzigues, D.K. Lubensky, D.R. Nelson, M. Prentiss, *Proc. Natl. Acad. Sci. USA* **100** 1694 (2003).

[46] C. Danilowicz, Y. Kafri, R.S. Conroy, V.W. Coljee, J. Weeks, M. Prentiss, *Phys. Rev. Lett.* **93** 078101 (2004).

[47] J.D. Moroz, P. Nelson *Proc. Natl. Acad. Sci. USA* **94**, 14418 (1997); P. Nelson, *Biophys. J* **74**, 2501 (1998); J.D. Moroz, P. Nelson, *Macromolecules* **31** 6333 (1998).

[48] T.C. Boles, J.H. White, N.R. Cozzarelli *J Mol Biol.* **213**(1990).

[49] J.F. Marko, *Physica A* **296**, 289 (2001); J.F. Marko, *Physica A* **244**, 263 (1997).

[50] J.F. Marko, E.D. Siggia, *Science* **265**, 506 (1994).

[51] J.F. Marko, E.D. Siggia, *Phys. Rev. E* **52**, 2912 (1995).

[52] C. Bouchiat, M. Mezard, *Phys. Rev. Lett.* **80** 1556 (1998); C. Bouchiat, M. Mezard, Eur. Phys. J. **E2**, 377 (2000).

[53] A.E.H. Love, *A treatise on the mathematical theory of elasticity*, 4th edition (Dover, New York, 1944).

[54] V. Rossetto, *Europhys. Lett.* **69**, 142 (2005).

[55] A. Sarkar, J.-F. Leger, D. Chatenay, J.F. Marko, *Phys. Rev. E* **63**, 051903 (2001).

[56] V.V. Rybenkov, C. Ullsperger, A.V. Vologodskii, N.R. Cozzarelli, *Science* **277**, 648 (1997).

[57] D.W. Sumners, S.G. Whittington, *J. Phys. A Math. Gen.* **21** 1689 (1988).

[58] K. Koniaris and M. Muthukumar, *Phys. Rev. Lett.* **66** 2211 (1991).

[59] S.Y. Shaw, J.C. Wang, *Science* **260**, 533 (1993).

[60] V.V. Rybenkov, N.R. Cozzarelli, A.V. Vologodskii, *Proc. Natl. Acad. Sci. USA* **11** 5307 (1993).

[61] J.F. Marko, E.D. Siggia, *Mol. Biol. Cell* **8** 2217 (1997).

[62] H.C. Berg, *Random Walk in Biology* (Princeton, Princeton, 1993)

[63] O.G. Berg, R.B. Winter, P.H von Hippel, *Biochemistry* **20** 6929 (1981); R.B. Winter, P.H. von Hippel, (1981) *Biochemistry* **20** 6948 (1981); R.B. Winter, O.G. Berg, P.H. von Hippel, *Biochemistry* **20** 6961 (1981).

[64] S.E. Halford, J.F. Marko *Nucl. Acids Res.* **32**, 3040 (2004).

[65] B. van den Broek, M.C. Noom, G.J. Wuite, Nucl. Acids Res. **33**, 2676 (2005).

[66] J.F. Marko, E.D. Siggia *Biophys. J.* **73** 2173 (1997).

[67] L. Finzi, G. Gelles *Science* **267** 378 (1995).

[68] G. Lia, D. Bensimon, V. Croquette, J.-F. Allemand, D. Dunlap, D.E. Lewis, S. Adhya, L. Finzi *Proc. Natl. Acad. Sci. USA* **20** 11373 (2003).

[69] M.L. Bennink, S.H. Leuba, G.H. Leno, J. Zlatanova, B.G. de Grooth, J. Greve, *Nat. Struct. Biol.* **8** 606 (2001).

[70] B.D. Brower-Toland, C.L. Smith, R.C. Yeh, J.T. Lis, C.L. Peterson, M.D. Wang *Proc. Natl. Acad. Sci. USA* **99** 1960 (2002).

[71] J. Liphardt, B. Onoa, S.B. Smith, I.T. Tinoco, C. Bustamante *Science* **292** 733 (2001).

[72] S. Sankararaman, J.F. Marko, *Phys. Rev. E* **71** 021911 (2005).

[73] J. Yan, R. Kawamura, J.F. Marko, *Phys. Rev. E* in press (2005).

[74] J. Yan, D. Skoko, J.F. Marko, *Phys. Rev. E* **70**, 011905 (2004).

[75] K.K. Swinger, P.A. Rice, *Curr. Opin. Struct. Biol.* **14**, 28 (2004).

[76] J.O. Thomas, A.A. Travers, *Trends. Biochem. Sci.* **26**, 167 (2001).

[77] N. Trun, J.F. Marko, *AMS News* **64**, 276 (1998).

[78] B.M.J. Ali, R. Amit, I. Braslavsky, A.B. Oppenheim, O. Gileadi, J. Stavans, *Proc. Natl. Acad. Sci. USA* **98** 10658 (2001).

[79] J. van Noort, S. Verburgge, N. Goosen, C. Dekker, R.T. Dame, *Proc. Natl. Acad. Sci. USA* **101** 6969 (2004).

[80] D. Skoko, J. Yan, J.F. Marko, *Biochemistry* **43** 13867 (2004).

[81] P. Cizeau, J.-L. Viovy, *Biopolymers* **42** 383 (1997).

[82] J. Rudnick, R. Bruinsma, *Biophys. J.* **76**, 1725–33 (1999).

[83] C. Storm, P.C. Nelson, *Phys. Rev. E* **67** 051906 (2003).

[84] H. Diamant, D. Andelman, *Phys. Rev. E* **61**, 6740 (2000).

[85] J. Robert, L. Bourdieu, D. Chatenay and J. Marko, *Proc. Natl. Acad. Sci. USA* **95**, 12295 (1998).

[86] B. Maier, D. Bensimon, V. Croquette, *Proc. Natl. Acad. Sci. USA* **97**, 12002 (2000).

Course 8

THE ANALYSIS OF REGULATORY SEQUENCES

J. van Helden

SCMBB Université Libre de Bruxelles, Campus Plaine, CP 263, Boulevard du Triomphe, B-1050 Bruxelles, Belgium. E-mail: jvanheld@ucmb.ulb.ac.be

D. Chatenay, S. Cocco, R. Monasson, D. Thieffry and J. Dalibard, eds.
Les Houches, Session LXXXII, 2004
Multiple aspects of DNA and RNA: from Biophysics to Bioinformatics

Contents

1. Forewords

1.1. Scope of the course

This course will deal with different aspects of the analysis of regulatory sequences.

The first lesson will consist of a general presentation of the different type of questions that can be asked about regulatory sequences, and the different approaches that can be envisaged to answer these questions (pattern-discovery, pattern matching). The second lesson will be dedicated to string-based approaches, and the third lesson to matrix-based approaches. The theoretical concepts will mainly be illustrated by concrete examples from the yeast *Saccharomyces cerevisiae*.

1.2. Web site and practical sessions

The tools developed by Jacques van Helden are available for academic users via their web interface (http://rsat.scmbb.ulb.ac.be/). During the practical session, student will apply the concepts seen during the course, and test different approaches to detect putative regulatory signals in non-coding sequences. The main resources available on the web (databases and specific sequence analysis programs) will be presented.

2. Transcriptional regulation

In this chapter we briefly describe some fundamental aspects of transcriptional regulation that are relevant for the analysis of regulatory sequences. Our purpose is minimalist, and we do not pretend to review, even partially, the huge and complex field of transcriptional regulation.

2.1. The non-coding genome

Traditionally, sequence analysis and genomics have mainly been focussed on coding sequences. These sequences however represent only a fraction of the information contained in the genome. As shown in table 1, the proportion of coding sequences decreases with evolution.

Table 1

The non-coding genome

Organism	Year	Size	Genes	coding	non-coding
		Mb		%	%
Mycoplasma genitalium	1995	0.6	481	90	10
Haemophilus influenzae	1995	1.8	1717	86	14
Escherichia coli	1997	4.6	4289	87	13
Saccharomyces cerevisiae	1996	12	6286	72	28
Arabidiopsis thaliana	2001	120	27000	30	70
Caenorhabditis elegans	1998	97	19000	27	73
Drosophila melanogaster	2000	165	16000	15	85
Homo sapiens	2000	3000	50000	3	97

2.2. Transcriptional regulation

Non-coding sequences play an essential role in all cellular processes, since they mediate transcriptional regulation. Transcriptional regulation ensures the temporal and spatial specificity of expression for each gene of a genome. The level of expression is not only determined independently for each gene, but in addition, it can vary in response to a variety of signals: presence or absence of metabolites in the extra-cellular medium, inter-cellular communication, temperature, ... These signals generally provide information about the conditions outside the cell.

Transcription factors

Transcriptional regulation is mediated by classes of proteins, called *transcription factors*. These proteins interact with the general transcription machinery (RNA polymerase) in a way that either enhances (activation) or reduces (repression) the level of transcription. The same transcription factor are called dual, because they combine both effects: activate the expression of some genes while repressing the expression of other genes.

Transcriptional activators (Figure 1A) generally contain a domain that binds DNA in a sequence-specific manner (DNA-binding domain, and a domain that interacts with the RNA polymerase (activation domain). Repression encompasses a variety of mechanisms by which the transcription factor (repressor) reduces the expression level of a gene. Some repressors bind DNA in close vicinity (or downstream) of the transcription, and directly prevent RNA polymerase from starting transcription (Figure 1B). Another mechanism of repression is to compete with a transcriptional activator for the occupancy of the same site on DNA (Figure 1C). Some transcriptional repressors do not bind DNA at all, but rather their function

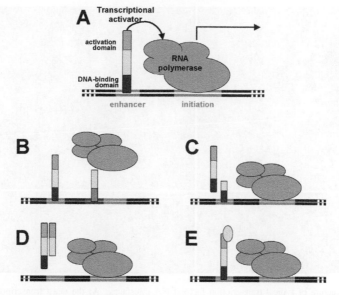

Fig. 1. Schematic representation of transcriptional regulation. **A:** transcriptional activation. **B:** transcriptional repression.

is mediated by direct protein–protein interaction with a transcriptional activator. In some cases, this interaction prevents the activator from binding DNA (Figure 1D). In other cases, the repressor forms a complex with the activation domain of the transcription activator, thereby preventing its interaction with RNA polymerase (Figure 1E).

Protein-DNA interfaces

Transcription factor-DNA interfaces are generally restricted to a very few amino acids and bases. Figure 2 shows the tri-dimensional structure of some transcription factor-DNA complexes, as determined by X-ray crystallography.

Many transcription factors are active in the form of dimers, two polypeptides forming a non-covalent complex via a dimerization domain. The dimer acts on DNA like tweezers (Figure 2A,C). Each monomeric unit enters in contact with a very limited number of nucleotides (typically 3–4). In some cases, the two contact points are adjacent (Figure 2B). Several classes of transcription factors (Helix-turn-helix in bacteria, Zinc cluster proteins in fungi) contain an intermediate domain that imposes spacing between the two contact points (Figure 2D).

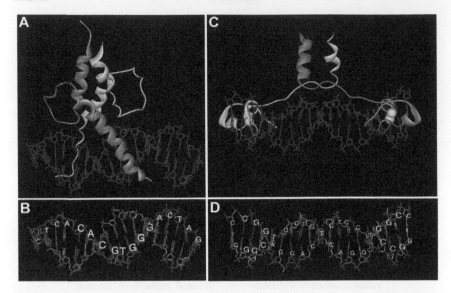

Fig. 2. Structure of typical transcription factor-DNA interfaces. **A:** the yeast transcription factor Pho4p forms a homodimer, which enters in contact with a set of contiguous nucleotides. **B:** sequence of nucleotides on Pho4p DNA binding site. **C:** the yeast Gal4p protein forms a homodimer, which binds a spaced pair of trinucleotides. **D:** sequence of nucleotides in the DNA binding of Gal4p.

Regulatory elements

The site on DNA where a transcription activator binds is denoted by different terms (depending on the biological field) : the yeast community favours *upstream activating sequence* (UAS), in higher organisms one speaks about *enhancers*, . . . The site where a repressor binds on DNA is often called *operator* (by bacteriologists), *upstream repressing sequence* (URS, in the yeast community), *silencer* (by drosophilist), ... The generic terms *cis-acting element* or *regulatory site* are used to denote the locations where transcription factors bind DNA, irrespective of their positive or negative effect on the level of expression.

Regulatory elements are very short sequences (between 5 and 30 bp) of highly conserved nucleotides. One class of regulatory element consists of a highly conserved core of 5–8 base pairs (bp), flanked by a few partly conserved bases. Another type of regulatory sites consists of a pair of very short conserved oligonucleotides (typically 3 bases) separated by a region of fixed width but variable content.

Gene	Site Name	Sequence	Affinity
PHO5	UASp2	`---aCtCaCACACGTGGGACTAGC-`	high
PHO84	Site D	`---TTTCCAGCACGTGGGGCGGA--`	high
PHO81	UAS	`----TTATGGCACGTGCGAATAA--`	high
PHO8	Proximal	`GTGATCGCTGCACGTGGCCCGA---`	high
PHO5	UASp3	`--TAATTTGGCATGTGCGATCTC--`	low
PHO84	Site C	`-----ACGTCCACGTGGAACTAT--`	low
PHO84	Site A	`-----TTTATCACGTGACACTTTTT`	low
group 1	consensus	`--------gCACGTGggac-----`	high-low
PHO5	UASp1	`--TAAATTAGCACGTTTTCGC----`	medium
PHO84	Site E	`----AATACGCACGTTTTTAATCTA`	medium
PHO84	Site B	`-----TTACGCACGTTGGTGCTG--`	low
PHO8	Distal	`---TTACCCGCACGCTTAATAT---`	low
group 2	consensus	`--------cgCACGTTt--------`	med-low
Degenerate consensus		`--------GCACGTKKk------`	

Fig. 3. Binding sites for the Pho4p transcription factor (Oshima et al. 1996).

3. Representations of regulatory elements

3.1. String-based representations

Different types of experiments provide primary information about the binding specificity of a DNA-binding protein. Collections of experimentally proven binding sites are stored in specialized databases such as TRANSFAC (Wingender 2004; Wingender et al. 1996), RegulonDB (Huerta et al. 1998; Salgado et al. 2001), SCPD (Zhu and Zhang 1999). These databases provide valuable information for the development and assessment of pattern detection algorithms.

Figure 3 displays a collection of binding sites for the yeast transcription factor Pho4p (Oshima et al. 1996). The table displays a qualitative estimation of the factor's binding affinity for different sequence fragments. The comparison of these sequences shows that the high affinity binding site share a "core" motif CACGTG, usually followed by a two or three cytosines (C) or guanines (G). The core CACGTG is however not sufficient to confer a high affinity: the protein does not bind to the sequences tCACGTGa or cCACGTGgaa. The lower part of the table shows two sites bound with a medium affinity, and showing a variation in the core (CACGTT) and followed by a few additional tyrosines (T). Despite the medium affinity, these sites have been shown to be actively involved in the regulation of the genes PHO5 and PHO84.

The collection of binding sites can be summarized with *consensus* strings such as CACGTGGG (high affinity) of CACGTTT (medium affinity). The two types of binding sites can even be represented in a more compact way, with a *degenerate consensus* CACGTKKK, where K denotes "either T of G", according to the IUPAC convention on ambiguous nucleotide code (Table 4). This representation is however an over-simplification, and suffers from several weaknesses.

1. By merging the letters G and T into the degenerate code K, we give the same weight to these letters, and we thus loose the concept that CACGTGGG is bound with a higher affinity than CACGTTT.

2. Several high affinity binding sites from Figure 3 do not match this consensus, and would thus be missed in a string-based search based on this pattern.

3. The degenerate consensus fails to indicate the *dependencies between successive residues*: in the collection of binding sites, the high affinity core CACGTG is usually followed by a few Gs or Cs, and the medium affinity core CACGTT by a few Ts. However, the pattern CACGTKKK would as well match sequences like CACGTGTT, CACGTGTG, CACGTTGG, which were never observed in the initial collection.

The two first limitations can be solved by using Position-Specific Scoring Matrices (PSSM), as will be shown in the next chapter. Higher-order dependencies can be treated with some more complex PSSM, or with Hidden Markov Models (HMM).

3.2. Matrix-based representation

Position-specific scoring matrices (PSSM)
A position-specific scoring matrix (PSSM) represents the binding specificity at each position of the DNA binding site for a transcription factor. The matrix is build from an alignment of a collection of binding sites.

Each row of the matrix represents one letter of the alphabet (in this case the 4 nucleotides A, C, G and T), and each column one position of the sequence alignment. The simplest representation is a *occurrence matrix* (Table 2A), where the values in the cells indicate the absolute frequency of each residue (letter) at each position in the multiple alignment.

The weight matrix
The absolute frequency is generally not very indicative of the significance of a residue. Indeed, a general observation is most non-coding sequences are AT-rich.

For instance, in the yeast *Saccharomyces cerevisiae*, the average composition of intergenic sequences is $F(A) = F(T) = 0.325$, $F(C) = F(G) = 0.175$. This intergenic composition can be used to estimate prior probabilities $p_A = p_T \sim 0.325$; $p_C = p_G \sim 0.175$.

$$W_{i,j} = \ln\left(\frac{f'_{i,j}}{p_i}\right) \qquad f'_{i,j} = \frac{n_{i,j} + p_i k}{\sum_{i=1}^{A} n_{i,j} + k} \tag{1}$$

Where

A	alphabet size (4 for nucleic acids, 20 for peptides)
w	matrix width (=12 in the TRANSFAC matrix \$PHO4_01)
$n_{i,j}$	occurrences of residue i in column j of the matrix
p_i	prior residue probability for residue i
$f_{i,j}$	relative frequency of residue i at position j
k	pseudo weight (arbitrary, 1 in our example)
$f'_{i,j}$	corrected frequency of residue i at position j

Differences in residue composition can be taken into account by calculating a *weight* ($W_{i,j}$), which represents log ratio of observed frequency ($f_{i,j}$) and prior residue probability (p_i). In addition, a *pseudo-weight* (k) can be introduced to obtain a *corrected frequency f'ij* (Hertz and Stormo 1999). The reason for introducing a pseudo-weight is that the collections of known sites used to build the matrix are generally small. For example, the TRANSFAC matrix F\$PHO4_01 (Table 2A) was calculated from no more than 8 binding sites. At some positions of the matrix, some residues have a frequency of 0 (for example the T at position 4), using a (uncorrected) frequency of 0 would give a weight of $-\infty$, which amounts to consider as completely impossible for the factor to bind at such a position. However, the absence of this residue in our data set could either indicate that this residue hinders the factor binding, our that our current collection does not yet contain this variant for a simple reason of insufficient sampling. The introduction of the pseudo-weight resolves this problem pragmatically, since corrected frequencies cannot be null, and the weight can thus not be infinitely negative anymore. The problem is of course to estimate the importance assigned to the pseudo-weight (k) relative to the observed sites (n). A *weight matrix* (Table 2B) is derived from the occurrence matrix by calculating the weight of each residue at each position of the alignment. The weight matrix is used to assign, at each position of a sequence, a score reflecting the likelihood for the transcription factors to bind there (see chapter on *Pattern Matching*).

Table 2

A: occurrence matrix representing the binding specificity of the Pho4p transcription factor from *Saccharomyces cerevisiae* (source TRANSFAC F$PHO4_01). **B:** frequencies (corrected with a pseudo-weight of 1). **C:** Weights. Positive values are shadowed. **D:** information content. Positive values are shadowed

A: occurrences (counts)

Prior	Pos	1	2	3	4	5	6	7	8	9	10	11	12
0.325	A	1	3	2	0	8	0	0	0	0	0	1	2
0.175	C	2	2	3	8	0	8	0	0	0	2	0	2
0.175	G	1	2	3	0	0	0	8	0	5	4	5	2
0.325	T	4	1	0	0	0	0	0	8	3	2	2	2
1	Sum	8	8	8	8	8	8	8	8	8	8	8	8

B: frequencies

Prior	Pos	1	2	3	4	5	6	7	8	9	10	11	12
0.325	A	0.15	0.37	0.26	0.04	0.93	0.04	0.04	0.04	0.04	0.04	0.15	0.26
0.175	C	0.24	0.24	0.35	0.91	0.02	0.91	0.02	0.02	0.02	0.24	0.02	0.24
0.175	G	0.13	0.24	0.35	0.02	0.02	0.02	0.91	0.02	0.58	0.46	0.58	0.24
0.325	T	0.48	0.15	0.04	0.04	0.04	0.04	0.04	0.93	0.37	0.26	0.26	0.26
1	Sum	1	1	1	1	1	1	1	1	1	1	1	1

C: weights

Prior	Pos	1	2	3	4	5	6	7	8	9	10	11	12
0.325	A	-0.79	0.13	-0.23	-2.20	1.05	-2.20	-2.20	-2.20	-2.20	-2.20	-0.79	-0.23
0.175	C	0.32	0.32	0.70	1.65	-2.20	1.65	-2.20	-2.20	-2.20	0.32	-2.20	0.32
0.175	G	-0.29	0.32	0.70	-2.20	-2.20	-2.20	1.65	-2.20	1.19	0.97	1.19	0.32
0.325	T	0.39	-0.79	-2.20	-2.20	-2.20	-2.20	-2.20	1.05	0.13	-0.23	-0.23	-0.23
1	Sum	-0.37	-0.02	-1.02	-4.94	-5.55	-4.94	-4.94	-5.55	-3.08	-1.13	-2.03	0.186

D: information content

Prior	Pos	1	2	3	4	5	6	7	8	9	10	11	12
0.325	A	-0.12	0.05	-0.06	-0.08	0.97	-0.08	-0.08	-0.08	-0.08	-0.08	-0.12	-0.06
0.175	C	0.08	0.08	0.25	1.50	-0.04	1.50	-0.04	-0.04	-0.04	0.08	-0.04	0.08
0.175	G	-0.04	0.08	0.25	-0.04	-0.04	-0.04	1.50	-0.04	0.68	0.45	0.68	0.08
0.325	T	0.19	-0.12	-0.08	-0.08	-0.08	-0.08	-0.08	0.97	0.05	-0.06	-0.06	-0.06
1	Sum	0.111	0.087	0.356	1.294	0.803	1.294	1.294	0.803	0.609	0.392	0.465	0.037

Information content

The information content (Hertz and Stormo 1999) is obtained by multiplying the weight by the frequency (corrected by the pseudo weight).

$$I_{i,j} = f'_{i,j} \ln\left(\frac{f'_{i,j}}{p_i}\right) \qquad I_j = \sum_{i=1}^{A} I_{i,j} \qquad I_{matrix} = \sum_{j=1}^{w} \sum_{i=1}^{A} I_{i,j} \qquad (2)$$

The information content can be calculated for each cell of the matrix, and then summed over rows and column to obtain I_{matrix}, the total information content of the matrix. The total information content represents the discrimination between a binding site (represented by the matrix) and the background model. Pattern discovery programs such as *consensus* (Hertz et al. 1990) select a matrix by optimizing the information content.

The information content also provides an estimate for the upper limit of the expected frequency of the binding sites in random sequences (Hertz and Stormo 1999).

$$P(site) \leq e^{-I_{matrix}} \tag{3}$$

4. Pattern discovery

4.1. Introduction

The application of pattern discovery to predict regulatory motifs can be formulated in the following way: *given a set of functionally related genes, can we detect exceptional motifs in their upstream regions, which could be responsible for their co-regulation?* This problem became very popular during the last years, due to the increasing amount of data about functional grouping of genes. A first domain of application was for the interpretation of microarray data (DeRisi et al. 1997): starting from clusters of co-expressed genes, try to predict cis-acting elements potentially responsible for their co-regulation. The same approach can be applied to other data types such as protein complexes (Gavin et al. 2002; Ho et al. 2002), genes with similar phylogenetic profiles (Pellegrini et al. 1999), pairs of genes detected by the analysis of fusions/fission (Marcotte et al. 1999a; Marcotte et al. 1999b) (Enright et al. 1999).

The pattern discovery problem can be addressed by a variety of algorithmic approaches and statistical models. We will describe here some of these approaches, and illustrate them with selected test cases.

4.2. Study cases

A simple way to evaluate a pattern discovery software is to submit a set of sequences S which contain some known motif M_{known}. The sequence is given as input for the pattern discovery program, which returns a predicted motif M_{pred}. We then compare the predicted (M_{pred}) and known (M_{known}) motifs.

As test cases, we selected the target genes of a few transcription factors from the yeast *Saccharomyces cerevisiae* (van Helden et al. 1998).

Table 3

Test cases for pattern discovery: list of target genes for some well-characterized transcription factors from the yeast *Saccharomyces cerevisiae*

Set name	Transcription factor	Regulated genes	# genes	Description
PHO	Pho4p	PHO5; PHO8; PHO11; PHO84; PHO81	5	Activated under phosphate stress conditions
NIT	Gln3p	DAL5; GAP1; MEP1; MEP2; MEP3; PUT4; DAL80	7	Activated in response to some sources of nitrogen.
MET	Met4p	MET1; MET2; MET3; MET6; MET14; MET19; MET25; MET30; MUP3; SAM1; SAM2	11	Activated when methionine concentration is low.
GAL	Gal4p	GAL1; GAL2; GAL7; GAL80; MEL1; GCY1	6	Expressed when the yeast is fed with galactose.

5. String-based pattern discovery

5.1. Analysis of word occurrences

We saw in the chapter 0 that the consensus of the transcription factor Pho4p consists in a short sequence of conserved residues (CACGTKKK). This is also the case for many (but not all) other transcription factors: their binding sites share a common core, consisting in a set of 5–10 contiguous residues. Starting from this observation, a simple conceptual approach to pattern discovery is to analyze the occurrences of oligonucleotides in order to detect those having an exceptionally high frequency in this input set, by comparison with some background model.

We will illustrate this approach with the test groups described above. Results obtained with some additional data sets are described in the original publication (van Helden et al. 1998).

Estimation of expected frequencies

Expected occurrences were calculated on the basis of intergenic frequencies.

$$E(W) = F_{bg}(W) * T; T = s * (L - k + 1)$$

$E(W)$ expected number of occurrences for word W

$F_{bg}(W)$ background frequency of word W. This frequency is estimated by the intergenic frequencies of the same word.

W a given word (oligonucleotides)

k	word length (6 for hexanucleotides)
S	number of sequences in the set (5 in this case)
L	length of each sequence in the input set
T	possible positions for a k-letter word in the sequence set

Comparison of expected and observed frequencies

Figure 4 compares the expected (abscissa) and observed (ordinate) occurrences for hexanucleotides in the upstream sequences of the PHO genes.

Hexanucleotides were grouped by pairs of reverse complements, because in yeast, cis-acting elements are generally strand-insensitive. Each dot represents one pair of reverse complements (e.g. GATAAG|CTTATC for the NIT genes).

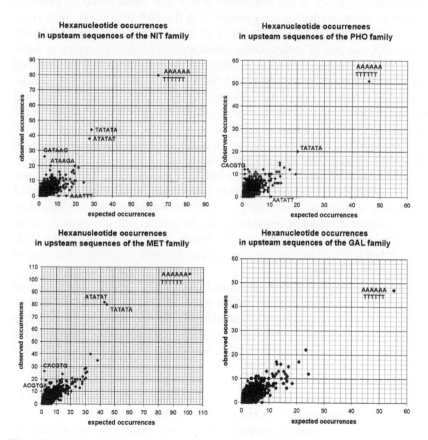

Fig. 4. Comparison of observed and expected hexanucleotide frequencies in the upstream sequences of groups of co-regulated genes. **A:** NIT group. **B:** PHO group. **C:** MET group. **D:** GAL group.

On these plots, most words align more or less on the diagonal, with some fluctuations. The fluctuations are more important for small groups (e.g. PHO, which contains 5 genes) than for larger groups (e.g. MET, 10 genes).

On each of these plots, the most frequent pair of words is AAAAAA|TTTTTT. The next most frequent words are usually TATATA and ATATA. These words cannot be considered as over-represented, since their observed and expected occurrences are similar. This illustrates the essential difference between frequent words and over-represented words: since these frequent words are the same for all the groups, their high frequency reflects some general property of yeast upstream sequences rather than the presence of group-specific regulatory signals.

Interesting words are thus not the most frequent ones, but those which are found more frequently in the considered group than what would be expected by chance, given our background model. On the plot (Figure 4), such over-represented words appear on the top left of the diagonal. For the NIT family (Figure 4A), one pair of reverse-complementary words clearly appears as separated from the diagonal: GATAAG|CTTATC. This hexanucleotide is the so-called GATA-box, which is bound by the GATA factors, involved in nitrogen regulation. For the MET family (Figure 4C), another hexanucleotide is clearly separated from the diagonal: CACGTG, a reverse-palindrome which corresponds to the consensus of the Met4p transcription factor. For the PHO family (Figure 4B), the plot is less obvious to interpret, due to the wider overall dispersion of the cloud around the diagonal. This lower signal-to-noise ratio is due to the small number of genes in the PHO family (5 members only). However, some words seem reasonably separated from the main diagonal. In particular, CACGTG is found in 12 occurrences, whereas no more than 2 occurrences would be expected according to the background model. Consistently, this hexanucleotide corresponds to the core of the high-affinity binding sites for Pho4p. For the last group, the GAL genes, all hexanucleotides seem to align on the diagonal, suggesting that none of them is over-represented.

The graphical representation shown in Figure 4 is useful to get an intuition about the principle of word-based pattern discovery, but the simple visual comparison of observed and expected frequencies is not very accurate for selecting over-represented patterns. We saw that the hexanucleotides discarding from the diagonal correspond to regulatory signals, but where should the limit be placed?

Measuring over-representation with a P-value
We proposed a very simple probabilistic model to calculate the statistical significance of over-representation (van Helden et al. 1998).

The sequence S of length L is considered as a succession of T positions from which starts a substring of size k (word length). Since the sequence is generally linear, the number of positions for a word W of length k is smaller than L, since the last $k - 1$ positions do not contain a full k-letter word.

$$T = L - k + 1$$

For circular sequences (e.g. plasmids, bacterial chromosomes) the end of the string is continuous with its beginning and that a substring can be extracted from each position, so that $T = L$.

If we focus on a given word W, we can consider the sequence as a series of T trials, each of which can either result in a success (the word found at this position is W) or in a failure (the word found at this position is not W). The probability to observe at least x successes (occurrences of the word W) in a succession of T trials can be calculated with the binomial probability.

$$Pvalue = P(X \geq x) = \sum_{i=x}^{T} \frac{T!}{i!(T-i)!} p^i (1 - p)^{T-i}$$

Assumptions for the binomial distributions
The binomial distribution assumes that the successive trials are independent from each other and that the probability to find a word is constant over the sequence. This assumption is not properly verified, since the presence of a word of length k depends on words found at the $k - 1$ preceding positions, and affects those found at the $k + 1$ successive positions. For example, if the word GATAAG is found at position i of sequence S, the only words that can be found at position $i + 1$ are ATAAGA, ATAAGC, ATAAGG and ATAAGT. There are thus short-term dependencies between successive words. However, when the sequence is much larger than the pattern length, and when the pattern is not self-overlapping, the hypothesis of independent positions is reasonably verified.

A notable exception to this assumption of independence is the case of self-overlapping words, like GGGGGG, TATATA, TAGTAG. Indeed, the first occurrence of such word will strongly increase the probability to find another occurrence at the following position (GGGGGG), or two (TATATA) or three (TAG-TAG) positions further. This problem has been addressed by several statisticians and several corrections have been proposed. For instance, Pevzner (Pevzner et al. 1989) defined a self-overlap coefficient, which can be used to correct the estimation of variance in Gaussian models. This model relies on a normality assumption, which is verified only if the expected number of occurrences is large ($\gg 10$). In our conditions, the expectation is typically small (often smaller than 1) and Gaussian models should be avoided. Schbath (Reinert and Schbath 1998; Schbath et al. 1995) uses a compound Poisson distribution to model occurrences of clumps of words (the first occurrence being followed by overlapping occurrences of the same word).

Another way to circumvent this problem is to exclude overlapping occurrences from the counts. When the word W is found at position i of sequence S, the next occurrences of W are ignored for positions $i + 1$ to $i + k - 1$. The binomial schema has to be corrected accordingly: if x occurrences of word W are found, the number of possible positions for this word become   .

$$T = L - k + 1 - x(k - 1) = L - (x + 1)(k - 1)$$

This counting mode might look like a tricky way to circumvent the problem of overlap, but it has some biological justification: the binding interface between the transcription factor and the DNA covers the whole word, and no other protein can bind simultaneously on the overlapping positions, even though the same word can be found in our string representation. We adopted this exclusion of mutually overlapping occurrences as default counting mode for web interface of the program *oligo-analysis* (van Helden 2003), but overlapping occurrences can also be counted if the user finds it appropriate according to his/her biological model.

From P-value to E-value

Another important issue is the number of words considered in a single analysis. Since the same test is simultaneously applied to all the words of the same size, the *P-value* has to be corrected for multi-testing. The number of considered words depends on the word length, and on the counting mode (regrouping or not the pairs of reverse complements). When occurrences are counted in a strand-sensitive way, there are $D = 4^k$ possible words of length k. For hexanucleotides, this makes $D = 4^6 = 4096$ possibilities. If occurrences are counted in a strand-insensitive way, each word is regrouped with its reverse complement. For odd values of k, the number of patterns is simply divided by two: $D = 4^k/2$. There are thus $4^5/2$ pairs of reverse-complementary pentanucleotides. For even values of k, the count of D is slightly more complicated. Indeed, reverse-palindromic words (e.g. CACGTG) will not be regrouped with another word. There are $4^{k/2}$ reverse-palindromes of size k (the second half of the word is determined by the first half). The total number of patterns is thus $D = (4^k - 4^{k/2})/2 + 4^{k/2} = (4^k + 4^{k/2})/2$.

A simple way to take multi-testing into account is to multiply the P-value by the number of tests (D), in order to obtain an E-value.

$$Evalue = Pvalue * D$$

The interpretation of the E-value is straightforward: it represents the expected number of false positive, given the P-value considered. For example, if we analyze hexanucleotides grouped by pairs of reverse complements and select a P-value threshold of 0.01, the *E-value* is $E = 2080 * 0.01 = 20.8$, indicating that we should expect 21 false positives. This level of false positive can be easily verified by submitting random sequences to the program.

The significance score

A significance score can further be calculated from the E-value.

$$sig = -\log_{10}(Evalue)$$

This significance is convenient to interpret the over-representation: the larger is the significance, the more over-represented is the pattern. When the threshold of significance is set to 0, one expects on the average one false positive among all the words analyzed for a sequence set. With a threshold of $sig = 1$, a false positive is expected every 10 sequence sets. With a threshold of $sig = s$, a false positive is expected every 10^s sequence sets.

Over-represented hexanucleotides in upstream sequences of the MET genes

Figure 5 shows the result returned by *oligo-analysis* for upstream sequences of the MET genes. Among the 2080 possible pairs of hexanucleotides, no more than 8 are statistically over-represented ($sig > 0$). The most significant word (CACGTG) corresponds to the core of the consensus for Met4p, the main regulatory of methionine metabolism in yeast. Among the 10 upstream sequences of the MET family, 9 contain at least one occurrence of this word (column *matching sequences*). In addition, some sequences contain multiple occurrences of this word, leading to a total count of 13 occurrences. The expected frequency, calculated on the whole set of yeast upstream sequences, is $F(W) = 0.000164$ occurrences/positions. This word has a very high significance ($sig = 5.08$), corresponding to a very low expected number of false positives (E-$value = 8.4e$-06).

Word pair	*F(W)*	*Match. Seq.*	*occ*	*E (W)*	*P-value*	*E-value*	*sig*	*Overlaps (discarded)*	*Rank*
CACGTG | CACGTG	0.000164	9	13	1.42	4e-09	8.4e-06	5.08	0	1
CCACAG | CTGTGG	0.000265	8	11	2.30	3e-05	6.2e-02	1.21	0	2
ACGTGA | TCACGT	0.000368	9	13	3.19	3e-05	6.3e-02	1.20	6	3
AACTGT | ACAGTT	0.000610	10	17	5.28	3.8e-05	8.0e-02	1.10	0	4
ACTGTG | CACAGT	0.000374	9	12	3.24	0.00015	3.0e-01	0.52	0	5
GCTTCC | GGAAGC	0.000421	7	12	3.65	0.00042	8.6e-01	0.06	0	6
GCCACA | TGTGGC	0.000307	7	10	2.66	0.00045	9.4e-01	0.03	0	7
AGTCAT | ATGACT	0.000489	8	13	4.24	0.00046	9.6e-01	0.02	0	8

Fig. 5. Significant hexanucleotides in the upstream sequences of PHO genes.

The other selected words are much less significant, but we will see that another criterion suggest that they might be relevant.

Assembling words to describe more complex patterns

The 8 words selected in Figure 5 present some relationships, because some of them are mutually overlapping. For example, CACGTG can be assembled with ACGTGA to form the heptanucleotide CACGTGA. This heptanucleotide can in turn be assembled with TCACGT (the reverse complement of ACGTGA), to form the octanucleotide TCACGTGA. Among the remaining words, we also find another group of mutually overlapping words: CCACAG, CACAGT (reverse complement of ACTGTG), ACAGTT, . . .

The program called *pattern-assembly (van Helden 2003)* automatically assembles this type of patterns. The result of this assembly is shown in Figure 6.

The assembly of the 8 hexanucleotides returns two larger patterns. The first pattern (TCACGTGA, a reverse palindrome) corresponds to the binding site of Met4p, the main regulator of methionine metabolism in the yeast *Saccharomyces cerevisiae* (Thomas and Surdin-Kerjan 1997). The second pattern (GC-CACAGTT|AACTGTGGC) is bound by a pair of homologous transcription factors, Met31p and Met32p, also involved in the regulation of methionine (Blaiseau et al. 1997). The two last hexanucleotides, GCTTCC and AGTCAT, cannot be included in an assembly. Given their low level of significance, these words are likely to be false positive.

```
;cluster # 1        seed: CACGTG      3 words      length
TCACGT..        ..ACGTGA    1.20
.CACGTG.        .CACGTG.    5.08
..ACGTGA        TCACGT..    1.20
TCACGTGA        TCACGTGA    5.08  best consensus


;cluster # 2        seed: CCACAG      4 words      length 8
GCCACA...       ...TGTGGC   0.03
.CCACAG..       ..CTGTGG.   1.21
..CACAGT.       .ACTGTG..   0.52
...ACAGTT       AACTGT...   1.10
GCCACAGTT       AACTGTGGC   1.21  best consensus


; Isolated patterns: 2
GCTTCC          GGAAGC      0.06
AGTCAT          ATGACT      0.02
```

Fig. 6. Assembly of the significant hexanucleotides selected from the MET upstream sequences.

5.2. Analysis of dyad occurrences (spaced pairs of words)

A frustrating case: the GAL regulon

In this chapter, we only discussed a few examples, but the same analysis has been performed for other groups of co-regulated genes with similar results. Despite its conceptual simplicity, the program *oligo-analysis* was shown to return remarkably good results with most (but not all) yeast regulons (van Helden et al. 1998). However the analysis of oligonucleotides fails to detect the binding motif for Gal4p, and returns a negative answer: on the observed/expected frequency plot (Figure 4D), all the words align onto the diagonal. Consistently, the binomial test indicates that none of the 2080 words (grouped by pairs of reverse complement) is significantly over-represented. The failure of the program to detect the GAL-specific binding motif is particularly frustrating, since Gal4p is one of the bet characterized transcription factors in the yeast. The reason for this failure is pretty trivial: Gal4p forms a dimer, and each unit enters in contact with DNA over a few nucleotides (Figure 2C,D). The two contact points are separated by a spacing of fixed width (11bp for Gal4p), but variable content. The binding specificity is restricted to 3–4 nucleotides on each side of the spacing. One possibility would be to reduce the size of oligonucleotides, but the random expectation of trinucleotides is already quite high, so that the trinucleotides involved in the contact points of the binding sites will not be detected as significant. Another approach has been to develop a specific approach to detect over-represented pairs as a whole, as explained in the next chapter.

Analysis of spaced patterns with dyad-analysis

Spaced patterns are commonly found in transcription factor binding sites. This type of motifs are typical of some families of transcription factors, for example the fungal Zinc cluster proteins or the bacterial Helix-Turn-Helix (HTH) factors. As discussed above, word-based pattern discovery fails to detect such patterns (Figure 4D). This represents a serious inconvenient, since no less than 56 Zinc cluster proteins have been identified in the yeast genome, and in the bacteria *Escherichia coli*, most transcription factor belong to the HTH family.

In order to directly address this type of motifs, we developed a specific program, *dyad-analysis* (van Helden et al. 2000), which counts the number of occurrences of all possible spaced pairs, and compares expected and observed. Figure 7 shows the comparison of observed and expected frequencies for all pairs of trinucleotides, with all possible spacings between 0 and 16, in upstream sequences of the GAL genes. Expected frequencies were estimated as above, by counting dyad frequencies in the whole set of yeast upstream sequences (background model). As in the previous plots, most dots are more or less aligned onto the diagonal, but one dyad ($CCGn_{11}CGG$) appears clearly separated. This dyad

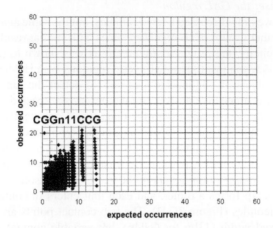

Fig. 7. Observed versus expected dyads in upstream sequences of the GAL genes.

dyad_identifier	F(W)	Occ	E(W)	P-value	E-value	sig	Rank	Ovl
CGGn$_{11}$CCG | CGGn$_{11}$CCG	0.0000662	20	0.60	2e-12	8.9e-08	7.05	1	2
CGGn$_{12}$CGA | TCGn$_{12}$CCG	0.0000621	10	0.58	8.6e-10	3.7e-05	4.43	2	2
CGGn$_{10}$TCC | GGAn$_{10}$CCG	0.0000687	10	0.64	2.2e-09	9.8e-05	4.01	3	3
CCGn$_{01}$GCG | CGCn$_{01}$CGG	0.0000533	6	0.50	1.6e-05	6.8e-01	0.17	4	0
CCGn$_{12}$CCG | CGGn$_{12}$CGG	0.0000545	6	0.51	1.8e-05	7.7e-01	0.11	5	0
AGAn$_{05}$CCG | CGGn$_{05}$TCT	0.0001153	8	1.08	2e-05	8.8e-01	0.06	6	0

Fig. 8. Statistically significant dyads in upstream sequences of the GAL genes.

corresponds to the two contact points of the interface between the Gal4p protein and its binding site (Figure 2).

We can now apply the binomial statistics as we did above for hexanucleotides. Figure 8 shows the statistically significant spaced pairs returned by the program *dyad-analysis*. In this analysis, we considered all possible pairs of trinucleotides separated by a spacing comprised between 0 and 20. In total, the number of possible dyads is $D = 21 * 4^3 * 4^3 = 86,016$, but we regrouped them by pairs of reverse complements, so that the total number is $D = 43,680$ (taking into account the number of reverse palindromes as above). Among these, no more than 6 dyads are significantly over-represented ($sig > 0$).

The most significant pattern is CGGn$_{11}$CCG|CGGn$_{11}$CCG, which appeared as the dot most distant from the diagonal in the observed/expected plot (Figure 8), and corresponds to the core of the Gal4p binding site. Several of the

```
;cluster # 1      seed: CGGnnnnnnnnnnnCCG 5 words      length
;           alignt                     rev_cpl      score
CCGnnnnnnnnnnnnCCG.      .CGGnnnnnnnnnnnnnCGG        0.11
TCGnnnnnnnnnnnnCCG.      .CGGnnnnnnnnnnnnnCGA        4.43
.CGGnnnnnnnnnnnnCCG.     .CGGnnnnnnnnnnnnCCG.        7.05
.CGGnnnnnnnnnnnnCGA      TCGnnnnnnnnnnnnnCCG.        4.43
.CGGnnnnnnnnnnTCCu..     ..GGAnnnnnnnnnnnCCG.        4.01
..GGAnnnnnnnnnnCCG.      .CGGnnnnnnnnnnnTCC..        4.01
TCGGAnnnnnnnnnTCCGA     TCGGAnnnnnnnnnnTCCGA        7.05   best consensus

; Isolated patterns: 2
;     alignt            rev_cpl score
CCGnGCG     CGCnCGG       0.17
AGAnnnnnCCG        CGGnnnnnTCT 0.06
```

Fig. 9. Assembly of the statistically significant dyads detected in upstream sequences of the GAL genes.

other selected dyads strongly overlap with this pattern. One can for example assemble $CGGn_{11}CCG$, $CGGn_{10}TCC$ and $CGGn_{12}CGA$ to form a larger pattern $CGGn_{12}TCCGA$. In addition, the core of the motif is reverse palindromic, and the reverse complements of the additional dyads can be included in the assembly as well (Figure 9). The resulting consensus is **TCGGAn$_8$TCCGA**.

We should keep in mind that the assembled motif is a simplification, compared to the collection of dyads. Indeed, the central dyad $CCGn_{11}CCG$ is more significant than the overlapping ones, suggesting that this might be the core of the binding interface. Searching for the complete consensus $TCGGAn_8TCCGA$ would result in the loss of some functionally active sites, because the flanking bases (T before CGG and A after it) may be present in some cis-acting elements, but absent in other ones. In order to predict the location of putative binding site, we will thus keep the collection of patterns (words or dyads) and the score associated to each of these, as illustrated in the chapter on string-based pattern matching. Besides the 3 dyads involved in the assembly, two isolated dyads are also selected. Their level of significance is however very low (0.17 and 0.06, respectively) and these are likely to be false positive.

5.3. Strengths and weaknesses of word- and dyad-based pattern discovery

Advantages

1. **Computational efficiency.** The computation time increases linearly with size of the input set. It can thus be applied to large sequence sets (e.g. complete genomes can be analyzed in a few minutes).

2. **Detection of under-represented patterns.** The same type of statistics can be applied to detect under-represented motifs, which can reveal a selective

pressure for the avoidance of some functional elements. Mathias Vandenbogaert (Vandenbogaert and Makeev 2003) applied word-counting approaches to detect under-represented hexanucleotides in different bacterial genomes, and showed that the most significantly under-represented motifs correspond to restriction sites.

3. **Exhaustivity.** Given the relatively small number of possible solutions ($D_W = 4^k$ for oligonucleotides of size k, $D_D = (s + 1) * 4^{2k}$ for dyads of length k with spacings between 0 and s), it is easy to calculate the P-value for each of these, and to systematically return all the over- or under-represented patterns.

4. **Ability to return negative answers.** The calculation of the P-value and, even better, of the E-value, allows to define significance thresholds and interpret these thresholds in terms of expected rate of false positive.

Weaknesses

1. **Treatment of variable residues.** A classical criticism addressed to string-based pattern discovery is that the resulting patterns (words and dyads) poorly reflect the degeneracy of the motif. In some cases (such as the PHO family above), the set of words partly reflects the degeneracy of the motif (it contains both the CACGTG and CACGTT words, as well as their surroundings). However, this is a case where the motif has two clearly distinct variants. Some motifs with a higher degree of degeneracy can be missed by the method, because none of the possible variants is significant alone.

2. **Pattern matching.** Pattern discovery is generally followed by pattern matching, i.e. trying to identify the positions of the discovered patterns, in order to predict putative regulatory elements. It is easy to detect the positions of the significant words and dyads obtained by the above methods, but most of their occurrences will not really correspond to motifs. Indeed, each word or dyad generally reflects only a fragment of the motif, but it is also expected to occur in other places of the sequence.

6. String-based pattern matching

A simple string-based pattern matching generally gives poor predictions for transcription binding sites, for the obvious reasons that a single string-based representations fails to capture the probabilistic aspect of binding site variability, as discussed above.

The results can however be improved by matching a collection of mutually overlapping patterns (word or regular expressions), instead of a single regular expression. Multiple patterns can be used to represent overlapping fragments of a larger binding site, or the variants arising from the degeneracy of the consensus. Collections of mutually overlapping patterns can also be used to match complex motifs with higher order dependencies between neighbouring positions. For example, the following combination of words: CACGTG, ACGTGG, CGTGGG, CACGTT and ACGTTT, would capture the two variants of Pho4p binding sites (CACGTGGG and CACGTGTTT), but not the mixtures of G and T after the binding core. Such collections of mutually overlapping words are typically detected with string-based pattern discovery approaches, as we will see below. The matching can also be improved by assigning a weight to different patterns of a collection. This allows one to distinguish the strongly constrained core of the binding site (e.g. CACGTG, CACGTT) from the flanking positions, which are more degenerated (CACGTGgg, CACGTTtt). The result of such a search can be represented graphically on a feature-map (Figure 10). Annotated binding sites

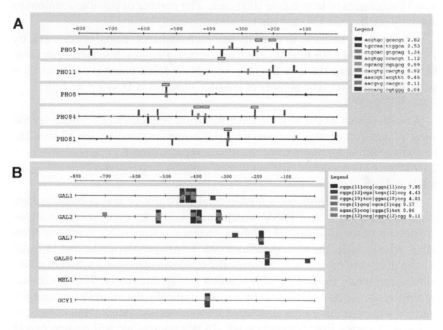

Fig. 10. Feature-map of pattern matching with a collection of words (A) and dyads (B). A specific weight was assigned to each pattern according to its significance in pattern discovery. **A:** over-represented hexanucleotides in upstream sequences of the PHO genes. The wider grey boxes above and below the maps indicate experimentally proven binding sites for the factor Pho4p. **B:** over-represented dyads in upstream sequences of the GAL genes.

(green horizontal boxes) are generally denoted by a clump of mutually overlapping hexanucleotides belonging to the collection of predicted patterns.

Another possible refinement of string-based pattern matching is to allow a certain number of substitutions (mismatches). This possibility is however generally not recommended, since it would consider as equivalent any substitution at any position of the pattern. This does not correspond to the typical DNA-protein interfaces, which impose some strong constraints on specific positions, whereas other positions may show some variability. This type of position-specific variability is typically treated by matrix-based pattern matching.

7. Matrix-based pattern discovery

Let us consider a simple case: we want to build a matrix of width $w = 10$ with $n = 12$ sequences of length $L = 1000$ each. The number of possible solutions to this very small-sized test case can be estimated easily.

A first option would be to consider that each sequence should contain exactly 1 site, on either of both strands (direct or reverse). From each sequence, we need to select one among the $T = 2 * (L - w + 1) = 1,982$ possible positions for a substring of size 10. The number of possible matrices is $D = T^n = 1,982^{12} = 3.67e + 39$.

Another option would be to consider that some sequences can contain several sites, whereas other might not contain a single site. In this case, the 12 sites can be chosen within the whole set of sequences, representing $T = 2n(L - w + 1) = 23,784$ possible positions. The number of possible matrices is $C_T^n = C_{23,784}^{12} = 6.82e + 43$.

This estimation illustrates a fundamental difficulty of matrix-based pattern discovery: the number of PSSM which could be made, even from a small sequence set, raises astronomical numbers, so that it is impossible to analyze them all in order to select the most significant one. Consequently, all the matrix-based pattern discovery programs are intrinsically condemned to scan a subset of possibilities, and return the best possible solution among this subset. The "goodness" of a matrix is generally estimated by a score (typically the information content). Various strategies have been developed to optimize the information content of a matrix extracted from a sequence set. In this course, we will present two of these strategies: a greedy algorithm developed by Hertz and Stormo (Hertz et al. 1990; Hertz and Stormo 1999; Stormo and Hartzell 1989), and a gibbs sampling algorithm originally developed by Newald and Lawrence (Lawrence et al. 1993; Neuwald et al. 1995; Neuwald et al. 1997).

7.1. Consensus: a greedy approach

A greedy algorithm has been implemented by Jerry Hertz (Hertz et al. 1990; Hertz and Stormo 1999; Stormo and Hartzell 1989) in a program named *consensus*. The principle is to start the matrix with two sequences only, and to incorporate the other sequences one by one. At each step, a subset of matrices with the highest information content are retained for the next iteration.

If the sequences have a length of, say $L = 1000$ and a matrix of width $w = 10$, there are $T = L - w + 1 = 991$ possible sites in each sequence, and thus $T^2 = 982,081$ possible matrices made of one site from the first sequence and one site from the second sequence.

Figure 11 illustrates the result returned by the program *consensus* with upstream sequences of the PHO genes. The three top motifs are actually very similar to each other, and they match the high-affinity binding site of Pho4p (CACGTGGG). The program failed to detect medium affinity variants (CACGTTtt). An important feature of *consensus* is that a P-value and an E-value (expected frequency) are calculated for each matrix. The E-value is very informative, since it corrects the P-value for multi-testing (as discussed above), by taking into account the number of matrices analyzed. The E-value indicates the number of false positives expected for a given P-value. For the top motif (described under MATRIX 1 in Figure 11), the P-value is very low (4.03e-18) but the E-value is 0.02 indicating that such a level of significance would be expected 2 times out of 100 random analyses. In this case, the E-value is still low, and the motif can be considered as significant.

7.2. Gibbs sampling

The *gibbs* program was initially developed to discover motifs in sets of unaligned protein sequences (Lawrence et al. 1993; Neuwald et al. 1995; Neuwald et al. 1997). In short, the *gibbs* sampler is a stochastic version of the Expectation-Maximization (EM) algorithm. To initialize the program, a PSSM is built from a set of random sites collected from the input sequence. At this stage, the matrix is thus not expected to contain any specific information. After this initialization, the program iterates between a *sampling* step and a *predictive update*. During the *sampling* step, a score is assigned to each position of the input set. A random position is selected at random, with probabilities proportional to the score. During the *predictive update* step, the selected site is integrated in the matrix, from which another site is removed.

Since the initial positions were chosen at random, the initial matrix is not supposed to contain any information. During the subsequent sampling step, the scores are thus not very informative, and the selection of the next site is mainly random. During a certain number of iterations, the information content of the

```
MATRIX 1
number of sequences = 5
unadjusted information = 12.264
sample size adjusted information = 28.1942
ln(p-value) = -40.0503    p-value = 4.03996E-18
ln(expected frequency) = -3.91122   expected frequency = 0.0200161
A |    1    2    0    5    0    0    0    0    0    0
C |    3    0    5    0    5    0    0    0    1    2
G |    0    3    0    0    0    5    0    5    4    3
T |    1    0    0    0    0    0    5    0    0    0
   1|1   :    1/546    CACACGTGGG
   2|2   :    2/516    CACACGTGGG
   3|5   :   -3/265    TGCACGTGGC
   4|3   :    4/385    AGCACGTGGG
   5|4   :   -5/455    CGCACGTGCC

MATRIX 2
number of sequences = 5
unadjusted information = 12.2136
sample size adjusted information = 28.1438
ln(p-value) = -39.6863    p-value = 5.81381E-18
ln(expected frequency) = -3.54722   expected frequency = 0.0288047
A |    0    2    0    5    0    0    0    0    0    0
C |    4    0    5    0    5    0    0    0    1    2
G |    0    3    0    0    0    5    0    4    4    3
T |    1    0    0    0    0    0    5    1    0    0
   1|1   :    1/546    CACACGTGGG
   2|2   :    2/516    CACACGTGGG
   3|5   :   -3/265    TGCACGTGGC
   4|3   :    4/212    CGCACGTTGG
   5|4   :   -5/455    CGCACGTGCC

MATRIX 3
number of sequences = 5
unadjusted information = 12.0546
sample size adjusted information = 27.9848
ln(p-value) = -38.5478    p-value = 1.8151E-17
ln(expected frequency) = -2.40873   expected frequency = 0.0899295
A |    1    2    0    5    0    0    0    0    0    0
C |    2    0    5    0    5    0    0    0    1    1
G |    1    3    0    0    0    5    0    5    4    4
T |    1    0    0    0    0    0    5    0    0    0
   1|1   :    1/546    CACACGTGGG
   2|2   :    2/516    CACACGTGGG
   3|5   :   -3/265    TGCACGTGGC
   4|3   :    4/385    AGCACGTGGG
   5|4   :    5/455    GGCACGTGCG
```

Fig. 11. The 3 matrices with the highest information content detected by the program consensus in upstream sequences of the PHO genes.

matrix remains thus quite low. However, if, by chance, an occurrence of the motif is incorporated in the matrix during a sampling step, it will slightly bias the next sampling step in favour of other occurrences of the same motif. And if, due to this slight bias, a second occurrence is incorporated, the bias will be reinforced. The sampler thus tends to incorporate a third, then a fourth, . . . occurrence of the motif, and the sampler rapidly converges towards a PSSM with high information content.

Although the original gibbs sampler (Lawrence et al. 1993; Neuwald et al. 1995; Neuwald et al. 1997) was already able to analyze DNA sequences, it had not been optimized for this task. Given the remarkable results obtained with this approach on proteins and DNA sequences, several other groups implemented their own version of a DNA-dedicated *gibbs* sampler, with various improvements:

1. Possibility to search patterns on boths strands.

2. Possibility to search multiple motifs, with iterative masking (sites used in a motif cannot be re-used for a subsequent motif).

3. Calculation of additional scores (information content, MAP, . . .)

4. Background models based on Markov chains of arbitrary order.

Figure 12 illustrates the result obtained with Gert Thijs' MotifSampler (Thijs et al. 2001) on upstream sequences of the PHO genes. For this analysis, we used a Markov chain of order 5. Actually, this is equivalent to a calibration of expected frequencies based on hexanucleotides frequencies. Motifs were searched on both strands, with a width of 10 bp. For each motif, the program returns the consensus, followed by a frequency matrix (the frequency matrix is presented vertically: rows correspond to positions, columns to residues). The top motif (consensus ACGTGCnnmn) matches the PHO4p consensus (CACGTKkk), but it is shifted rightwards, so that the beginning of the motif is missing. The second motif (CsCACGTknk) has a weaker score, but it is better centred, and it reflects the degeneracy of the right side of the Pho4p consensus (CACGTG or CACGTK).

7.3. Strengths and weaknesses of matrix-based pattern discovery

Matrix-based pattern discovery presents the advantage of returning a probabilistic description of motif degeneracy: the matrix indicates the frequency of each residue at each position of the motif. The main difficulty is in the choice of appropriate parameters: most programs require for the user to specify the matrix width, and the expected number of site occurrences. Since this information is typically

```
#INCLUSive Motif Model v1.0
#
#ID = box_1_1_ACGTGCnnmn
#Score = 41.4
#W = 10
#Consensus = ACGTGCnnmn
0.980384              0.0053098         0.00515859        0.00914785
0.00933481            0.976359          0.00515859        0.00914785
0.00933481            0.0053098         0.976208          0.00914785
0.138808              0.134783          0.00515859        0.72125
0.00933481            0.0053098         0.976208          0.00914785
0.00933481            0.717412          0.264105          0.00914785
0.268281              0.0053098         0.523051          0.203358
0.527228              0.0053098         0.328842          0.138621
0.591964              0.393729          0.00515859        0.00914785
0.203545              0.0053098         0.199368          0.591777


#ID = box_1_2_CsCACGTknk
#Score = 28.7803
#W = 10
#Consensus = CsCACGTknk
0.205241              0.773606          0.00762748        0.013526
0.109522              0.390728          0.390505          0.109245
0.0138024             0.965044          0.00762748        0.013526
0.970995              0.00785106        0.00762748        0.013526
0.0138024             0.965044          0.00762748        0.013526
0.0138024             0.00785106        0.964821          0.013526
0.0138024             0.00785106        0.00762748        0.970719
0.0138024             0.00785106        0.486224          0.492123
0.0138024             0.19929           0.199066          0.587842
0.0138024             0.19929           0.390505          0.396403


#ID = box_1_3_GCTGnTnTTs
#Score = 9.30447
#W = 10
#Consensus = GCTGnTnTTs
0.0152634             0.0086821         0.961097          0.0149577
0.121115              0.855493          0.00843485        0.0149577
0.0152634             0.0086821         0.00843485        0.96762
0.0152634             0.0086821         0.961097          0.0149577
0.332817              0.537939          0.00843485        0.120809
0.0152634             0.0086821         0.00843485        0.96762
0.226966              0.114533          0.643543          0.0149577
0.121115              0.0086821         0.00843485        0.861768
0.0152634             0.0086821         0.00843485        0.96762
0.0152634             0.432087          0.43184           0.120809
```

Fig. 12. 3 top motifs discovered in upstream sequences of the PHO genes with MotifSampler. The Markov order of order 5 was generated with all the yeast upstream sequences. The program was used with the following options: MotifSampler –f PHO_up800.fasta –b mkv5_yeast_allup800_noorf.txt –s 1 –n 3 –w 10 –x 1 –r 1.

not provided, the user has to make guesses, or to try various possibilities and select the most convincing result.

The greedy approach, implemented in the program *consensus,* returns good results (at least with microbial data sets used in our tests), but is sensitive to the order of the sequences in the input set. If, for some reason, the first sequence does not contain any occurrence of the motif, the program will not be able to recover it subsequently.

One advantage of the gibbs sampler is time efficiency: large sequence sets can be treated in a few seconds. In comparison to the EM algorithm, the gibbs sampler shows a better ability to avoid suboptimal solutions (local optima), due to the stochastic sampling. A drawback of this is that independent runs of the program are expected return different motifs, even if the same input sequence has been analyzed with the same parameters. The program can easily be stuck in suboptimal solutions, like AT-rich motifs. The choice of a higher order Markov model is essential to reduce this effect.

8. Concluding remarks

The aim of this chapter was to give a short introduction to the prediction of regulatory signals in non-coding sequences. This introduction is incomplete and biased. Incomplete because a whole book would be necessary to describe the multitude of approaches developed to detect motifs in biological sequences. Biased because I deliberately placed a stronger emphasis on string-based pattern discovery approaches, firstly because these are conceptually simpler and secondly because, as developer of two of them, I know them better.

Since a few years, the decryption of regulatory signals has been recognized as a major challenge to interpret genome information, and many researchers have joined the field. Besides the methodological issues (which algorithm should be chosen, with which parameters, etc.), the availability of an increasing number of genomes has opened the door to a perspective which was out of reach no more than 5 years ago: applying comparative genomics to understand the evolution of gene regulation. This perspective is particularly exciting for higher organisms, since morphological differences are probably to be found in gene regulation rather than in protein structures themselves. But we are far from there: if some pattern discovery methods return decent results with sets of co-regulated genes from microbial organisms, the detection of signals in mammalian genomes is still in its infancy, and the rates of false positives are currently so high that the results are barely interpretable. There is no doubt that the future will be paved of exciting developments and discoveries for bioinformaticians willing to face this challenge.

9. Practical sessions

A series of tutorials and exercises can be found at http://rsat.scmbb.ulb.ac.be/rsat/.

10. Appendices

10.1. IUPAC ambiguous nucleotide code

Table 4

IUPAC AMBIGUOUS NUCLEOTIDE CODE

A	A	Adenine
C	C	Cytosine
G	G	Guanine
T	T	Thymine
R	A or G	puRine
Y	C or T	pYrimidine
W	A or T	Weak hydrogen bonding
S	G or C	Strong hydrogen bonding
M	A or C	aMino group at common position
K	G or T	Keto group at common position
H	A, C or T	not G
B	G, C or T	not A
V	G, A, C	not T
D	G, A or T	not C
N	G, A, C or T	aNy

References

Blaiseau, P.L., A.D. Isnard, Y. Surdin-Kerjan, and D. Thomas. 1997. Met31p and Met32p, two related zinc finger proteins, are involved in transcriptional regulation of yeast sulfur amino acid metabolism. *Mol Cell Biol* 17: 3640–3648.

DeRisi, J.L., V.R. Iyer, and P.O. Brown. 1997. Exploring the metabolic and genetic control of gene expression on a genomic scale. *Science* 278: 680–686.

Enright, A.J., I. Iliopoulos, N.C. Kyrpides, and C.A. Ouzounis. 1999. Protein interaction maps for complete genomes based on gene fusion events. *Nature* 402: 86–90.

Gavin, A.C., M. Bosche, R. Krause, P. Grandi, M. Marzioch, A. Bauer, J. Schultz, J.M. Rick, A.M. Michon, C.M. Cruciat, M. Remor, C. Hofert, M. Schelder, M. Brajenovic, H. Ruffner, A. Merino, K. Klein, M. Hudak, D. Dickson, T. Rudi, V. Gnau, A. Bauch, S. Bastuck, B. Huhse, C. Leutwein,

M.A. Heurtier, R.R. Copley, A. Edelmann, E. Querfurth, V. Rybin, G. Drewes, M. Raida, T. Bouwmeester, P. Bork, B. Seraphin, B. Kuster, G. Neubauer, and G. Superti-Furga. 2002. Functional organization of the yeast proteome by systematic analysis of protein complexes. *Nature* 415: 141–147.

Hertz, G.Z., G.W.d. Hartzell, and G.D. Stormo. 1990. Identification of consensus patterns in unaligned DNA sequences known to be functionally related. *Comput Appl Biosci* 6: 81–92.

Hertz, G.Z. and G.D. Stormo. 1999. Identifying DNA and protein patterns with statistically significant alignments of multiple sequences. *Bioinformatics* 15: 563–577.

Ho, Y., A. Gruhler, A. Heilbut, G.D. Bader, L. Moore, S.L. Adams, A. Millar, P. Taylor, K. Bennett, K. Boutilier, L. Yang, C. Wolting, I. Donaldson, S. Schandorff, J. Shewnarane, M. Vo, J. Taggart, M. Goudreault, B. Muskat, C. Alfarano, D. Dewar, Z. Lin, K. Michalickova, A.R. Willems, H. Sassi, P.A. Nielsen, K.J. Rasmussen, J.R. Andersen, L.E. Johansen, L.H. Hansen, H. Jespersen, A. Podtelejnikov, E. Nielsen, J. Crawford, V. Poulsen, B.D. Sorensen, J. Matthiesen, R.C. Hendrickson, F. Gleeson, T. Pawson, M.F. Moran, D. Durocher, M. Mann, C.W. Hogue, D. Figeys, and M. Tyers. 2002. Systematic identification of protein complexes in Saccharomyces cerevisiae by mass spectrometry. *Nature* 415: 180–183.

Huerta, A.M., H. Salgado, D. Thieffry, and J. Collado-Vides. 1998. RegulonDB: a database on transcriptional regulation in Escherichia coli. *Nucleic Acids Res* 26: 55–59.

Lawrence, C.E., S.F. Altschul, M.S. Boguski, J.S. Liu, A.F. Neuwald, and J.C. Wootton. 1993. Detecting subtle sequence signals: a Gibbs sampling strategy for multiple alignment. *Science* 262: 208–214.

Marcotte, E.M., M. Pellegrini, H.L. Ng, D.W. Rice, T.O. Yeates, and D. Eisenberg. 1999a. Detecting protein function and protein–protein interactions from genome sequences. *Science* 285: 751–753.

Marcotte, E.M., M. Pellegrini, M.J. Thompson, T.O. Yeates, and D. Eisenberg. 1999b. A combined algorithm for genome-wide prediction of protein function. *Nature* 402: 83–86.

Neuwald, A.F., J.S. Liu, and C.E. Lawrence. 1995. Gibbs motif sampling: detection of bacterial outer membrane protein repeats. *Protein Sci* 4: 1618–1632.

Neuwald, A.F., J.S. Liu, D.J. Lipman, and C.E. Lawrence. 1997. Extracting protein alignment models from the sequence database. *Nucleic Acids Res* 25: 1665–1677.

Oshima, Y., N. Ogawa, and S. Harashima. 1996. Regulation of phosphatase synthesis in Saccharomyces cerevisiae–a review. *Gene* 179: 171–177.

Pellegrini, M., E.M. Marcotte, M.J. Thompson, D. Eisenberg, and T.O. Yeates. 1999. Assigning protein functions by comparative genome analysis: protein phylogenetic profiles. *Proc Natl Acad Sci USA* 96: 4285–4288.

Pevzner, P.A., M. Borodovsky, and A.A. Mironov. 1989. Linguistics of nucleotide sequences. I: The significance of deviations from mean statistical characteristics and prediction of the frequencies of occurrence of words. *J Biomol Struct Dyn* 6: 1013–1026.

Reinert, G. and S. Schbath. 1998. Compound Poisson and Poisson process approximations for occurrences of multiple words in Markov chains. *J Comput Biol* 5: 223–253.

Salgado, H., A. Santos-Zavaleta, S. Gama-Castro, D. Millan-Zarate, E. Diaz-Peredo, F. Sanchez-Solano, E. Perez-Rueda, C. Bonavides-Martinez, and J. Collado-Vides. 2001. RegulonDB (version 3.2): transcriptional regulation and operon organization in Escherichia coli K-12. *Nucleic Acids Res* 29: 72–74.

Schbath, S., B. Prum, and E. de Turckheim. 1995. Exceptional motifs in different Markov chain models for a statistical analysis of DNA sequences. *J Comput Biol* 2: 417–437.

Stormo, G.D. and G.W.d. Hartzell. 1989. Identifying protein-binding sites from unaligned DNA fragments. *Proc Natl Acad Sci USA* 86: 1183–1187.

Thijs, G., M. Lescot, K. Marchal, S. Rombauts, B. De Moor, P. Rouze, and Y. Moreau. 2001. A higher-order background model improves the detection of promoter regulatory elements by Gibbs sampling. *Bioinformatics* 17: 1113–1122.

Thomas, D. and Y. Surdin-Kerjan. 1997. Metabolism of sulfur amino acids in Saccharomyces cerevisiae. *Microbiol Mol Biol Rev* 61: 503–532.

van Helden, J. 2003. Regulatory sequence analysis tools. *Nucleic Acids Res* 31: 3593–3596.

van Helden, J., B. Andre, and J. Collado-Vides. 1998. Extracting regulatory sites from the upstream region of yeast genes by computational analysis of oligonucleotide frequencies. *J Mol Biol* 281: 827–842.

van Helden, J., A.F. Rios, and J. Collado-Vides. 2000. Discovering regulatory elements in non-coding sequences by analysis of spaced dyads. *Nucleic Acids Res* 28: 1808–1818.

Vandenbogaert, M. and V. Makeev. 2003. Analysis of bacterial RM-systems through genome-scale analysis and related taxonomy issues. *In Silico Biol* 3: 127–143.

Wingender, E. 2004. TRANSFAC, TRANSPATH and CYTOMER as starting points for an ontology of regulatory networks. *In Silico Biol* 4: 55–61.

Wingender, E., P. Dietze, H. Karas, and R. Knuppel. 1996. TRANSFAC: a database on transcription factors and their DNA binding sites. *Nucleic Acids Res* 24: 238–241.

Zhu, J. and M.Q. Zhang. 1999. SCPD: a promoter database of the yeast Saccharomyces cerevisiae. *Bioinformatics* 15: 607–611.

Course 9

A SURVEY OF GENE CIRCUIT APPROACH APPLIED TO MODELLING OF SEGMENT DETERMINATION IN FRUIT FLY

M.G. Samsonova[1], A.M. Samsonov[2], V.V. Gursky[2] and
C.E. Vanario-Alonso[3]

[1]*Department of Computational Biology, Center for Advanced Studies, St. Petersburg State Polytechnic University, St. Petersburg, 195259 Russia*
[2]*Theoretical Department, The Ioffe Physico-Technical Institute of the Russian Academy of Sciences, St. Petersburg, 194021 Russia*
[3]*Department of Applied Mathematics and Statistics, and Center for Developmental Genetics, Stony Brook University, Stony Brook, NY 11794-3600, U.S.A.*

D. Chatenay, S. Cocco, R. Monasson, D. Thieffry and J. Dalibard, eds.
Les Houches, Session LXXXII, 2004
Multiple aspects of DNA and RNA: from Biophysics to Bioinformatics

Contents

Contents

1. Preamble

In this Chapter we demonstrate the ability of gene circuit method to interpret and predict regulatory mechanisms using as an example the segment determination system in fruit fly Drosophila. We have selected this process because of its biological importance, and also because computational investigations of gene regulation can be done in the exceptionally accurate way in this system. We show that this method can predict experimental results as well as solve certain problems better than standard experimental methods. The utility of the gene circuit method was previously discussed [6]. In this review we restrict ourselves to consideration of new results obtained since 1998 and not discussed in the last review.

2. Introduction

In modern genetics the regulatory interactions in multicellular organisms are inferred by a comparisons of mutant and wild type phenotypes. An important example of use of this method is the deduction that virtually all of the pair-rule class of segmentation genes in the fruit fly *Drosophila melanogaster* are regulated by members of gap gene class [1,2]. This conclusion follows directly from observations of segmentation gene expression patterns at gastrulation in a variety of gap and pair-rule mutants.

In spite of yielding considerable information the inference of regulatory interactions from qualitative mutant expression data remains a highly nontrivial task in all but the simplest cases. These are the problems of consistency, uniqueness and completeness in interpretation of mutant expression patterns that make our knowledge on regulatory interactions incomplete.

To solve these problems we need a method that allows us to reconstitute wild-type gene expression patterns *in silico*, infer underlying regulatory interactions from these wild type patterns, and keep track of all regulatory interactions in all nuclei at all times.

The gene circuit method provides such an approach [3–5]. The gene circuit is a data driven mathematical modelling method, whose main aim is to extract information about dynamical regulatory interactions between transcription factors from given gene expression patterns (Figure 1A). This is achieved in four steps: (1) formulation of a mathematical modelling framework, (2) collection of gene expression data, (3) fitting of the model to expression data to obtain regulatory parameters, and (4) biological analysis of the resulting gene circuits.

3. The biology of segment determination

The process of segment determination occurs during the syncytial blastoderm stage, which starts with the completion of nuclear migration to the periphery of the embryo at early cleavage cycle 10 and ends at the end of cleavage cycle 14A [7]. At syncitial blastoderm stage cell membranes do not form and each nucleus is surrounded by an island of cytoplasm, called an energid. During cellularization which starts in mid cleavage cycle 14A cell membranes begin to invaginate progressively engulfing the blastoderm nuclei. Cellularization is complete by the end of cycle 14A [7, 8].

Cells in an embryo organize themselves into appropriate structures at correct locations by interpreting positional information encoded by chemical signals, denoted as morphogens (reviewed in [9]). In application to segment determination these signals are the protein gradients formed by products of the maternal co-ordinate genes, namely anterior gradient of proteins Bicoid (Bcd) and Hunchback (Hb), as well as posterior gradient of Caudal (Cad) [10–13]. The positional information provided by morphogenetic gradients is interpreted by zygotic genes [14, 15]. A large majority of the segmentation genes encode transcription factors that form a multilayered network of gene regulatory interactions. The segmentation gene network consists of three classes of zygotic segmentation genes distinguished by the nature of their mutant phenotypes and expression patterns. The developmental function of gap and pair-rule genes is to establish expression of the segment-polarity genes, especially *wingless* and *engrailed*, whose transcripts first appear during late cycle 14A and which are thought to constitute the final segmentation prepattern [14, 15].

The *Drosophila* blastoderm permits exceptionally precise modelling, since pattern formation is a consequence of regulatory interactions. In a typical developmental process, well characterized genetics alone does not provide enough information to model and understand the system behavior. The blastoderm is a very important exception of this rule. Segmentation gene mutations do not cause any morphological defects before the onset of gastrulation [16, 17]. Thus, the internal state of each blastoderm nucleus can be described by concentration levels of transcription factors encoded by segmentation genes. The segmentation genes have been cloned, and hence their level of expression can be monitored by antibody methods. In addition, there is no tissue growth, and we do not have to consider intercellular signaling since nuclei are not yet surrounded by membranes during the syncytial blastoderm stage [8]. These properties allow to understand important aspects of developmental genetics in unprecedented detail.

4. Method description

4.1. The gene circuit modelling framework

The gene circuits modelling framework has been described in detail in [3] and [4]. In the presumptive segmented germ band, segmentation gene expression is exclusively a function of A-P position to a very good level of approximation, and therefore the model considers the one dimensional line of nuclei running laterally along the A-P axis. In most cases, each nucleus was explicitly represented and numbered with an index i from anterior to posterior. The model has three rules governing the behavior of nuclei in time t: (1) interphase, (2) mitosis, and (3) division. Rules 1 and 2 describe the dynamics of protein synthesis and decay within a nucleus and protein diffusion between nuclei. Rule 3 is discrete and describes how each nucleus is replaced by its two daughter nuclei upon division. The schedule for these rules is based on [7] and is summarized in Figure 1B.

The internal state of nucleus i is described by concentrations v_i^a of transcription factors encoded by segmentation genes denoted by index a. The change in transcription factor concentration over time dv_i^a/dt depends on three processes during interphase: (1) protein synthesis, (2) protein diffusion and (3) protein decay, represented by the summation terms on the right hand side of the equation below:

$$\frac{dv_i^a}{dt} = R^a g \left(\sum_{b=1}^{N} T^{ab} v_i^b + m^a v_i^{\text{Bcd}} + h^a \right) +$$
$$+ D^a(n) \left[(v_{i-1}^a - v_i^a) + (v_{i+1}^a - v_i^a) \right] - \lambda^a v_i^a, \tag{4.1}$$

where $a = 1, ..., N$, and N is the total number of zygotic genes in the model.

In equation (4.1), T^{ab} represents a matrix of regulatory coefficients where each coefficient T^{ab} characterizes the regulatory effect of the product of gene b on the expression of gene a (Figure 1D). This matrix is independent of i reflecting the fact that each nucleus contains a copy of the same genome. v_i^{Bcd} is the concentration of Bcd protein in nucleus i. Bcd is exclusively maternal and its concentration is constant in time. The regulatory effect of Bcd on gene a is represented by the parameter m^a. Parameter h^a is a threshold representing regulatory contributions of uniformly expressed maternal transcription factors. The relative rate of protein synthesis is then given by the sigmoid regulation-expression function $g(u^a) = \frac{1}{2} \left[\left(u^a / \sqrt{(u^a)^2 + 1} \right) + 1 \right]$, where $u^a = \sum_{b=1}^{N} T^{ab} v_i^b + m^a v_i^{\text{Bcd}} + h^a$ is the total regulatory input on gene a (Figure 1C). The maximum synthesis rate for the product of gene a is given by R^a. The diffusion parameter $D^a(n)$ depends on the number of nuclear divisions n that have taken place before the current time t. The diffusion coefficient is assumed to vary inversely with the square of the distance between neighboring nuclei and this distance is halved upon nuclear division. λ^a is the decay rate of the product of gene a and is related to the protein half life $\tau_{1/2}^a$ of the product of gene a by $\tau_{1/2}^a = \ln 2/\lambda^a$.

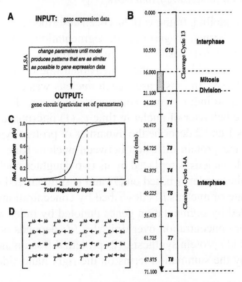

Fig. 1. The gene circuit method. (A) The basic principle. Regulatory interactions are inferred from wild type expression patterns by fitting gene circuit models to quantitative data. (B) Time schedule for gap gene circuits. The model spans the time from the onset of cycle 13 (0.0 min) to the onset of gastrulation at the end of cycle 14A (71.1 min). The three rules of the model (interphase, mitosis and nuclear division) are shown to the right. There is one time class in cycle 13 (C13), and eight time classes (T1–T8) in cycle 14A. Time points used for comparison of model output to data for time classes C13 and T1–T8 are indicated. (C) The regulation-expression function $g(u)$. Total regulatory input u is shown on the horizontal axis. Corresponding relative activation of protein synthesis $g(u)$ is shown on the vertical axis. $g(u)$ rapidly approaches saturation for values of u above 1.5, and rapidly approaches zero for values of u below −1.5 (dashed lines). (D) Regulatory interactions within a gene circuit are represented by the genetic interconnection matrix T (shown here for interactions of *hb*, *Kr*, *gt* and *kni*). See text for details.

The only regulatory molecules considered in (4.1) are proteins synthesized by segmentation genes. RNA is not included because there is no evidence for a direct role of RNA in the regulation of zygotic segmentation genes.

The equation (4.1) keeps track of individual nuclei. The nuclear structure can be mathematically abolished by rewriting the equations for a spatial continuum, giving the PDE form of (4.1), which is written

$$\frac{\partial v^a(x,t)}{\partial t} = R^a(t)g(u^a) - \lambda^a v^a(x,t) + D^a \frac{\partial^2 v^a(x,t)}{\partial x^2}, \quad a = 1, ..., N, \quad (4.2)$$

where x is the spatial coordinate varying along the A-P axis of the embryo. The maximum synthesis rate $R^a(t)$ now depends on time to capture the effects of mitosis. Possible specific forms of this dependence are discussed in Section 6.

The PDEs can be used in two ways. First, with PDEs it is possible to probe the role of nuclear structure in the segment determination (see Section 6). Second, PDEs can be solved analytically and exact solutions can be compared with numerical results.

4.2. Quantitative expression data

Images of expression patterns of segmentation genes in fruit fly were obtained by immunostaining whole mount embryos using either fluorescence microscope [4] or confocal scanning microscope as described in [18]. Two datasets were generated from these images.

The first dataset (low-resolution dataset) was generated from fluorescence double antibody stained embryos. The expression levels in this dataset were estimated by eye from Kodachrome slides and image registration was performed by hand using transparencies. The data described a row of 32 nuclei running along the lateral equator in an A-P direction, extending from the middle of *even-skipped* (*eve*) stripe 1 to the interstripes between *eve* stripes 5 and 6. Data were obtained for genes *eve*, *Krüppel* (*Kr*), *giant* (*gt*), *knirps* (*kni*), *hunchback* (*hb*), and *bicoid* (*bcd*) at 4 time points, corresponding to early cleavage cycles 13 and 14, middle cleavage cycle 14 and late cleavage cycle 14. The *eve* expression data in cleavage cycle 13 and early cleavage cycle 14A [7] was spatially uniform. In the late cleavage cycle 14A data, *eve* is approximately periodic, with the spatial period of 7 nuclei mentioned above.

In the second dataset (named as high-resolution dataset) obtained with confocal scanning microscope expression levels were normalized per gene to a relative fluorescence intensity range of 0–255 based on the most intensely fluorescent pattern on each slide with multiple embryos. This dataset was generated from 2862 images of expression patterns by applying a five step data pipeline [18–29]. The dataset contains quantitative data on expression of gap and pair-rule genes, as well as maternal genes *caudal* (*cad*) and *bcd* at nuclear resolution and for a period spanning 71 minutes of development (cleavage cycles 13 and 14). The temporal resolution of data at cleavage cycle 14 is about 6.5 minutes of development.

4.3. Optimization by Parallel Lam Simulated Annealing (PLSA) and Optimal Steepest Descent Algorithm (OSDA)

PLSA was used as described in [4] and [30]. The set of ordinary differential equations (4.1) was solved numerically using either forward Euler method, with 10–20% residual error (in numerical experiments with the first dataset) or a Bulirsch–Stoer adaptive step size solver scheme taken from [31]. In the second case equations were solved to a relative accuracy of 0.1%, and solutions were tested for numerical stability. The parameters R^a, T^{ab}, m^a, h^a, D^a and λ^a were adjusted to minimize the following cost function, representing the difference between a model solution and data:

$$E = \sum \left(v_i^a(t)_{\text{model}} - v_i^a(t)_{\text{data}} \right)^2. \tag{4.3}$$

Summation is performed over the total number of data points N_d, i.e., the number of protein measurements across all genes a, nuclei i and time classes t.

Parameter search spaces were defined by explicit search limits for R^a, D^a and λ^a and a collective penalty function for T^{ab}, m^a, h^a as described in [4]. Parameters h^a for *Kr*, *kni*, *gt* and *hb* were fixed to negative values representing a constitutive 'off' state of the gene. This accelerated the annealing process considerably and slightly improved annealing results while not altering the overall quality of the resulting gene circuits.

OSDA is described in [29]. This algorithm is based on a Lagrangian approach, in which the function (4.3) is minimized subject to the constraints that (4.1) is satisfied and that the parameters lie within their search space. Search space constraints, initially expressed as inequalities, are transformed into additional constraint equations. An expanded cost function (Lagrangian) is constructed by adding each constraint equation multiplied by its Lagrange multiplier to (4.3), and minimized by steepest descent.

4.4. Selection of gene circuits

The root mean square (rms) score

$$\text{rms} = \sqrt{\frac{E}{N_d}}$$

was used as a measure for the quality of a gene circuit. The rms represents the average absolute difference between protein concentrations in model and data.

PLSA is a stochastic optimization method yielding gap gene circuits of varying quality. Gene circuits most faithfully reproducing gap gene expression were selected as follows: First, only circuits with an rms of less than 12.0 were considered. All gap gene circuits with an rms of more than 12.0 showed obvious pattern defects, some of them severe such as displaced or missing expression boundaries. Second, each of the selected circuits was carefully tested for patterning defects by visual inspection and plotting of squared differences between model and data for each protein and time class.

4.5. Software and bioinformatics

Simulator and optimization code were implemented in C, gene circuit analysis and plotting tools were implemented in Perl and Java. Software and gene circuit files are available at

> http://flyex.ams.sunysb.edu/lab/gaps.html.

Data of the second dataset can be downloaded from the FlyEx database at

> http://urchin.spbcas.ru/flyex, or

> http://flyex.ams.sunysb.edu/flyex.

5. Analysis of regulatory mechanisms controlling segment determination

5.1. Regulatory interactions in gap gene system

Using the high-resolution dataset, we report in [33] a new features of the gap system behavior: Spatial positions of domains of gap gene expression on the anterior-posterior axis of the embryo ("gap domains"), which thought to be almost stationary during the domain formation, are in fact substantially shifting towards anterior pole during the cleavage cycle 14A (Figure 2).

We model gene expression in the gap gene network comprising 6 zygotically expressed genes: *Kr*, *gt*, *kni*, *hb*, *cad*, and *tailless* (*tll*), as (4.1) [33, 34]. Concentrations of proteins produced by maternally expressed *hb* and *cad* ("maternal gradients") are used as initial conditions at cleavage cycle 13 for related zygotic gene products. Another maternal gradient, produced by *bcd*, comes into model equations as an external input.

All parameter values are found by fitting solution of the model to the expression data at 8 consecutive time points from cycle 13 to the end of cycle 14A. Optimal parameter values found by PLSA predict the activation of gap genes by maternal factors and gap–gap cross-repression, which is consistent with results of qualitative studies of mutant gene expression patterns. A solution related to the found optimal parameter values mimic data at high accuracy and temporal resolution, including the described shifts of gap domain boundaries.

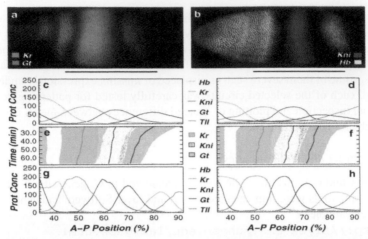

Fig. 2. Dynamical shifts in gap gene domains are reproduced by g ap gene circuits. a,b, Drosophila melanogaster blastoderm stage embryos at late cleavage cycle 14A (time class: T8), immunostained for a, Kr and Gt (FlyEx database embryo name: rge9), and b, Kni and Hb (rb8). Anterior is to the left, dorsal is up. Black bars indicate the region included in gap gene circuits. c,g, gene expression data and d,h, gap gene circuit model output at early (c,d, T1) and late (g,h, T8) cycle 14A. Vertical axes represent relative protein concentrations, horizontal axes represent position along the anteroposterior (A-P) embryo axis (where 0% is the anterior pole). There are no Tll data for T1 (c). e,f, gap domain shifts for *Kr, kni* and *gt* covering the time between patterns shown in c,d and g,h. Solid dark coloured lines indicate position of maximum concentration for each domain. Lighter coloured areas cover regions in which protein concentration is above half maximum value. Positional values for data were obtained using interpolation with quadratic splines.

Main results of the analysis can be formulated as the following five points:

(1) The gap gene system is reconstructed *in silico*, with the help of the novel PLSA technique. The patterns in the model are of excellent agreement with the data (Figure 2).

(2) The gradients of maternal genes alone are not sufficient for positioning of gap gene domains and hence do not qualify as morphogens in a strict sense. At cycle 14A the shifts of these domains and, hence, their positions are determined by gap–gap cross-regulation (Figure 3).

(3) Regulatory loops of mutual repression create positive regulatory feedback between complementary gap genes providing a straightforward mechanism for their mutually exclusive expression patterns. This mechanism is com-

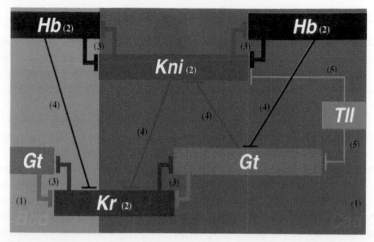

Fig. 3. Overview of the gap gene network. Expression domains of *hb*, *kni*, *gt*, *Kr*, and *Tll* are shown schematically as black boxes. Anterior is to the left. Repressive interactions are represented by T-bar connectors. Background shading represents main maternal activating inputs by Bcd (dark) and Cad (light). The gap gene network consists of five basic regulatory mechanisms: (1) Activation of gap genes by Bcd and/or Cad, (2) autoactivation, (3) strong repression between mutually exclusive gap genes, (4) repression between overlapping gap genes, (5) repression by Tll.

plemented by repression among overlapping gap genes. Overlap in expression patterns of two repressors imposes a limit on the strength of repressive interactions between them. Accordingly, repression between neighboring gap genes is generally weaker than between complementary ones (Figure 3). Moreover, repression among overlapping gap genes is asymmetric, centered on the *Kr* domain (see Figure 2). Posterior of this domain, only posterior neighbors contribute functional repressive inputs to gap gene expression, while anterior neighbors do not. This asymmetry is responsible for anterior shifts of posterior gap gene domains during cycle 14A [34].

(4) The diffusion of gap proteins is present in both embryo and gap gene circuits, however it does not have a significant role in shifting gap domain boundaries.

(5) Positional information in the blastoderm embryo can no longer be seen as a static coordinate system imposed on the embryo by maternal morphogens. Rather, it needs to be understood as the dynamic process underlying the positioning of expression domain boundaries, which is based on both external inputs by morphogens and tissue-internal feedback among target genes.

5.2. Stripe forming architecture of the gap gene system

A critical step in the determination of the periodic segments in the fruit fly *Drosophila* is the transformation of aperiodic positional information encoded by gap domains and maternal gradients into the periodic pattern of pair-rule gene expression [15]. This transformational step is modulated by pair-rule cross regulation [1, 35].

An important question in segment determination is the identification of those features of pair-rule expression patterns that are controlled by the gap gene system as opposed to those that are determined by pair-rule cross regulation.

The difficulty of this problem arises because pair-rule genes have many inputs. Their combined effect leads to a complex but precisely positioned set of overlapping patterns [36]. The correct set of overlaps is absolutely required for embryonic viability: The pair-rule genes *fushi-tarazu* and *eve* are normally expressed in separate, complementary stripes at gastrulation, but any overlap in expression at this time results in lethality [37].

An analysis of the relative contributions of gap and pair-rule genes to the pair-rule patterns would begin by identifying which periodic patterns can be generated by the gap gene system alone. This question could be answered by a simple experiment: Monitor the expression of each pair-rule gene in embryos mutant for all of the seven other pair-rule genes. Such an experiment is probably not feasible by genetic methods, because the animal would die as a heterozygote.

The difficulty of constructing a multiple mutant genetic stock may be circumvented by applying the gene circuit method. Using the low resolution dataset in [5] Reinitz et al. approximated the stripe pattern of a variety of pair-rule genes by the expression of a single pair-rule gene, shifted along the anteroposterior axis by one or more nuclei to investigate whether or not the gap gene system can encode each of these periodic patterns. Such analysis makes it possible to study the stripe forming capabilities of the gap gene system in the complete absence of pair-rule cross regulation.

A set of fits performed with gap gene parameters described elsewhere [4] demonstrated that due to the architecture of the gap domain system the gap genes encode only one set of pair-rule stripes in the native *eve* position.

6. Pattern formation and nuclear divisions are uncoupled in Drosophila segmentation

In [38], we have used the PDE model (4.2) to study relation between nuclear structure of the embryo and expression pattern formation by segmentation genes. Nuclei are replaced in this model by a continuum, and we try to find out if such system is capable of correct pattern formation using the low-resolution dataset. The model is formulated in the spatial domain on the A-P axis which covers a central part of the embryo including *eve* stripes 2–5, and in time interval from cleavage cycle 11 to cycle 14A.

The choice of time-dependence of $R^a(t)$ in the equations is related to the way that mitosis is represented in the model, which is a nontrivial theoretical issue. Actual nuclei in the embryo divide at the end of each cleavage cycle; therefore, at the subsequent cleavage cycle we have the doubled number of gene copies, and, hence, we might expect the doubled potential for protein synthesis. If the nuclei were smaller in size than any spatial scale in which gene expression changes, and if macromolecular synthesis took place in a region of infinitesimal spatial extent, writing (4.2) would be a straightforward exercise in taking concentrations, and $R^a(t)$ would double in each successive cleavage cycle. In fact, this approximation does not hold. This is both because nuclei are large compared to the scale of spatial variation and because the actual process of protein synthesis takes place in a volume larger than a nucleus, since RNA must be transcribed and processed in the nucleus and transported to the cytoplasm for translation. Newly translated protein returns to the nucleus to bind to chromatin and regulate other genes. A central problem in modelling biological systems is the extent to which various mechanisms must be incorporated into a model of a particular process in order to correctly understand its behavior. Here we are concerned with the spatial part of this coarse-graining problem, which is to ask which (if any) formulations of $R^a(t)$ allow the blastoderm to be well represented by PDEs.

We consider three possible formulations of $R^a(t)$, representing different approximations of this dependence, and in order of increasing complexity they are as follows:

(A) $R^a(t) = R^a$, a constant.

(B) $R^a(t) = 0$ during mitosis and has a positive value $R^a(t) = R^a$ during interphase. This approximation takes into account the specific fact that there is no synthesis during a short time period right before a cleavage takes place.

(C) $R^a = 0$ during mitosis and $R^a(t) = 2^{C-14}R^a$ during interphase, where C is a number of cleavage cycle and R^a is the cycle 14 synthesis rate. This is the same as B, but $R^a(t) \mapsto 2R^a(t)$ in each successive cleavage cycle.

The mitosis schemes A–C reflect different ways of incorporating mitosis into the model. Their ability to reproduce expression patterns (or lack thereof) allows us to draw conclusions about the importance of mitosis to the pattern formation process.

Parameter values are found for all three continuum models by applying OSDA [32, 38]. This optimization algorithm is used to minimize (in parameter space) the functional representing the spatially continuous extension of the cost function (4.3) with the data from the low-resolution dataset.

We obtained correct pattern dynamics from all of the models (see Figure 4), as well as from the model with explicit nuclear structure [4]. The sets of parameter values in models A, B, C, and in the model from [4] are quantitatively different from each other, but are qualitatively equivalent. Therefore, they represent the same genetic regulatory system (or the same gene network topology) which is independent of the representation of subcellular structure and the implementation of mitosis in the model. This leads us to conclude that *nuclear divisions* are *not coupled to pattern formation* and serve only to populate the blastoderm with nuclei.

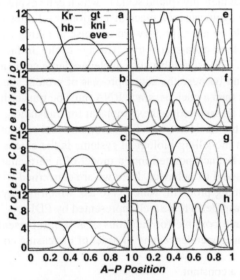

Fig. 4. Segmentation gene expression patterns: comparison between data and continuum models A–C. Protein concentration profiles are shown at early (*a–d*) and late (*e–h*) cleavage cycle 14A: (*a* and *e*) Data from the low-resolution dataset, (*b* and *f*) model A, (*c* and *g*) model B, (*d* and *h*) model C. The horizontal axis represents the rescaled spatial domain (covering the middle 32% of the A-P axis of the embryo), the vertical axis represents protein concentrations in conventional units.

7. Conclusions

In conclusion we summarize the ways in which gene circuits can be used to solve a number of problems about the mechanisms of gene regulation.

Being *in silico* method gene circuits enable us to predict the results of experiments that have not been done, or which are quite difficult to carry out, as well as to prove the sufficiency of the inferred mechanisms without reconstructing the system *ab initio*. Simulations probing the stripe forming architecture of the gap gene system, as well as the necessity of nuclear divisions for pattern formation would be examples of the former, while the analysis of regulatory interactions in the gap gene system would be an example of the latter.

Another important property of the gene circuits is their ability to keep track of all regulatory inputs to a specific gene in the intact and complete developmental system. This cannot be done in genetics, where functional information comes from removing genes one at a time from a complete system via mutation, and the regulatory structure of the wild type network must be assembled on the basis of evidence from different experiments.

Additional power of gene circuits resides in its support of the quantitative reasoning about the dynamics of living systems. As applied to the *Drosophila* segmentation system this property allows to reveal the role of autoactivation in sharpening the gap domain boundaries, as well as to explain the mechanism governing the posterior domain shifts during cycle 14A.

Acknowledgments

We would like to thank our colleagues who participate in the work described in this overview: M. Blagov, J. Jaeger, H. Janssens, D. Kosman, K. N. Kozlov, Manu, E. M. Myasnikova, A. Pisarev, E. Pustel'nikova, J. Reinitz, D. Sharp, S. Surkova.

References

[1] S.B. Carroll and M.P. Scott, Cell **45** (1986) 113.

[2] D. Tautz, Nature **332** (1988) 281.

[3] E. Mjolsness, D.H. Sharp and J. Reinitz, J. Theor. Biol. **152** (1991) 429.

[4] J. Reinitz and D.H. Sharp, Mech. Dev. **49** (1995) 133.

[5] J. Reinitz, D. Kosman, C.E. Vanario-Alonso and D.H. Sharp, Dev. Gen. **23** (1998) 11.

[6] J. Reinitz and D.H. Sharp, in: *Integrative Approaches to Molecular Biology*, editors: J. Collado, B. Magasanik, and T. Smith, Ch. 13 (MIT Press, Cambridge, Massachusetts, USA, 1996), p. 253.

[7] V.E. Foe and B.M. Alberts, J. Cell Sci. **61** (1983) 31.

[8] J.A. Campos-Ortega and V. Hartenstein, *The Embryonic Development of Drosophila melanogaster* (Springer, Germany, 1985).

[9] D. St Johnston and C. Nüsslein-Volhard, Cell **68** (1992) 201.

[10] T. Berleth, M. Burri, G. Thoma, D. Bopp, S. Richstein, G. Frigerio, M. Noll and C. Nüsslein-Volhard, The EMBO J. **7** (1988) 1749.

[11] S.K. Chan and G. Struhl, Nature **388** (1997) 634.

[12] R. Lehmann and C. Nüsslein-Volhard, Dev. **112** (1991) 679.

[13] R. Rivera-Pomar, D. Niessing, U. Schmidt-Ott, W.J. Gehring and H. Jackle, Nature **379** (1996) 746.

[14] M. Akam, Dev. **101** (1987) 1.

[15] P.W. Ingham, Nature **335** (1988) 25.

[16] P.T. Merrill, D. Sweeton and E. Wieschaus, Dev. **104** (1988) 495.

[17] E. Wieschaus and D. Sweeton, Dev. **104** (1988) 483.

[18] D. Kosman, J. Reinitz, and D. H. Sharp, in: *Proceedings of the 1998 Pacific Symposium on Biocomputing*, editors: R. Altman, K. Dunker, L. Hunter, and T. Klein (World Scientific Press, Singapore, 1997), p. 6, also available in: http://www.smi.stanford.edu/projects/helix/psb98/kosman.pdf.

[19] H. Janssens, D. Kosman, C.E. Vanario-Alonso, J. Jaeger, K.N. Kozlov, M.G. Samsonova and J. Reinitz, Dev. Gen. Evol. (2005), in press.

[20] D. Kosman, S. Small and J. Reinitz, Dev. Gen. Evol. **208** (1998) 290.

[21] E.M. Myasnikova, M.G. Samsonova and J. Reinitz, Dev. Gen. Evol. (2005), in press, DOI: 10.1007/s00427-005-0472-2.

[22] I. Aizenberg, C. Butakoff, E.M. Myasnikova, M.G. Samsonova, and J. Reinitz, in: *SPIE Proceedings*, **4668**, editors: N.M. Nasrabadi and A.K. Katsaggelos (SPIE, San Jose, CA, USA, 2002), p. 10.

[23] I. Aizenberg, E.M. Myasnikova, M.G. Samsonova and J. Reinitz, Math. Biosci. **159** (2002) 145.

[24] E.M. Myasnikova, A.A. Samsonova, M.G. Samsonova and J. Reinitz, Bioinformatics (Suppl.) **18** (2002) S87.

[25] E.M. Myasnikova, A.A. Samsonova, K.N. Kozlov, M.G. Samsonova and J. Reinitz, Bioinformatics **17** (2001) 3.

[26] E.M. Myasnikova, A.A. Samsonova, M.G. Samsonova and J. Reinitz, Molekulyarnaya Biologiya **35** (2001) 1110, in Russian.

[27] K.N. Kozlov, E.M. Myasnikova, A.S. Pisarev, M.G. Samsonova and J. Reinitz, In Silico Biology **2** (2002) 125.

[28] E.M. Myasnikova, D. Kosman, J. Reinitz, and M.G. Samsonova, in: *Proceedings of the Seventh International Conference on Intelligent Systems for Molecular Biology*, editors: T. Lengauer, R. Schneider, P. Bork, D. Brutlag, J. Glasgow, H.W. Mewes, and R. Zimmer (AAAI Press, Menlo Park, California, 1999), p. 195.

[29] K.N. Kozlov, E.M. Myasnikova, M.G. Samsonova, J. Reinitz and D. Kosman, Comp. Technol. **5** (2000) 112.

[30] K.W. Chu, Y. Deng and J. Reinitz, J. Comput. Phys. **148** (1999) 646.

[31] W.H. Press, S.A. Teukolsky, W.T. Vetterling and B.P. Flannery, *Numerical Recipes in C* (Cambridge University Press, UK, 1992).

[32] K.N. Kozlov and A.M. Samsonov, Tech. Phys. **48** (2003) 6.

[33] J. Jaeger, S. Surkova, M. Blagov, H. Janssens, D. Kosman, K.N. Kozlov, Manu, E.M. Myasnikova, C.E. Vanario-Alonso, M.G. Samsonova, D.H. Sharp and J. Reinitz, Nature **430** (2004) 368.

[34] J. Jaeger, M. Blagov, D. Kosman, K.N. Kozlov, Manu, E.M. Myasnikova, S. Surkova, C.E. Vanario-Alonso, M.G. Samsonova, D.H. Sharp and J. Reinitz, Genetics **167** (2004) 1721.

[35] S. Baumgartner and M. Noll, Mech. Dev. **33** (1990) 1.

[36] S.B. Carroll, A. Laughon and B.S. Thalley, Gen. Dev. **2** (1988) 883.

[37] M. Frasch, R. Warrior, J. Tugwood and M. Levine, Gen. Dev. **2** (1988) 1824.

[38] V.V. Gursky, J. Jaeger, K.N. Kozlov, J. Reinitz and A.M. Samsonov, Physica D **197** (2004) 286.

Course 10

MODELING, ANALYSIS, AND SIMULATION OF GENETIC REGULATORY NETWORKS: FROM DIFFERENTIAL EQUATIONS TO LOGICAL MODELS

Hidde de Jong and Denis Thieffry

Institut National de Recherche en Informatique et en Automatique (INRIA)
Unité de recherche Rhône-Alpes
655 avenue de l'Europe, Montbonnot, 38334 Saint Ismier Cedex, France
Email: Hidde.de-Jong@inrialpes.fr

Institut de Biologie de Développement de Marseille (IBDM)
Laboratoire de Génétique et Physiologie du Développement (LGPD), CNRS UMR 6545
Luminy Campus, CNRS Case 907, 13288 Marseille Cedex 9, France
Email: thieffry@ibdm.univ-mrs.fr

D. Chatenay, S. Cocco, R. Monasson, D. Thieffry and J. Dalibard, eds.
Les Houches, Session LXXXII, 2004
Multiple aspects of DNA and RNA: from Biophysics to Bioinformatics
© *2005 Elsevier B.V. All rights reserved*

Contents

Abstract

Genetic regulatory networks, consisting of genes, proteins, small molecules, and their mutual interactions, control the functioning and differentiation of cells. Given the large number of components of most networks of biological interest, connected by positive and negative feedback loops, an intuitive comprehension of the dynamics of the system is often difficult, if not impossible to obtain. As a consequence, mathematical and computational approaches are indispensable for gaining a comprehension of the functioning of complex networks. In this chapter, we review three approaches towards the modeling, analysis, and simulation of genetic regulatory networks, based on ordinary differential equations, piecewise-linear differential equations, and logical models, respectively. We discuss the strengths and weaknesses of these formalisms, and illustrate their application to the study of a variety of prokaryotic and eukaryotic model systems.

1. Introduction

A remarkable development in molecular biology today is the upscaling to the genomic level of its experimental methods. Hardly imaginable only 20 years ago, the sequencing of complete genomes has become a routine job, highly automated and executed in a quasi-industrial environment. The miniaturization of techniques for the hybridization of labeled nucleic acids in solution to DNA molecules attached to a surface has given rise to DNA microarrays, tools for measuring the level of gene expression in a massively parallel way [39]. The development of proteomic methods based on two-dimensional gel electrophoresis, mass spectrometry, and the double-hybrid system allows the identification of proteins and their interactions at a genomic scale [46].

These novel methods in genomics produce enormous amounts of data about different aspects of the cell. On one hand, they allow the identification of interactions between the genes of an organism, its proteins, metabolites, and other small molecules, thus mapping the structure of its interaction networks. On the other hand, they enable biologists to measure the evolution of the state of the cell, that is, the temporal variation of the concentration and the localization of the different molecular components, in response to changes in the environment. The challenge of *systems biology* consists in relating these structural and functional data, in order to arrive at an integrated representation of the functioning

329

of the organism [28, 34]. This amounts to predicting and understanding how the observed behavior of the organism – the adaptation to its environment, the differentiation of its cells during development, even its evolution on a longer time-scale – emerges from the networks of molecular interactions.

The molecular interactions in the cell are quite heterogeneous in nature. They concern the transcription and translation of a gene, the enzymatic conversion of a metabolite, the phosphorylation of a regulatory protein, *etc.* While studying a cellular process, it is often sufficient, at least to a first approximation, to focus on a part of the interaction network, dominated by a particular type of interaction. In this chapter, we focus on *genetic regulatory networks*, which mainly concern interactions between proteins and nucleic acids, controlling the transcription and translation of genes. Genetic regulatory networks play an important role in the functioning and differentiation of cells. Moreover, a large part of the experimental data available today, notably transcriptome data, concern these networks. Not withstanding their importance, one should bear in mind that genetic regulatory networks are integrated in the cell with other types of networks, sometimes to the point that they become difficult to separate.

In addition to high-throughput experimental methods, mathematical and computational approaches are indispensable for the analysis of genetic regulatory networks. Given the large number of components of most networks of biological interest, often connected by positive and negative feedback loops, an intuitive comprehension of the dynamics of the system is difficult, if not impossible to obtain. *Mathematical modeling* supported by *computer tools* can contribute to the analysis of a regulatory network by allowing the biologist to focus on a restricted number of plausible hypotheses. The formulation of a mathematical model requires an explicit and non-ambiguous description of the hypotheses being made on the regulatory mechanisms under study. Furthermore, its simulation by means of the model yields predictions on the behavior of the cell that can be verified experimentally.

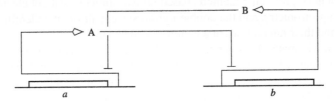

Fig. 1. Example of a simple genetic regulatory network, composed of two genes *a* and *b*, the proteins A and B, and their regulatory interactions.

In the last forty years, a large number of approaches for the dynamic modeling genetic regulatory networks have been proposed in the literature [4, 10, 19, 26, 53,

61]. The aim of this chapter is to review three modeling formalisms in some detail: ordinary differential equations, piecewise-linear differential equations, and logical models. The three formalisms will be introduced and compared by means of a simple network of two genes (Figure 1). Each of the genes encodes a regulatory protein that inhibits the expression of the other gene, by binding to a site overlapping the promoter of the gene. Simple as it is, this *cross-inhibition network* is a basic component of more complex, real networks and allows the analysis of some characteristic aspects of cellular differentiation [44, 59]. The application of the three formalisms to actual genetic regulatory networks will be illustrated by means of examples taken from a variety of prokaryotic and eukaryotic systems.

2. Ordinary differential equations

2.1. Models and analysis

Ordinary differential equations (ODEs) are probably the most-widespread formalism for modeling genetic regulatory networks. They represent the concentration of gene products – mRNAs or proteins – by continuous, time-dependent variables. The variables take their values from the set of non-negative real numbers $\mathbb{R}_{\geq 0}$, reflecting the constraint that a concentration cannot be negative. In order to model the regulatory interactions between genes, functional and differential relations are used.

More precisely, gene regulation is modeled by a system of ordinary differential equations having the following form:

$$\frac{dx_i}{dt} = f_i(\boldsymbol{x}), \; i \in [1, \ldots, n], \tag{2.1}$$

where $\boldsymbol{x} = (x_1, \ldots, x_n)' \in \Omega$ is a vector of cellular concentration variables, $\Omega \subset (R)_{\geq 0}^n$ is a bounded n-dimensional phase space box, and the function $f_i : \mathbb{R}_{\geq 0}^n \to \mathbb{R}$, usually highly nonlinear, represents the regulatory interactions. The system of equations (2.1) describes how the temporal derivative of the concentration variables depends on the values of the concentration variables themselves. In order to simplify the notation, we can write (2.1) as the vector equation

$$\frac{d\boldsymbol{x}}{dt} = \boldsymbol{f}(\boldsymbol{x}), \tag{2.2}$$

with $\boldsymbol{f} = (f_1, \ldots, f_n)'$.

An ordinary differential equation model of the cross-inhibition network in Figure 1 is shown in Figure 2(a). The variables x_a and x_b represent the concentration of the proteins A and B, encoded by the genes a and b, respectively. The temporal derivative of x_a is the difference between the *synthesis term* $\kappa_a\, h^-(x_b, \theta_b, m_b)$ and the *degradation term* $\gamma_a\, x_a$. The first term expresses that the rate of synthesis of protein A depends on the concentration of protein B and is described by the function $h^- : \mathbb{R}_{\geq 0} \times \mathbb{R}^2_{\geq 0} \to \mathbb{R}_{\geq 0}$. This so-called *Hill function* takes the value 1 for $x_b = 0$, and monotonically decreases towards 0 for $x_b \to \infty$. It is characterized by a threshold parameter θ_b and a cooperativity parameter m_b (Figure 2(b)). For $m_b > 1$, the Hill function has a sigmoidal form that is often observed experimentally [47, 65]. The synthesis term $\kappa_a\, h^-(x_b, \theta_b, m_b)$ thus means that, for low concentrations of the protein B, gene a is expressed at a rate close to its maximum rate κ_a ($\kappa_a > 0$), whereas for high concentrations of B, the expression of the gene is almost completely repressed. The second term of the differential equation, the degradation term, expresses that protein A disappears at a rate proportional to its own concentration x_a, where $\gamma_a > 0$. This may be due to degradation of the proteins or a consequence of growth dilution. The differential equation for x_b has an analogous interpretation.

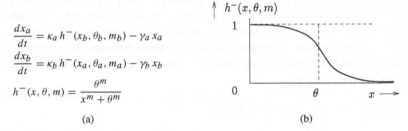

$$\frac{dx_a}{dt} = \kappa_a\, h^-(x_b, \theta_b, m_b) - \gamma_a\, x_a$$

$$\frac{dx_b}{dt} = \kappa_b\, h^-(x_a, \theta_a, m_a) - \gamma_b\, x_b$$

$$h^-(x, \theta, m) = \frac{\theta^m}{x^m + \theta^m}$$

(a)   (b)

Fig. 2. (a) Nonlinear ordinary differential equation model of the mutual-inhibition network (figure 1). The variables x_a and x_b correspond to the concentrations of proteins A and B, respectively, the parameters κ_a and κ_b to the synthesis rates of the proteins, the parameters γ_a and γ_b to the degradation rates, the parameters θ_a and θ_b to the threshold concentrations, and the parameters m_a and m_b to the degree of cooperativity of the interactions. All parameters are positive. (b) Graphical representation of the characteristic sigmoidal form, for $m > 1$, of the Hill function $h^-(x, \theta, m)$.

Because of the nonlinearity of the functions f, the solutions of the system of ordinary differential equations (2.2) cannot generally be determined by analytical means. However, because the model of the two-gene network (Figure 2) has only two variables, we can obtain a qualitative understanding of the dynamics of the network, by applying phase-plane analysis tools [31, 55].

The phase portrait in Figure 3(a) shows that the system is *bistable*, in the sense that it possesses two asymptotically stable equilibrium points, at which either protein A or protein B is present at a high concentration. The third equilibrium point, characterized by intermediate concentrations for proteins A and B, is unstable and has no biological significance. The phase-plane analysis also reveals that the system exhibits *hysteresis*. If one strongly perturbs the system from one of its stable equilibria – for instance, by provoking the degradation of the protein present at a high concentration – the other equilibrium can be reached (Figure 3(b)). From then onwards, even if the source of strong degradation has disappeared, the system will remain at the new equilibrium. In other words, the analysis suggests that a simple molecular mechanism may allow the system to switch from one functional state to another.

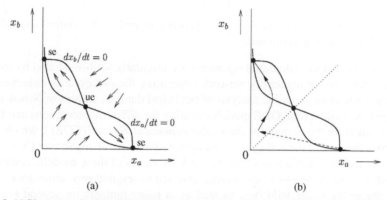

(a)   (b)

Fig. 3. (a) Phase portrait of the differential equation model of the cross-inhibition network (Figure 2). The system has two asymptotically stable equilibrium points (se) and one unstable equilibrium point (ue). The equilibria lie at the intersection of the nullclines of x_a and x_b (drawn curves annotated by $dx_a/dt = 0$ and $dx_b/dt = 0$). (b) Hysteresis effect, resulting from a transient perturbation of the system (dashed line with arrow).

It is important to remark that the above analysis is not just a theoretical exercise. In fact, the properties of the mutual inhibition network revealed by the analysis – bistability and hysteresis – have been experimentally investigated. The novelty of the study by Gardner *et al.* [18] is that the network of Figure 1 has been reconstructed in *Escherichia coli* cells by cloning the genes on a plasmid. The genes have been chosen such that the activity of the corresponding proteins can be regulated by external signals. In addition, a reporter gene has been added to allow the state of the system to be measured. The resulting mutual-inhibition network functions independently from the rest of the cell, like a 'genetic applet', in the words of the authors. Carefully-chosen experiments have shown that the system is bistable and can switch from one equilibrium to the other following chemical or heat induction.

Generally, for networks having more than two genes, an analysis in the phase plane is no longer possible. In certain cases, one can reduce the dimension of the system by simplifying the model, but most of the time, numerical techniques become necessary. *Numerical simulation* approximates the exact solution of the system of equations, by computing approximate values $x_0, \ldots, x_m$ for x at consecutive time-points $t_0, \ldots, t_m$ (see [36] for an introduction). Many computer tools for numerical simulation have been developed, some specifically adapted to networks of molecular interactions. Well-known examples of the latter are GEPASI [41], DBsolve [24], XPPAUT [15], and Ingeneue [40]. Recently, formats for the exchange of models between different simulation tools have appeared [27].

2.2. Analysis of regulatory networks involved in cell-cycle control, circadian rhythms, and development

The use of ordinary differential equations is particularly well-illustrated by modeling studies of the regulatory network controlling the *cell cycle* in eukaryotes. On the basis of an extensive analysis of published data, the groups of Novak and Tyson have built several ODE models covering the different interactions and factors controlling the activity of the *cyclin-dependent kinases (CDKs)*, which are found at the core of the cell-cycle regulatory networks in all eukaryotes [62]. The analysis and the simulations of the network by means of these models, comprising up to several dozens of equations, have led to testable prediction about the cell phenotype in the wild type as well as in many mutants for several organisms, including yeast and mammals [7, 45]. One of these predictions has been experimentally corroborated, whereas others appear to be in contradiction with available data, pointing toward the necessity to take into account new elements or interactions in the model [9].

Most ODE simulation and analysis techniques require precise numerical values for kinetic parameters and molecular concentrations, but unfortunately this information is rarely available. As an alternative, one could decide to explore the parameter space and check the behavior of the system. Although computationally intensive and hardly scalable, such an approach has been successfully applied to the modeling of a cross-regulatory module involved in the development of the fruit fly *Drosophila melanogaster* [64] (see also [40]). The module consists of the main segment-polarity genes responsible for the segmentation of the fly embryo. From their simulation study, von Dassow *et al.* concluded that this regulatory module is robust and can produce roughly correct gene expression patterns for substantial ranges of values for most parameters. The module is also robust with respect to the initial conditions, notably the gene expression pre-pattern.

Working on the same biological system, but concentrating on events occurring at an earlier developmental stage, the group of Reinitz has developed a reverse-engineering method for deriving the values of little-known kinetic parameters [48]. Starting with a differential equation model with unspecified parameter values, a simulated annealing and a gradient descent method have been used to fit the spatial-temporal behavior of the model with temporal series of digitalised embryo images. On the basis of this approach, Reinitz *et al.* have identified the mechanisms constraining the expression of the four gap genes across the trunk of the embryo, as well as emphasized specific dynamical properties of these expression patterns, notably spatial shifts, that had previously gone unnoticed [29, 30].

A final example of the application of ODE models concerns the analysis of the mammalian circadian clock by Leloup and Goldbeter [37] (see [23] for a review). The model is based on intertwined positive and negative feedback loops involving a number of genes identified in molecular studies of circadian rhythms. The model consists of some twenty equations with parameter values chosen semiarbitrarily in their physiological range, so as to satisfy constraints set by experimental observations (*e.g.*, a period of oscillations in continuous darkness close to 24 h). Analysis of the models uncovers the possible existence of multiple sources of oscillatory behavior. That is, in conditions where one negative feedback loop is inactive, a second negative feedback loop could take over. Another interesting suggestion is to use the model to explore syndromes or pathological conditions resulting from disorders of circadian rhythms. In fact, the model shows how changes in the value of certain control parameters can be related to perturbations of the human circadian clock.

3. Piecewise-linear differential equations

3.1. Models and analysis

Consider again the ordinary differential equation model of the cross-inhibition network in Figure 2. The model can be simplified by replacing the sigmoid Hill function h^- by a *step function* s^- : $S \times \mathbb{R}_{>0} \to \mathbb{R}_{\geq 0}$, $S \subset \mathbb{R}_{\geq 0}$, as shown in Figure 4. For concentrations x below the threshold θ, $s^-(x, \theta)$ equals 1, whereas for concentrations x above θ, the function evaluates to 0. The intuitive justification of this approximation is that as the sigmoid function becomes increasingly steep, it approaches the step function. The step-function approximation results in so-called piecewise-linear differential equation models which facilitate the qualitative analysis of the dynamics of large networks. This has obvious advantages at a time when reliable measurements of the kinetic constants are not available for most systems of biological interest (Section 2.2).

More precisely, the dynamics of genetic regulatory networks can be modeled by a class of *piecewise-linear (PL) differential equations*, originally proposed by Glass and Kauffman [21] (see also [43, 59]):

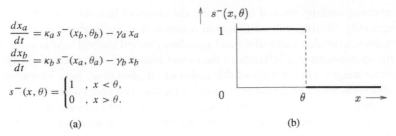

(a)   (b)

Fig. 4. (a) Piecewise-linear differential equation model of the mutual-inhibition network (Figure 1). The variables and the parameters have the same interpretation as in Figure 2. (b) Graphical representation of the step function $s^-(x, \theta)$.

$$\frac{dx}{dt} = h(x) = f(x) - g(x)\,x,\qquad\qquad (3.1)$$

where $x = (x_1, \ldots, x_n)' \in \Omega$ is a vector of cellular protein concentrations and $f = (f_1, \ldots, f_n)'$, $g = \mathrm{diag}(g_1, \ldots, g_n)$. The rate of change of each protein concentration x_i, $i \in [1, \ldots, n]$, is thus defined as the difference of the rate of synthesis $f_i(x)$ and the rate of degradation $g_i(x)\,x_i$ of the protein. The function $f_i : \Omega \to \mathbb{R}_{\geq 0}$ consists of a sum of step function expressions, each weighted by a rate parameter, which expresses the logic of gene regulation [43, 59]. The function $g_i : \Omega \to \mathbb{R}_{>0}$ is defined analogously. In our example, these functions have a simple form, for instance $f_a(x_b, x_b) = \kappa_a\, s^-(x_b, \theta_b^1)\, s^-(x_a, \theta_a^2)$ and $g_a = \gamma_a$ in the case of gene a (Figure 4). More complex expressions can represent the combined effects of several regulatory proteins.

The dynamical properties of the piecewise-linear models can be analyzed in Ω. Given that the protein encoded by gene i has p_i threshold concentrations, the $n - 1$-dimensional threshold hyperplanes $x_i = \theta_i^{k_i}$, $k_i \in [1, \ldots, p_i]$, partition Ω into (hyper)rectangular regions that are called *domains* [13]. Figure 5(a) shows the subdivision into domains of the two-dimensional phase space box of the cross-inhibition network. We distinguish between domains like D^4 and D^7, which are located on (intersections of) threshold planes, and domains like D^1, which are not. The former domains are called *switching* domains, whereas the latter are called *regulatory* domains. The phase space box in Figure 5(b) is partitioned into 4 regulatory and 9 switching domains.

When evaluating the step-function expressions of (3.1) in a regulatory domain D, we obtain a system of differential equations of a particularly simple form: the equations are linear and uncoupled. For such a system it is easy to show that all solution trajectories monotonically converge towards a so-called *focal point* $\phi(D)$ [21]. If $\phi(D) \in D$, then $\phi(D)$ is a stable equilibrium point of the system.

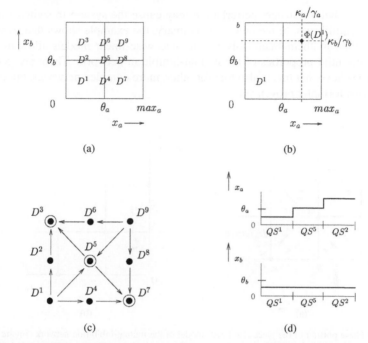

(a)   (b)

(c)   (d)

Fig. 5. (a) Partition of the phase space into regulatory and switching domains. (b) Analysis of the behavior of the system in regulatory domain D^1, under the assumption that $\theta_a < \kappa_a/\gamma_a < max_a$ and $\theta_b < \kappa_b/\gamma_b < max_b$. (c) Transition graph consisting of domains and transitions between domains. The domains containing equilibrium points have been circled. (d) Detailed description of the sequence of domains (D^1, D^4, D^7).

If not, then the solution trajectories will leave D at some point and enter another domain. For instance, in domain D^1 the piecewise-linear model of the mutual-inhibition network reduces to the equations $\dot{x}_a = \kappa_a - \gamma_a x_a$ and $\dot{x}_b = \kappa_b - \gamma_b x_b$, while the focal point is $\phi(D^1) = (\kappa_a/\gamma_a, \kappa_b/\gamma_b)'$ and $\phi(D^1) \notin D^1$ (Figure 5(b)). In the case of switching domains, where discontinuities can occur, the situation is more complicated. Gouzé and Sari [25] have shown that by extending the differential equations to differential inclusions, following an approach originally proposed by Filippov [17], the dynamics of the system in switching domains can be described in an analogous manner to the dynamics in regulatory domains.

The local analyses of the dynamics of the system in the different regions of the phase space can be combined into a global analysis, as illustrated in Figure 6(a). The predictions of the piecewise-linear model are qualitatively equivalent to those obtained by the nonlinear model (Section 2.1). The network has three equilibrium points, two of which are stable and one unstable. Part (b) of the figure shows that a transient perturbation may cause the system to switch from one stable equilibrium to the other. In summary, the example shows that, while facilitating the mathematical analysis, the piecewise-linear models allow us to preserve essential properties of the mutual-inhibition network. There are good reasons to believe that this is also true for other, more complex networks, but this has not been formally proven yet.

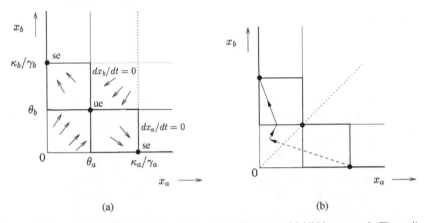

(a)   (b)

Fig. 6. (a) Phase portrait of the piecewise-linear model of the mutual-inhibition network (Figure 4). The system has two stable equilibrium points (se) and one unstable equilibrium point (ue). (b) Hysteresis phenomenon, following a transient perturbation of the system (broken line with arrow).

The analyse of the piecewise-linear model of the cross-inhibition network suggests a discrete, more compact representation of the dynamics of the system [13]. In fact, in every domain of the phase space the system behaves in a qualitatively homogeneous way. For instance, in domain D^1 all solution trajectories converge towards $\phi(D^1) = (\kappa_a/\gamma_a, \kappa_b/\gamma_b)'$ and the monotonicity of the solutions implies that the concentrations of x_a and x_b increase. This allows the definition of a discrete or qualitative abstraction of the system [1, 35], resulting in a *transition graph* consisting of domains and transitions between domains. Two contiguous domains are connected by a transition, if there exists a solution starting in the first domain that reaches the second domain, without passing through a third domain. This is the case for the solutions in D^1 which, while converging towards $\phi(D^1)$, reach D^5, D^6 or D^9. The transition graph obtained for the model of the cross-inhibition network is shown in Figure 5(c).

The discrete representation of the dynamics of the continuous system facilitates the analysis of its dynamics. For instance, the transition graph provides information on the reachability of an equilibrium point from a given initial domain. If the equilibrium point is reachable, there must exist a path in the graph going from the initial domain to the domain containing the equilibrium point. We have shown that the transition graph is invariant for certain inequality constraints on the parameters, which can be inferred from the experimental literature. In fact, it is not difficult to verify that for all parameter values $\theta_a < \kappa_a/\gamma_a < max_a$ and $\theta_b < \kappa_b/\gamma_b < max_b$, the transition graph in Figure 5(c) is obtained. That is, the properties of the graph represent qualitative properties of the dynamics of the system.

The transition graph can be computed by means of simple symbolic rules from the piecewise-linear model completed by inequality constraints. The algorithms have been implemented in the computer tool *Genetic Network Analyzer (GNA)* [12, 22], which allows the computation of all domains reachable from a given set of initial domains (*qualitative simulation*). In order to enable the analysis of large transition graphs in an efficient and reliable manner, the qualitative simulator has been connected to model-checking tools for the automatic verification of dynamic properties expressed in temporal logic [3,8]. GNA has been used to study various prokaryotic networks [11,50,63]. In the next section, we present the results of the qualitative simulation of the network controlling the initiation of sporulation in *Bacillus subtilis*.

3.2. Simulation of the initiation of sporulation in Bacillus subtilis

Under conditions of nutrient deprivation, the Gram positive soil bacterium *Bacillus subtilis* can abandon vegetative growth and form a dormant, environmentally-resistant spore instead [5]. During vegetative growth, the cell divides symmetrically and generates two identical cells. During sporulation, on the other hand, cell division is asymmetric and results in two different cell types: the smaller cell (the forespore) develops into the spore, whereas the larger cell (the mother cell) helps to deposit a resistant coat around the spore and then disintegrates (Figure 7).

The decision to abandon vegetative growth and initiate sporulation involves a radical change in the gene expression program of the cell. This switch is controlled by a complex genetic regulatory network integrating various environmental, cell-cycle, and metabolic signals. Due to the ease of genetic manipulation of *B. subtilis*, it has been possible to identify and characterize a large number of the genes, proteins, and interactions making up this network. Currently, more than 125 genes are known to be involved [16]. A graphical representation of the regulatory network controlling the initiation of sporulation is shown in Figure 8, displaying key genes and their promoters, proteins encoded by the genes, and the regulatory action of the proteins.

The network is centered around a *phosphorelay*, which integrates a variety of environmental, cell-cycle, and metabolic signals. Under conditions appropriate

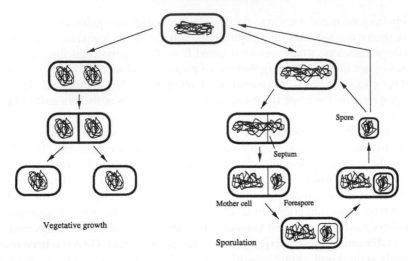

Fig. 7. Life cycle of *B. subtilis*: decision between vegetative growth and sporulation (adapted from [38]).

for sporulation, the phosphorelay transfers a phosphate to the SpoOA regulator, a process modulated by kinases and phosphatases. The phosphorelay has been simplified here by ignoring intermediate steps in the transfer of phosphate to SpoOA. However, this simplification does not affect the essential function of the phosphorelay: modulating the phosphate flux as a function of the competing action of kinases and phosphatases (here KinA and SpoOE). Under conditions conducive to sporulation, such as nutrient deprivation or high population density, the concentration of phosphorylated SpoOA (SpoOA~P) may reach a threshold value above which it activates various genes that commit the bacterium to sporulation. The choice between vegetative growth and sporulation in response to adverse environmental conditions is the outcome of competing positive and negative feedback loops, controlling the accumulation of SpoOA~P.

Not withstanding the enormous amount of work devoted to the elucidation of the network of interactions underlying the sporulation process, very little quantitative data on kinetic parameters and molecular concentrations are available. de Jong and colleagues have therefore used the qualitative simulation method introduced in Section 3.1 to analyze the network [11]. The objective of the study was to reproduce the observed qualitative behavior of wild-type and mutant bacteria from a model integrating data available in the literature. To this end, the graphical representation of the network has been translated into a piecewise-linear model supplemented by inequality constraints on the parameters. The resulting model

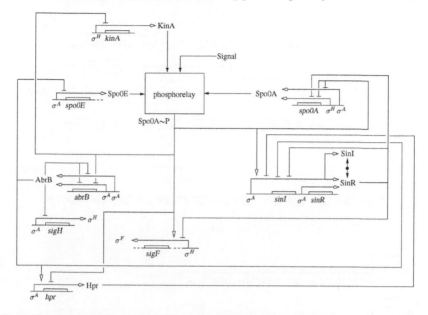

Fig. 8. Key genes, proteins, and regulatory interactions making up the network involved in the initiation of sporulation in *B. subtilis*. In order to improve the legibility of the figure, the control of transcription by the sigma factors σ^A and σ^H has been represented implicitly, by annotating the promoter with the corresponding sigma factor (figure reproduced from [11]).

consists of nine state variables and two input variables. The 48 parameters are constrained by 70 parameter inequalities, the choice of which is largely determined by biological data.

The tool GNA [12] has been used to simulate the response of a wild-type *B. subtilis* cell to nutrient depletion and high population density. Starting from initial conditions representing vegetative growth, the system is perturbed by a sporulation signal that causes KinA to autophosphorylate. Simulation of the network takes less than a few seconds to complete on a PC (500 MHz, 128 MB of RAM), and gives rise to a transition graph of 465 domains. Many of these domains are traversed instantaneously by the system and, since their biological relevance is limited, can be eliminated from the transition graph. This leads to a reduced transition graph with 82 domains.

The transition graph faithfully represents two possible responses to nutrient depletion that are observed for *B. subtilis*: either the bacterium continues vegetative growth or it enters sporulation. Sequences of domains typical for these two developmental modes are shown in Figure 9. The initiation of sporulation is determined by positive feedback loops acting through Spo0A and KinA, and a

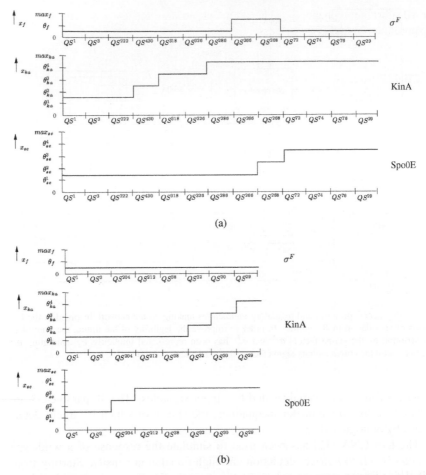

Fig. 9. (a) Temporal evolution of selected protein concentrations in a typical qualitative behavior corresponding to the *spo*+ phenotype. (b) Idem, but for a typical qualitative behavior corresponding to the *spo*⁻ phenotype (figure adapted from [11]).

negative feedback loop involving Spo0E. If the rate of accumulation of the kinase KinA outpaces the rate of accumulation of the phosphatase Spo0E, we observe transient expression of *sigF*, *i.e.* a *spo*+ phenotype (Figure 9(a)). The gene *sigF* is a sigma factor essential for the development of the forespore [54]. If the kinetics of these processes are inversed, *sigF* is never activated and we observe a *spo*⁻ phenotype (Figure 9(b)). Deletion or overexpression of genes in the network of Figure 8 may disable a feedback loop, leading to specific changes in the observed sporulation phenotype. The results of the simulation of a dozen examples of sporulation mutants are discussed in [11].

4. Logical models

4.1. Models and analysis

Starting in the early 1970s, several authors have proposed to simplify the description of regulatory networks even further through the use of Boolean algebra [32, 33, 57, 58]. In this context, a regulatory product is considered as *present* (*absent*) when its concentration or activity exceeds (remains below) a certain threshold level. Interactions between genes can then be formalized in terms of logical equations of the form

$$v_i(t + 1) = F_i(v(t)), \ i \in [1, \ldots, n], \tag{4.1}$$

where v_i is a Boolean variable representing the activity of gene i, $F_i : \{0, 1\}^n \to \{0, 1\}$ a Boolean function, and $v = (v_1, \ldots, v_n)'$ the variables associated with genes in the network.

Coming back to the example of the cross-inhibitory network of Figure 2, the Boolean description leads to the following *truth table*:

v_a	v_b	F_a	F_b
0	0	1	1
0	1	0	1
1	0	1	0
1	1	0	0

This table encompasses the whole dynamics of the Boolean model. Examination of the table leads to the following conclusions:
• the states 01 and 10 are stable, since, in both cases, each component of the variable vector $(v_a, v_b)'$ is equal to each corresponding component of the function vector $(F_a, F_b)'$. In order to indicate that they are stabe, 01 and 10 are written as [01] and [10];
• in the cases of states 00 and 11, both components of the variable vector differ from the corresponding components of the function vector. Consequently, both genes are called to change their levels, upwards in the first case, downwards in the second case.

Many authors have treated these equations under a *synchronicity* assumption [33], that is, the computation of $v_i(t + 1)$ in terms of $v(t)$ is carried out synchronously for all i. In our example, this leads to the state transition graph of Figure 10(a).

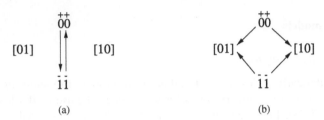

(a)   (b)

Fig. 10. State transition graphs for the two-gene cross-inhibition network of Figure 1. Each node represent a logical state of the system, and the Boolean components of these nodes represent the qualitative levels of the two regulatory products. The graphs in (a) and (b) are obtained by synchronous and asynchronous updating, respectively.

Such a synchronous treatment implies that the synthesis and the degradation of the different regulatory products occur at identical rates, which does not seem likely in most biological situations. More generally, the synchronous assumption often leads to simulation artefacts – notably, spurious dynamical cycles with multiple synchronous transitions (Figure 10(a)) – which can be avoided by considering specific time delays for each (upward or downward) change in value of a variable. As information about the relative values of transition delays is scarcely available, one often represents all alternative *asynchronous* pathways in the state transition graph. In the case of our two-gene toggle switch, this asynchronous assumption gives rise to the state transtion graph in Figure 10(b). We find the same stable states as under the synchronous updating assumption. However, the transitions from the other two states now lead to the two stable states, which fits the biological intuition much better.

In many situations, however, a Boolean representation oversimplifies the system being modeled, leading to the loss of important qualitative information, even sometime impeding the generation of biologically meaningful results. This realization led Thomas [59] to generalize the logical approach in order to:

- use multilevel variables whenever needed (*i.e.*, variables taking the values $0, 1, 2, \ldots$);
- define logical parameters to replace the logical operators;
- while still treating value changes under an asynchronicity assumption.

In the following section, this generalized logical formalism is illustrated in the context of the modeling of the regulatory network controlling the lysis-lysogeny decision in the bacteriophage lambda.

4.2. Modeling of the lysis-lysogeny decision during the infection of Escherichia coli by bacteriophage lambda

Bacteriophages are viruses infecting bacteria. Depending on the bacterium state, some bacteriophages, such as the phage lambda, can either multiply and kill the host bacterium quickly after the infection (*lysis*), or rather remain silent in the form of a piece of phage genome integrated into and replicated with the bacterial genome (*lysogeny*).

A number of bacterial and viral genes take part in the decision between lysis and lysogenization in temperate bacteriophages. In the case of the bacteriophage lambda, at least five viral regulatory products (CI, Cro, CII, N, and CIII) and several bacterial genes are involved (see [47] for an extensive overview). A schematic description of this network is given in Figure 11. Several attempts have been made to model this well-studied but yet relatively complex regulatory network, using discrete, differential or stochastic formalisms (see, *e.g.*, [2,49,56], and references therein).

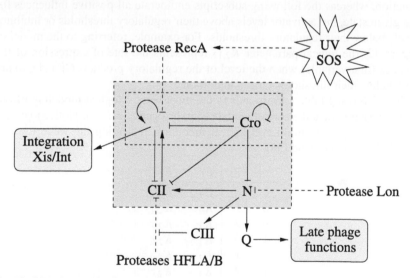

Fig. 11. Main interactions between the viral and bacterial regulatory genes controlling the lysis-lysogeny decision in bacteriophage lambda.

In order to illustrate the flexibility of the generalized logical formalism, we will first focus here on a two-variable model of the core of the lambda regulatory network, which consists of the cross-regulation between the regulatory genes *cI* (encoding the repressor) and *cro*. Figure 12 gives the regulatory graph of the central switch of the regulatory network controlling the lysis-lysogeny decision

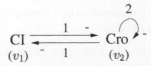

Fig. 12. Regulatory graph involving the genes *cI* and *cro* of bacteriophage lambda. The logical variable corresponding to each gene is indicated, as well as the type of interaction (−/inhibition) and the minimal value for which the interaction occurs.

in the bacteriophage lambda. Notice the similarity between the *cI-cro* network and the cross-inhibition network in Figure 1.

In the context of the generalized logical formalism, one can compute a general state table giving the values of the logical functions (rates of gene expression) corresponding to each value combination of the variables (concentration or activity levels of regulatory products), in terms of logical parameters (Ks). In the notation used here for these logical parameters, the first subscript identifies the function, whereas the following subscripts enumerate all positive influences for the given state, *i.e.* activator levels above their regulatory thresholds or inhibitor levels below their regulatory thresholds. For example, referring to the model of Figure 12, the logical parameter $K_{2.12}$ represents the rate of expression of the gene *cI* (first variable) when the level of the regulatory products CI and Cro are both below their thresholds for *cro* inhibition.[1]

The following table corresponds to the most natural logical modeling of our two-gene network at the core of lambda lysis-lysogeny decision. Notice that here F_1 and F_2 represent multilevel logical functions, indicating the value towards which the variables tend [58].

v_1	v_2	F_1	F_2
0	0	$K_{1.2}$	$K_{2.12}$
0	1	K_1	$K_{2.12}$
0	2	K_1	$K_{2.1}$
1	0	$K_{1.2}$	$K_{2.2}$
1	1	K_1	$K_{2.2}$
1	2	K_1	K_2

Note that a Boolean variable has been associated with the regulatory product CI, as it affects the expression of a single gene (*cro*) in this simple network, whereas a ternary variable has been associated with Cro. Furthermore, the highest level of Cro ($v_2 = 2$) is considered here as necessary for the auto-inhibition to occur.

---

[1] Another notation used in the literature consists in writing a subscript when the level of the regulatory product is over its corresponding interaction threshold, whatever the sign of the interaction.

Depending on the values given to the logical parameters, this table covers different dynamical situations. In the following table, we have selected the parameter set which best corresponds to the available experimental data.

v_1	v_2	F_1	F_2
0	0	1	2
0	1	0	2
0	2	0	0
1	0	1	0
1	1	0	0
1	2	0	0

For the same parameter values, Figure 13 shows all possible transitions in the form of a state transition graph.

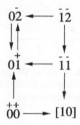

Fig. 13. State transition graph for the two-gene core regulatory network of bacteriophage lambda. Each node represent a logical state of the system, and the Boolean components of these nodes represent the qualitative levels of CI and Cro, in that order. The superscripts +/− denote the direction of change of each variable.

This graph encompasses a single stable state [10], which represent the lysogenic pathway (only the repressor CI is present). The lytic state is represented by oscillations between states 01 and 02. Accordingly, the level of Cro regulatory product is tightly regulated around its second threshold, due to the auto-inhibition of Cro at high levels of this protein. This would correspond to a stable equilibrium located on the threshold in the piecewise-linear description (Section 4). Note that both descriptions make some kind of simplification. In the first (logical) case, the thresholds are implicit, whereas in the second (piecewise-linear) case, intrinsic delays arising from protein synthesis are not taken into account, thereby allowing the system to reach and sometimes remain on threshold planes in the phase space. In the differential version of this model, one would also find a saddle point at the intersection of the lowest Cro and CI thresholds, located on the separatrix dividing the phase space into two basins of attraction. This saddle point is also implicit in the discrete formalism, although it is possible to qualitatively localize the separatrix as it involves the states which lie at the origin of alternative transitions (*i.e.*, states 00, 11, and 12).

Coming back to the regulatory network of Figure 11, one can derive more complex models, encompassing all the main viral regulatory genes, and perhaps even the bacterial regulators. For sake of space, we only provide here a brief overview of a four-gene multilevel model encompassing the main interactions between the viral genes *cI*, *Cro*, *cII*, and *N*. The corresponding regulatory graph is presented in Figure 14.

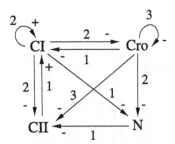

Fig. 14. Regulatory graph for the network involving the genes *cI*, *Cro*, *cII*, and *N* of bacteriophage lambda (adapted from [56]). The type of interaction (+/activation or −/inhibition) has been indicated, as well as the minimal value for the interaction to occur.

For proper parameter values, this model allows a qualitative reproduction of the dynamical behavior of the wild-type as well as of many mutant phages. Figure 15 shows the most relevant dynamical pathways leading to lysis and to lysogenesis for the wild-type phage.

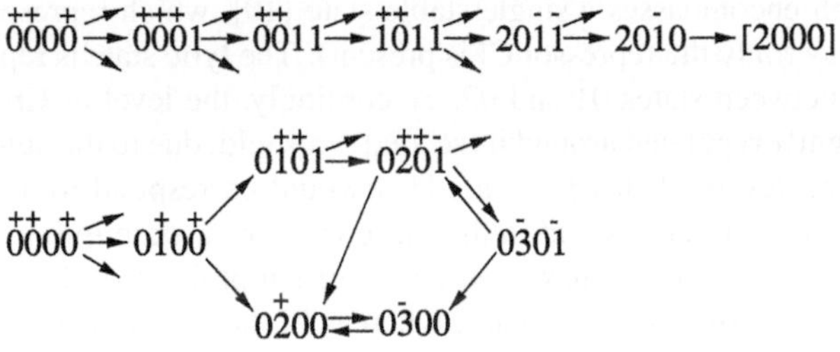

$$0\overset{++}{0}\overset{+}{0}0 \leftrightarrows 00\overset{+++}{0}1 \leftarrow 00\overset{++}{1}1 \leftarrow \overset{+}{1}0\overset{+}{1}1 \leftarrow 20\overset{-}{1}1 \leftarrow 2010 \rightarrow [2000]$$

$$0\overset{++}{0}\overset{+}{0}0 \leftrightarrows 0\overset{+}{1}0\overset{+}{0} \qquad 0\overset{++}{1}01 \leftarrow 0\overset{++}{2}0\overset{-}{1} \qquad 0\overset{-}{3}0\overset{-}{1}$$

$$0\overset{+}{2}00 \leftrightarrows 0\overset{-}{3}00$$

Fig. 15. Main dynamical pathways leading to lysis (bottom) *versus* lysogenesis (top) upon infection of *E. coli* by the bacteriophage lambda, obtained for the regulatory graph in Figure 14 (adapted from [56]). Each node represent a logical state of the system, and the discrete components of these nodes represent the qualitative levels of CI, Cro, CII, and N regulatory products, in that order. The superscripts +/− denote the direction of change of each variable.

The pathways of Figure 15 include the only two attractors of the system. The first of these attractors is the stable state [2000] corresponding to the lysogenic pathway, whereas the second attractor is a cycle between 0200 and 0300, thus corresponding to the homeostatic regulation of Cro after the onset of the lytic pathway. Note that *cII* and *N* are expressed only transiently shortly after the infection and do not play any role in later stages. By an large, this four-gene model has asymptotical dynamical properties very similar to those of our former, simpler two-gene model. In addition, it allows an explanation of the transient dynamical roles of the regulatory products of *cII* and *N*: the balance between the four regulatory products is crucial for the lysis-lysogeny decision. Furthermore, on the basis of this model, it become possible to simulate a much broader range of perturbations, in particular mutants.

4.3. *Extensions of logical modeling*

An important aspect of the generalized logical formalization lies in the possibility of computing specific parameter value constraints enabling any given feedback circuit (*i.e.*, a closed chain of interactions) to generate specific dynamical properties [59, 60]. Indeed, it is possible to show that, for proper parameter values, *positive circuits* (involving an even number of negative interactions) generate multistationarity, whereas *negative circuits* (involving an odd number of negative interactions) may generate homeostasis, possibly accompanied with damped or sustained oscillations. For the parameter values selected, the simulation of Figure 13 corresponds to a situation where both circuits (one positive two-gene circuit and one negative auto-inhibitory circuit) are functional. The positive circuit is at the origin of the coexistence of two attractors: one stable state and one cycle, whereas the negative circuit is at the origin of the homeostatic regulation of Cro (cycle between states 01 and 02). An easy way to check these relationships consists in changing the values of parameters, specifically affecting one or the two circuits. More generally, in the context of the generalized logical formalism, it is possible:

• for given parameter values, to check the potential dynamical role (or functionality) of each feedback circuit;

• to compute the parameter constraints to render any (combination) of circuit(s) simultaneously functional;

• to analytically compute the parameter constraints to have one specific (set of) stable state(s);

• for given parameter constrains, to compute all stable states of a complex multilevel logical model.

From a computational perspective, the logical approach has been implemented in a Java software suite called *GIN-sim* [6, 20]. Both synchronous and asynchronous simulation tools have been fully implemented. Starting with a set of initial conditions (*i.e.*, initial state(s) and parameter values), GIN-sim generates a state transition graph, which qualitatively represents all permitted state transitions corresponding to the network structure encoded in the original regulatory graph. The initial conditions and the parameter values can be defined by the user or by default, including the number of distinct levels for each regulatory product, and the qualitative weights of the different combinations of interactions on each gene (*i.e.* the values of the logical parameters).

In addition, we are implementing a series of graph analysis tools to delineate regulatory modules or interesting dynamical components (*i.e.*, subgraphs extracted from the state transition graph). Further tools being implemented are the feedback circuit analysis mentioned above, as well as a new symbolic computational approach allowing the analytic derivation of all stable states of a logical model, thereby avoiding the enumeration of all logical states in order to find those which correspond to interesting attractors [14].

This logical approach has been applied to various biological regulatory systems, in particular to the modeling of differentiation pathways during *Drosophila* development ([42,51,52,59].

5. Conclusions

In this chapter, we have discussed three formalisms for the modeling, analysis, and simulation of genetic regulatory networks: ordinary differential equations, piecewise-linear differential equations, and logical models. Ordinary differential equation models are able to make precise, quantitative predictions of the network dynamics. However, their practical use is often compromised by the general absence of quantitative information on kinetic parameters and molecular concentrations. Piecewise-linear differential equation and logical models trade quantitative precision for the ability to make predictions when only weak, qualitative information is available. Although these two qualitative approaches have been elaborated in different formal contexts, they are based on quite similar abstractions of the underlying biological processes. The application of the three approaches has been illustrated by means of several examples of prokaryote and eukaryote networks.

Acknowledgments

H. de Jong acknowledges financial support from the ARC initiative at INRIA (project GDyn), the ACI IMPBio initiative of the French Ministry for Research (project BacAttract), and the NEST Adventure programme of the European Commission (project Hygeia). The work of D. Thieffry has been supported by the ACI IMPBio initiative of the French Ministry for Research. The authors thank C. Chaouiya, J. Geiselmann, J.-L. Gouzé, M. Page, and T. Sari for contributions to the work reviewed in Sections 3 and 4 of this chapter.

References

[1] R. Alur, T.A. Henzinger, G. Lafferriere, and G.J. Pappas. Discrete abstractions of hybrid systems. *Proc. IEEE*, 88(7):971–984, 2000.

[2] A. Arkin, J. Ross, and H.A. McAdams. Stochastic kinetic analysis of developmental pathway bifurcation in phage λ-infected *Escherichia coli* cells. *Genetics*, 149(4):1633–1648, 1998.

[3] G. Batt, D. Ropers, H. de Jong, J. Geiselmann, R. Mateescu, M. Page, and D. Schneider. Validation of qualitative models of genetic regulatory networks by model checking: Analysis of the nutritional stress response in *Escherichia coli*. *Bioinformatics*, 2005. To appear.

[4] J.M. Bower and H. Bolouri, editors. *Computational Modeling of Genetic and Biochemical Networks*. MIT Press, 2001.

[5] W.F. Burkholder and A.D. Grossman. Regulation of the initiation of endospore formation in *Bacillus subtilis*. In Y.V. Brun and L.J. Shimkets, editors, *Prokaryotic Development*, pages 151–166. American Society for Microbiology, 2000.

[6] C. Chaouiya, E. Remy, B. Mossé, and D. Thieffry. Qualitative analysis of regulatory graphs: A computational tool based on a discrete formal framework. In *Positive Systems (POSTA 2003)*, volume 294 of *LNCIS*, pages 119–126. Springer-Verlag, 2003.

[7] K.C. Chen, L. Calzone, A. Csikasz-Nagy, F.R. Cross, B. Novak, and J.J. Tyson. Integrative analysis of cell cycle control in budding yeast. *Mol. Biol. Cell*, 15(8):3841–3862, 2004.

[8] E.M. Clarke, O. Grumberg, and D.A. Peled. *Model Checking*. MIT Press, 1999.

[9] F.R. Cross, V. Archambault, M. Miller, and M. Klovstadt. Testing a mathematical model of the yeast cell cycle. *Mol. Biol. Cell*, 13(1):52–70, 2002.

[10] H. de Jong. Modeling and simulation of genetic regulatory systems: A literature review. *J. Comput. Biol.*, 9(1):67–103, 2002.

[11] H. de Jong, J. Geiselmann, G. Batt, C. Hernandez, and M. Page. Qualitative simulation of the initiation of sporulation in *B. subtilis*. *Bull. Math. Biol.*, 66(2):261–299, 2004.

[12] H. de Jong, J. Geiselmann, C. Hernandez, and M. Page. Genetic Network Analyzer: Qualitative simulation of genetic regulatory networks. *Bioinformatics*, 19(3):336–344, 2003.

[13] H. de Jong, J.-L. Gouzé, C. Hernandez, M. Page, T. Sari, and J. Geiselmann. Qualitative simulation of genetic regulatory networks using piecewise-linear models. *Bull. Math. Biol.*, 66(2):301–340, 2004.

[14] V. Devloo, P. Hansen, and M. Labbé. Identification of all steady states in large networks by logical analysis. *Bull. Math. Biol.*, 65(6):1025–1052, 2003.

[15] B. Ermentrout. *Simulating, Analyzing, and Animating Dynamical Systems: A Guide to XPPAUT for Researchers and Students*. SIAM, 2002.

[16] P. Fawcett, P. Eichenberger, R. Losick, and P. Youngman. The trancriptional profile of early to middle sporulation in *Bacillus subtilis*. *Proc. Natl. Acad. Sci. USA*, 97(14):8063–8068, 2000.

[17] A.F. Filippov. *Differential Equations with Discontinuous Righthand Sides*. Kluwer Academic Publishers, 1988.

[18] T.S. Gardner, C.R. Cantor, and J.J. Collins. Construction of a genetic toggle switch in *Escherichia coli. Nature*, 403(6767):339–342, 2000.

[19] A. Gilman and A.P. Arkin. Genetic 'code': Representations and dynamical models of genetic components and networks. *Ann. Rev. Genom. Hum. Genet.*, 3:341–369, 2002.

[20] http://www.esil.univ-mrs.fr/~chaouiya/GINsim/.

[21] L. Glass and S.A. Kauffman. The logical analysis of continuous non-linear biochemical control networks. *J. Theor. Biol.*, 39(1):103–129, 1973.

[22] http://www-helix.inrialpes.fr/gna.

[23] A. Goldbeter. Computational approaches to cellular rhythms. *Nature*, 420(6912):238–245, 2002.

[24] I. Goryanin, T.C. Hodgman, and E. Selkov. Mathematical simulation and analysis of cellular metabolism and regulation. *Bioinformatics*, 15(9):749–758, 1999.

[25] J.-L. Gouzé and T. Sari. A class of piecewise linear differential equations arising in biological models. *Dyn. Syst.*, 17(4):299–316, 2002.

[26] J. Hasty, D. McMillen, F. Isaacs, and J.J. Collins. Computational studies of gene regulatory networks: *In numero* molecular biology. *Nat. Rev. Genet.*, 2(4):268–279, 2001.

[27] M. Hucka, A. Finney, H.M. Sauro, H. Bolouri, J.C. Doyle, H. Kitano, A.P. Arkin, B.J. Bornstein, D. Bray, A. Cornish-Bowden, A.A. Cuellar, S. Dronov, E.D. Gilles, M. Ginkel, V. Gor, I.I. Goryanin, W.J. Hedley, T.C. Hodgman, J.H. Hofmeyr, P.T. Hunter, N.S. Juty, J.L. Kasberger, A. Kremling, U. Kummer, N. Le Novere, L.M. Loew, D. Lucio, P. Mendes, E. Minch, E.D. Mjolsness, Y. Nakayama, M.R. Nelson, P.F. Nielsen, T. Sakurada, J.C. Schaff, B.E. Shapiro, T.S. Shimizu, H.D. Spence, J. Stelling, K. Takahashi, M. Tomita, J. Wagner, and J. Wang. The systems biology markup language (SBML): A medium for representation and exchange of biochemical network models. *Bioinformatics*, 19(4):524–531, 2003.

[28] T. Ideker, T. Galitski, and L. Hood. A new approach to decoding life: Systems biology. *Ann. Rev. Genom. Hum. Genet.*, 2:343–372, 2001.

[29] J. Jaeger, M. Blagov, D. Kosman, K.N. Kozlov, Manu, E. Myasnikova, S. Surkova, C.E. Vanario-Alonso, M. Samsonova, D.H. Sharp, and J. Reinitz. Dynamical analysis of regulatory interactions in the gap gene system of *Drosophila melanogaster. Genetics*, 167(4):1721–1737, 2004.

[30] J. Jaeger, S. Surkova, M. Blagov, H. Janssens, D. Kosman, K.N. Kozlov, Manu, E. Myasnikova, C.E. Vanario-Alonso, M. Samsonova, D.H. Sharp, and J. Reinitz. Dynamic control of positional information in the early *Drosophila* embryo. *Nature*, 430(6997):368–371, 2004.

[31] D. Kaplan and L. Glass. *Understanding Nonlinear Dynamics*. Springer-Verlag, 1995.

[32] S.A. Kauffman. Metabolic stability and epigenesis in randomly constructed genetic nets. *J. Theor. Biol.*, 22(3):437–467, 1969.

[33] S.A. Kauffman. *The Origins of Order: Self-Organization and Selection in Evolution*. Oxford University Press, 1993.

[34] H. Kitano. Systems biology: A brief overview. *Science*, 295(5560):1662–1664, 2002.

[35] B.J. Kuipers. *Qualitative Reasoning: Modeling and Simulation with Incomplete Knowledge*. MIT Press, 1994.

[36] J.D. Lambert. *Numerical Methods for Ordinary Differential Equations*. Wiley, 1991.

[37] J.-C. Leloup and A. Goldbeter. Toward a detailed computational model for the mammalian circadian clock. *Proc. Natl. Acad. Sci. USA*, 100(12):7051–7056, 2003.

[38] P.A. Levin and A.D. Grossman. Cell cycle and sporulation in *Bacillus subtilis. Curr. Opin. Microbiol.*, 1(6):630–635, 1998.

[39] D.J. Lockhart and E.A. Winzeler. Genomics, gene expression and DNA arrays. *Nature*, 405(6788):827–836, 2000.

[40] E. Meir, E.M. Munro, G.M. Odell, and G. Von Dassow. Ingeneue: A versatile tool for reconstituting genetic networks, with examples from the segment polarity network. *J. Exp. Zool.*, 294(3):216–251, 2002.

[41] P. Mendes. GEPASI: A software package for modelling the dynamics, steady states and control of biochemical and other systems. *CABOS*, 9(5):563–571, 1993.

[42] L. Mendoza, D. Thieffry, and E.R. Alvarez-Buylla. Genetic control of flower morphogenesis in *Arabidopsis thaliana*: A logical analysis. *Bioinformatics*, 15(7-8):593–606, 1999.

[43] T. Mestl, E. Plahte, and S.W. Omholt. A mathematical framework for describing and analysing gene regulatory networks. *J. Theor. Biol.*, 176(2):291–300, 1995.

[44] J. Monod and F. Jacob. General conclusions: Teleonomic mechanisms in cellular metabolism, growth, and differentiation. In *Cold Spring Harb. Symp. Quant. Biol.*, volume 26, pages 389–401, 1961.

[45] B. Novak and J.J. Tyson. A model for restriction point control of the mammalian cell cycle. *J. Theor. Biol.*, 230(4):563–579, 2004.

[46] A. Pandey and M. Mann. Proteomics to study genes and genomes. *Nature*, 405(6788):837–846, 2000.

[47] M. Ptashne. *A Genetic Switch: Phage λ and Higher Organisms*. Cell Press & Blackwell Science, 2nd edition, 1992.

[48] J. Reinitz, D. Kosman, C.E. Vanario-Alonso, and D.H. Sharp. Stripe forming architecture of the gap gene system. *Dev. Genet.*, 23:11–27, 1998.

[49] J. Reinitz and J.R. Vaisnys. Theoretical and experimental analysis of the phage lambda genetic switch implies missing levels of co-operativity. *J. Theor. Biol.*, 145:295–318, 1990.

[50] D. Ropers, H. de Jong, M. Page, D. Schneider, and J. Geiselmann. Qualitative simulation of the nutritional stress response in *Escherichia coli*. Technical Report RR-5412, INRIA, 2004.

[51] L. Sánchez and D. Thieffry. A logical analysis of the *Drosophila* gap-gene system. *J. Theor. Biol.*, 211(2):115–141, 2001.

[52] L. Sánchez, J. van Helden, and D. Thieffry. Establishment of the dorso-ventral pattern during embryonic development of *Drosophila melanogaster*: A logical analysis. *J. Theor. Biol.*, 189:377–389, 1997.

[53] P. Smolen, D.A. Baxter, and J.H. Byrne. Modeling transcriptional control in gene networks: Methods, recent results, and future directions. *Bull. Math. Biol.*, 62(2):247–292, 2000.

[54] P. Stragier and R. Losick. Molecular genetics of sporulation in *Bacillus subtilis*. *Ann. Rev. Genet.*, 30:297–341, 1996.

[55] S.H. Strogatz. *Nonlinear Dynamics and Chaos: With Applications to Physics, Biology, Chemistry, and Engineering*. Perseus Books, 1994.

[56] D. Thieffry and R. Thomas. Dynamical behaviour of biological networks: II. Immunity control in bacteriophage lambda. *Bull. Math. Biol.*, 57(2):277–297, 1995.

[57] R. Thomas. Boolean formalization of genetic control circuits. *J. Theor. Biol.*, 42:563–585, 1973.

[58] R. Thomas. Regulatory networks seen as asynchronous automata: A logical description. *J. Theor. Biol.*, 153:1–23, 1991.

[59] R. Thomas and R. d'Ari. *Biological Feedback*. CRC Press, 1990.

[60] R. Thomas, D. Thieffry, and M. Kaufman. Dynamical behaviour of biological regulatory networks: I. Biological role of feedback loops and practical use of the concept of the loop-characteristic state. *Bull. Math. Biol.*, 57(2):247–276, 1995.

[61] J.J. Tyson, C.C. Chen, and B. Novak. Sniffers, buzzers, toggles and blinkers: Dynamics of regulatory and signaling pathways in the cell. *Curr. Opin. Cell Biol.*, 15(2):221–231, 2003.

[62] J.J. Tyson, K. Chen, and B. Novak. Network dynamics and cell physiology. *Nat. Rev. Mol. Cell Biol.*, 2(12):908–916, 2001.

[63] A. Usseglio Viretta and M. Fussenegger. Modeling the quorum sensing regulatory network of human-pathogenic *Pseudomonas aeruginosa*. *Biotech. Prog.*, 20(3):670–678, 2004.

[64] G. von Dassow, E. Meir, E.M. Munro, and G.M. Odell. The segment polarity network is a robust developmental module. *Nature*, 406(6792):188–192, 2000.

[65] G. Yagil and E. Yagil. On the relation between effector concentration and the rate of induced enzyme synthesis. *Biophys. J.*, 11(1):11–27, 1971.